A TEXTBOOK OF

BIOTECHNOLOGY

A TEXTBOOK OF
BIOTECHNOLOGY

Dr. Rashmi Tyagi
Assistant Professor
Dept. of Applied Sciences & Humanities
Institute of Technology & Management
Gurgaon, Haryana

UNIVERSITY SCIENCE PRESS
(An Imprint of Laxmi Publications Pvt. Ltd.)
BANGALORE • CHENNAI • COCHIN • GUWAHATI • HYDERABAD
JALANDHAR • KOLKATA • LUCKNOW • MUMBAI • RANCHI
NEW DELHI

Published by :

UNIVERSITY SCIENCE PRESS

(An Imprint of Laxmi Publications Pvt. Ltd.)
113, Golden House, Daryaganj,
New Delhi-110002

Phone : 011-43 53 25 00
Fax : 011-43 53 25 28

www.laxmipublications.com
info@laxmipublications.com

First Edition : 2009

OFFICES

✆ **Bangalore**	080-26 61 15 61	✆ **Chennai**	044-24 34 47 26
✆ **Cochin**	0484-237 70 04, 405 13 03	✆ **Guwahati**	0361-254 36 69, 251 38 81
✆ **Hyderabad**	040-24 65 23 33	✆ **Jalandhar**	0181-222 12 72
✆ **Kolkata**	033-22 27 43 84	✆ **Lucknow**	0522-220 95 78
✆ **Mumbai**	022-24 91 54 15, 24 92 78 69	✆ **Ranchi**	0651-221 47 64

UBI-9350-195-ATB OF BIOTECHNOLOGY **C—**

Typeset at : Monu Printographics, Delhi printed at : Repro Knowledgecast Limited, Thane

CONTENTS

PREFACE

Biotechnology is a multidisciplinary subject and needs the knowledge of many areas, like, Botany, Zoology, Cytology, Genetics, Biochemistry, Microbiology, etc., where the microbial cells or cultured plant or animal cells or their products, like antibiotics, enzymes, various organic acids, etc., are produced industrially to be used practically for the welfare of human beings. Actually, the term Biotechnology, though, is of recent use, but it is a quite ancient technology and the man is exploiting unknowingly the services of various microorganisms, like yeast for alcoholic beverages, lactobacilli for curdling of milk, *Acetobacter* for vinegar production, etc. Nowadays, the technique is more efficiently used by modifying the genetic constitution of naturally occurring commercial strains of microorganisms, plants and animals, which are being used in the diverse fields of agriculture, medicine, pollution management, biofuels, food and other industries, etc.

The book includes 20 chapters and the various topics are written illustratively in a very simple way to make all the undergraduate B. E., B. Tech. and B. Sc. students understand the rather difficult subject easily and comfortably. The valuable suggestions and corrections in this first edition are highly appreciated from the readers.

I am especially thankful to my Head of the Department, Prof. A. P. Gupta, for encouraging me to write this book, without which the completion of this work was impossible. I also thank to Mr. Inderpal Singh for his help in operation of computer for various diagrams.

—AUTHOR

ACKNOWLEDGEMENT

I express my gratitude towards **Prof. A.P. Gupta,** Head of Department of Applied Science & Humanities, Institute of Technology and Management, Gurgaon, for inspiring me to write this book. I am also thankful to my colleague, **Mr. Inderpal Singh,** Senior in Physics, for his sincere assistance for diagrams in this book. Thanks are also due to **Institute ofTechnology & Management,** Gurgaon, where I am working as Assistant Professor, for providing all kinds of facilities, like library, laptop, internet, printing, scanning, photocopying, etc., for writing this book.

As it is the first edition of this book, I will appreciate all kinds of valuable suggestions from the readers for its improvement.

—AUTHOR

Chapter 1

INTRODUCTION TO CELL STRUCTURE AND VIRUSES

Robert Hook (1655) was the first person who observed cells of cork (dead cells) under the microscope, which appeared like honey comb. **Leeuwenhock** (1674) observed white blood cells with the help of a microscope with some organization inside. **Schleiden** and **Schwann** (1838–39) proposed the **cell theory**, according to which the cell is the **unit of structure** of living system (cell is the unit of life), *i.e.,* all living organisms are made up of cells (except for viruses, which are living but have acellular organization and are composed of only proteins and DNA/RNA, some viruses also contain membranes) and the cell **arises from the pre-existing cells.**

CELL STRUCTURE

The cellular organisms have been divided into two categories, **prokaryotes** (*e.g.*, archaebacteria, eubacteria, cyanobacteria, prochlorophytes) that are primitive and evolved around 3 billion years ago (Fig. 1.2); and more advanced **eukaryotes** that appeared around 1 billion years ago, *e.g.*, algae (except for the algae belonging to classes Cyanophyta and Prochlorophyta), fungi, bryophytes, pteridophytes, gymnosperms, angiosperms, unicellular and multicellular animals (Fig. 1.1).

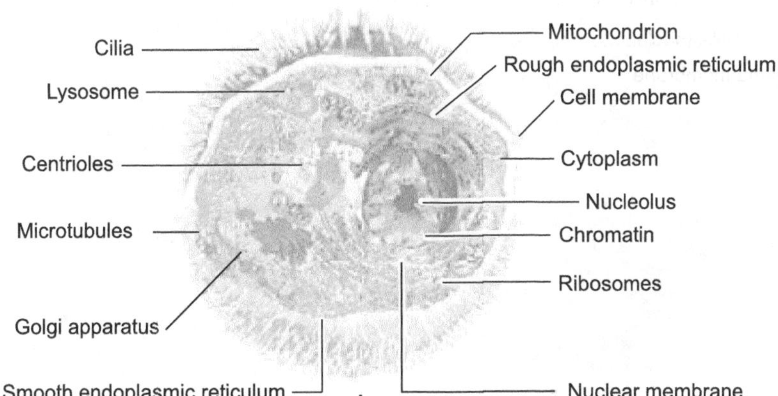

A

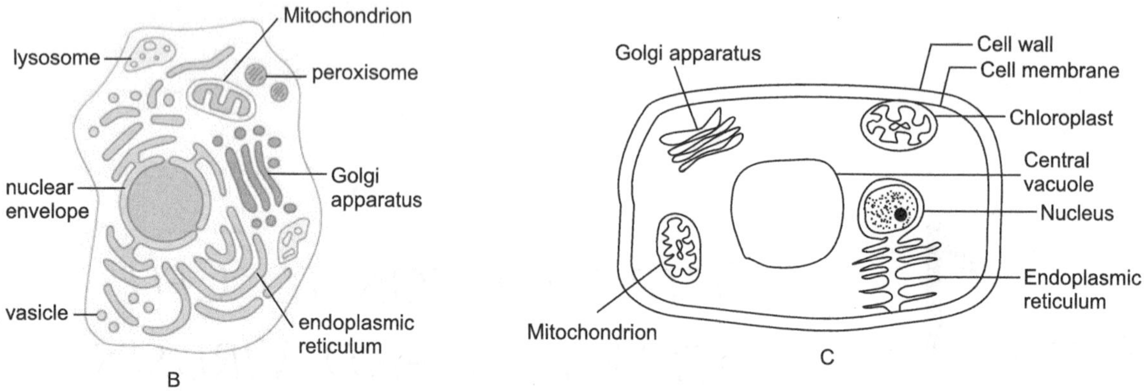

Fig. 1.1 The eukaryotic cells.

The cell **size** and **shape** varies from organism to organism. Some cells are visible to naked eyes whereas most of the cells can be observed through microscope. Generally, eukaryotic cells are 1–2 μm to 1 mm and prokaryoyic cells 0.15–2.0 μm in size. The largest cell found is the ostrich egg having 15–20 cm diameter. Regarding the **shape**, the cells may be spherical, cylindrical, irregular, etc. Some plant (*e.g.*, *Euglena*, *Dunaliella*, etc.) and animal (*e.g.*, *amoeba*, leucocytes, etc.) cells keep on changing their shape.

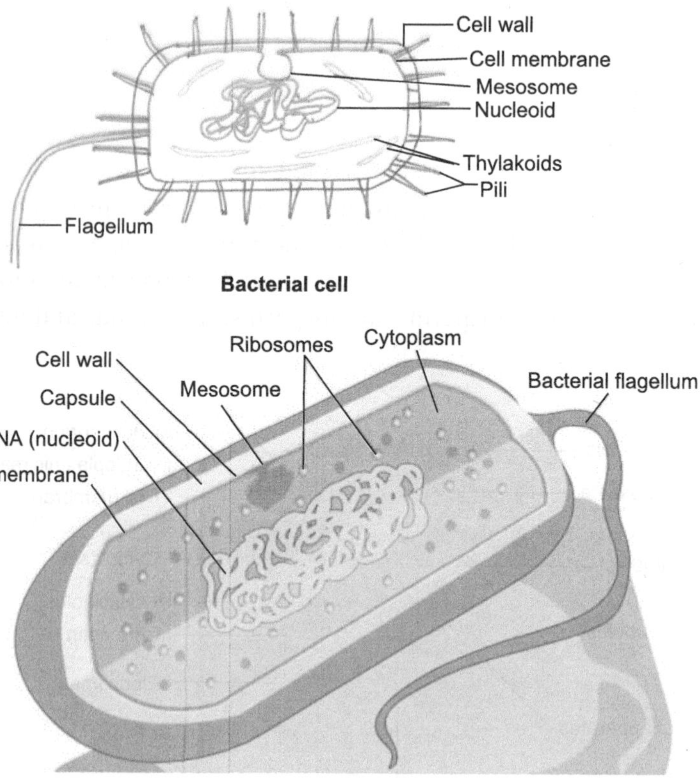

Fig. 1.2 The bacterial cells.

Various parts of a cell are:

The Cell Wall

The plant cells contain the outermost boundary, the cell wall (some unicellular phytoplankton lack a cell wall in their vegetative cells, *e.g.*, *Dunaliella, Euglena*, etc., but during their life cycle at sometime they form a cell wall, *e.g.*, during spore formation) that is lacking in animals. The rigid, porous (pores, called plasmodesmata, allow movement of substances between the adjacent cells in multicellular plants) cell wall providing shape and protection to the cell and prevents the cell from bursting. Some cells have a capsule (glycocalyx) around the cell wall that is made up of polysaccharides (*e.g.*, *Diplococcus pneumoniae*, *Streptococcus pneumoniae*), polypeptide (*e.g.*, *Bacillus anthracis*), mixture of both polysaccharides and proteins or waxy material (*e.g.*, *laprosy bacterium*).

The compositions of the cell wall of prokaryotic and eukaryotic cells are as fallows.

1. **Prokaryotic Cell Wall.** In some prokaryotes the vegetative cells are devoid of any cell wall, *e.g.*, mycoplasma. In most of the prokaryotic cells, the cell wall is made up of peptidoglycan, a combination of proteins and carbohydrates. The peptidoglycan consists of alternating N-acetyl glucosamine (NAG) and N-acetyl muramic acid (NAM) sugars linked by β-1,4-glycosidic bonds (repeated disaccharides). NAM molecules are attached with a tetrapeptide chain of L-alanine, D-glutamic acid, L-lysine and D-alanine amino acids (Fig. 1.7). In G^+ bacteria the peptidoglycan is quite thick and the two adjacent, parallel tetrapeptides are linked through a pentaglycine at the points of L-lysine and D-alanine (Fig. 1.3 and 1.6). In G^- bacteria the peptidoglycan is thinner, having an outer membrane above it and the pentaglycene is absent, so that the terminal D-Ala is attached directly to the neighbouring tetrapeptide through L-Lys (or L-Lys in some cases may be replaced by diaminopimelic acid that is lysine like amino acid) (Fig. 1.4 and 1.5).

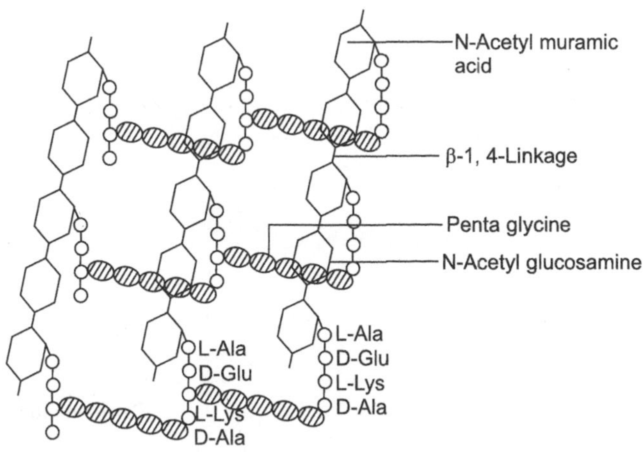

Fig. 1.3 Peptidoglycan of Gram positive bacteria (*e.g.*, *Staphylococcus*).

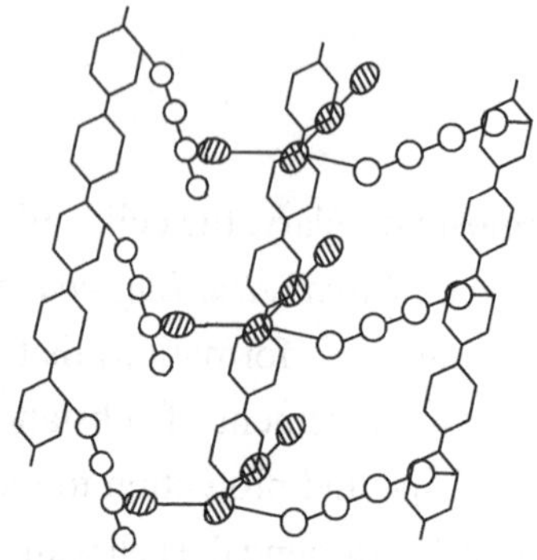

Fig. 1.4 Peptidoglycan of Gram negative bacteria.

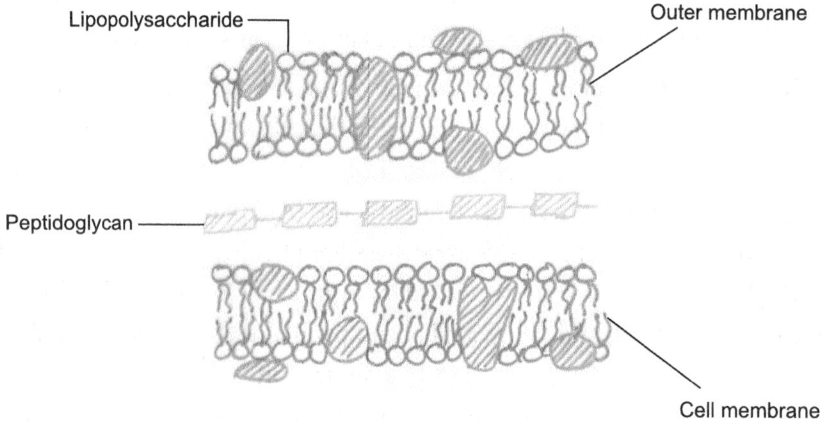

Fig. 1.5 Gram negative bacterial cell wall.

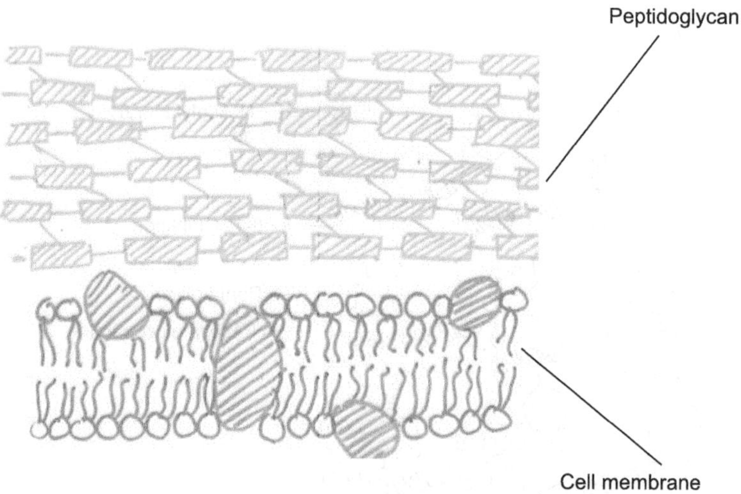

Fig. 1.6 Gram positive bacterial cell wall.

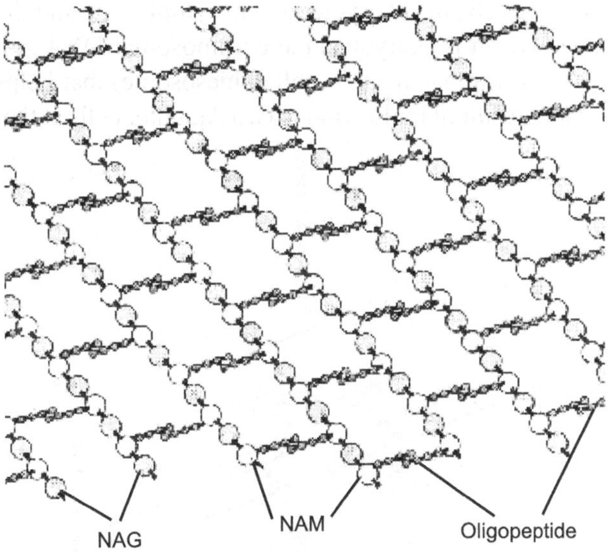

Fig. 1.7 Peptidoglycan.

2. **Eukaryotic Cell Wall.** Different groups of eukaryotic cells have different composition of the cell wall. In green algae and higher plants the cell wall is mainly made up of **cellulose** fibres that are embedded in a matrix of pectin, lignin and hemicellulose; the middle lamella found between the adjacent cells is made up of pectin. Sometimes, in higher plants, in addition to cellulose the outer cell wall also contains lignin (wood) and suberin (cork). Similarly, in various other groups of algae the cell walls have different compositions, *e.g.*, in diatoms it is mainly made up of **silica**. In fungi, the main component of the cell wall is **chitin**.

Plasma Membrane

The plasma membrane (cell membrane) lies just beneath the cell wall in plant cells, whereas it makes the outermost boundary in case of animal cells. It is **semi-permeable** (selectively permeable) as it facilitates the transportation of selected substances through it and prevents the escape of cellular materials outside the cells. **S. J. Singer** and **G. Nicholson** (1972) proposed **fluid mosaic model** for the biological membranes (cell membrane and organeller membranes) according to which the biological membranes (about 75 Å or 7.5 nm thick) are mainly composed of phospholipids and proteins that are arranged in a mosaic manner, and smaller amounts of oligosaccharides and other substances. A **phospholipid** molecule is made up of an alcohol (generally glycerol) attached with a phosphate group (hydrophilic polar head) and two fatty acids (hydrophobic non-polar tails) (Fig 1.9). The phospholipids (for details see the chapter 'Lipids') form a double layer structure in which the polar heads (phosphate groups) are directed towards the two surfaces (outer and inner) of the membrane and the two non-polar fatty acid tails on the inner side (Fig. 1.10). **Proteins** are found either at the surfaces (extrinsic proteins) or embedded (partially or completely) inside this phospholipid bilayer (intrinsic proteins) (Fig. 1.8). Some proteins serve as the **carriers** (intrinsic proteins) for the transportation of water-soluble materials through the membrane, whereas others function as

enzymes (*e.g.*, various respiratory and photosynthetic enzymes are found in mitochondrial and chloroplastic membranes, respectively, in eukaryotes and in cytoplasmic and thylakoids membranes, respectively, in prokaryotic cells). In prokaryotes, the chromosomal DNA is found to be attached with the plasma membrane (the invaginated site is called mesosome) that helps in the separation of two DNA molecules formed as a result of replication to two daughter cells at the time of cell division.

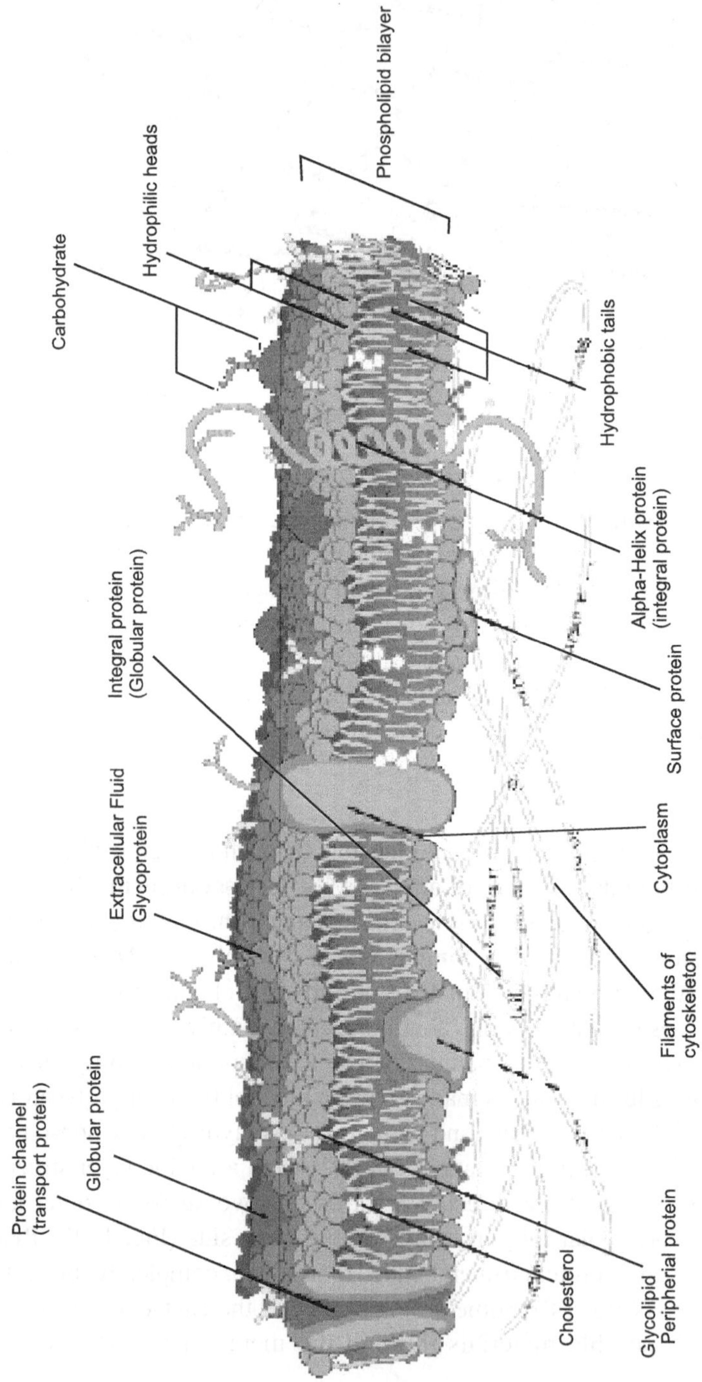

Fig. 1.8 Plasma Membrane (Cell membrane).

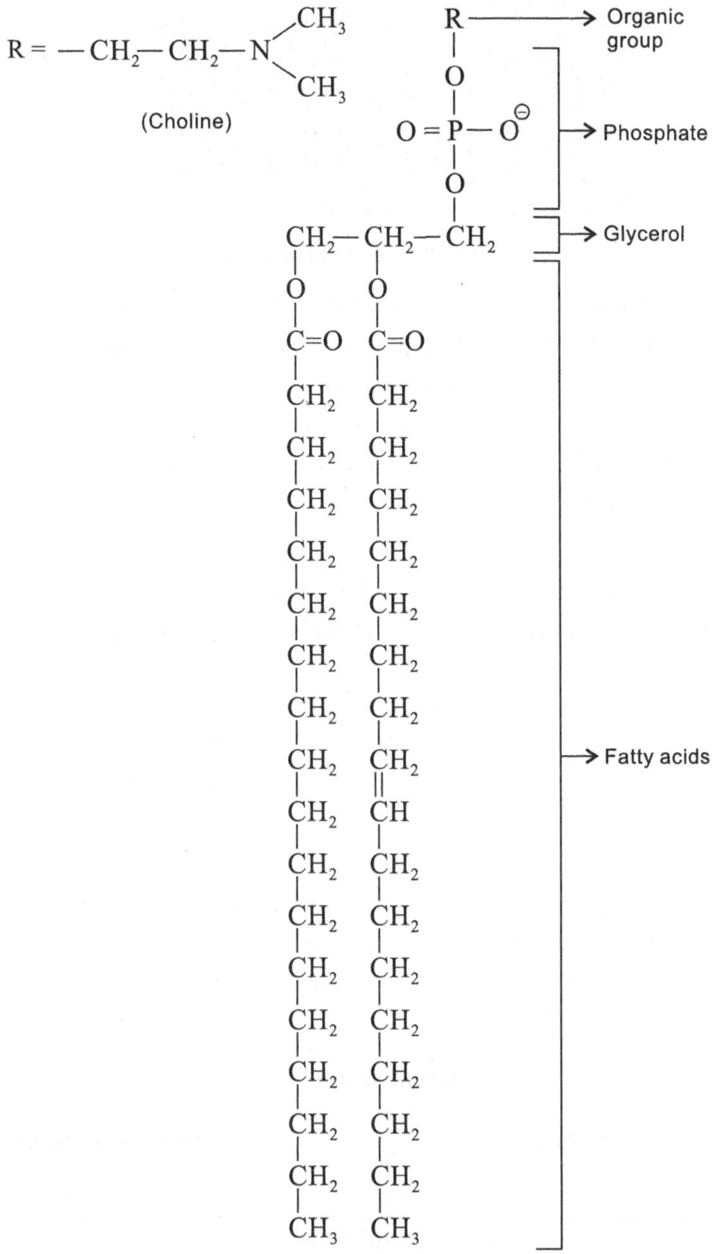

Fig. 1.9 Chemical structure of a phospholipid.

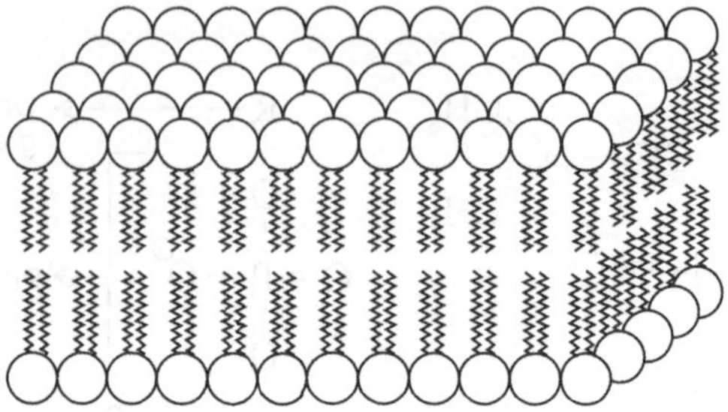

Fig. 1.10 Phospholipid bilayer.

The transportation through membranes can be of two types (Fig. 1.13):

1. Passive Transport. This transportation across the membrane involves the movement of solutes or gases (**diffusion**) and solvents (**osmosis**) (Fig. 1.11) along the concentration or pressure gradients (from a region of high concentration or pressure to a region of low concentration or pressure) till the concentration or pressure becomes equal on the two sides of the membrane and the molecules move equally in both the directions. This transportation does not require any energy.

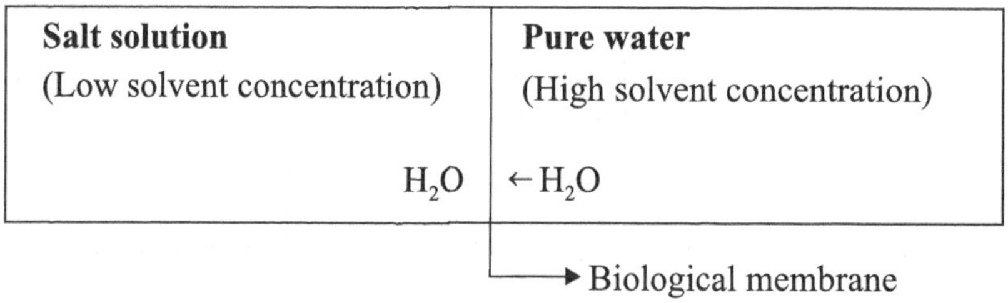

Fig. 1.11 Osmosis

2. Active Transport. This transportation of materials occurs against the concentration or electrochemical gradient (from a region of low concentration to a region of high concentration) and hence requires energy in the form of ATP, *e.g.*, movement of Na^+ and K^+ (Fig. 1.12).

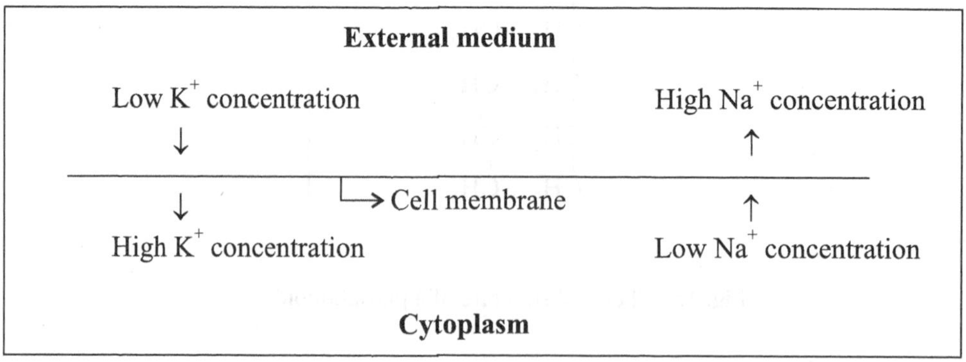

Fig. 1.12 Active transport of Na^+ and K^+ ions through cell membrane.

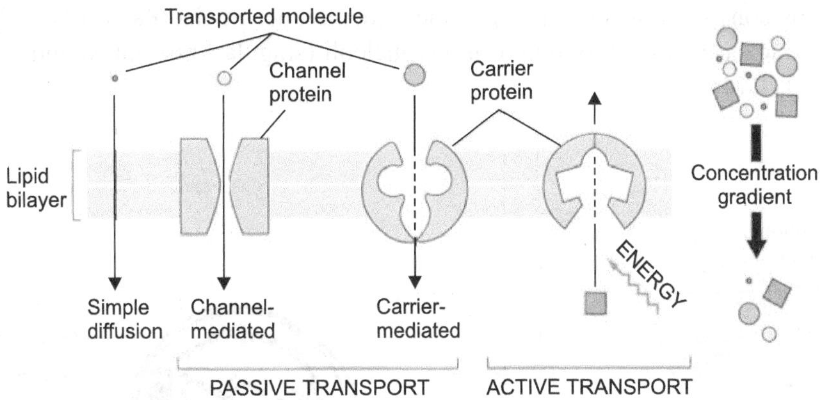

Fig. 1.13 Active and passive transports across the cell membrane.

Cytoplasm

Cytoplasm is the jelly-like fluid in between the plasma membrane and the nuclear membrane. It contains various organelles (like mitochondria, chloroplasts, Golgi bodies, endoplasmic reticulum in eukaryotes) or other membrane bound structures (like thylakoids, gas vacuoles in prokartotes), ribosomes, vitamins, various enzymes, proteins, etc.

Cyclosis is the cytoplasmic streaming found in eukaryotic cells that helps in the movement of organelles and other cell inclusions and in the intracellular distribution of nutrients, metabolites, enzymes, etc.

Nucleus

In **eukaryotic** cells the genetic material (several linear molecules of DNA) is found inside the nucleus and controls all the activities of a cell. Some cells are devoid of any nucleus, *e.g.,* mature sieve cells of vascular plants that help in the transportation of substances, red blood cells of mammals, etc. Generally, the cells contain a single nucleus (**uninucleate**), but in some cases they may contain more than one nuclei (**multinucleate** or coenocytic condition) resulting due to repeated nuclear divisions without any cytokinesis (division of cytoplasm), *e.g*., many algae (*Vaucheria*) and fungi (*Rhizopus*). Nucleus was discovered by **Robert Brown** (1833) and generally it is found in the centre of a cell, but in plant cells it may be located on one side due to the presence of a large central vacuole. The **shape** of the nucleus may be spherical, elliptical, elongated, kidney-shaped, lobed (in WBC), irregular, etc., and the size depends upon the type and activity of the cell.

The fact that nucleus contains **hereditary** information was shown by a Danish scientist **J. Hammerling** (1953). He performed transplantation experiment using two species of a green alga *Acetabularia crenulata* and *A. mediterrania* that contains a single nucleus in their rhizoids and differ with respect to their cap (reproductive body) shape.

Parts of Nucleus

The nucleus consists of a surrounding nuclear membrane, **nucleoplasm** (the jelly-like substance inside the nuclear membrane), one or more nucleoli (singular: nucleolus) and chromatin or chromosomes (Fig. 1.14).

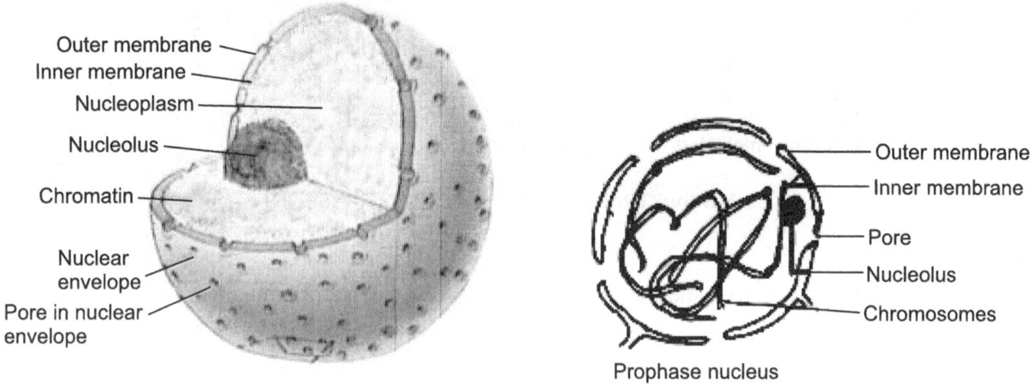

Fig. 1.14 Nucleus.

(*i*) ***Nuclear Membrane: (Nuclear envelope).*** It is made up of two membranes that merge at the pores. The pores (diameter 30–100 nm) allow the transfer of materials between the nucleus and the cytoplasm, *e.g.*, mRNA that is synthesized in nucleus and escapes into the cytoplasm through pores. In animal cells a diaphragm, called annulus, extends across each pore. The space between the two membranes is called perinuclear space (15 nm wide) that may be continuous with the lumen of the endoplasmic reticulum. The inner nuclear membrane is smooth, whereas the outer membrane may be attached with ribosomes at some locations to give a rough appearance.

(*ii*) ***Nucleolus:*** It is round or irregular in shape and found associated with a specific region (nucleolar-organizing region) of a chromosome called nucleolar-organizing chromosome that contains the gene for rRNA of which the nucleolus is made up of. The nucleolus and the nuclear membrane disappear at some stages (Metaphase and Anaphase) of the cell division.

(*iii*) ***Chromatin:*** The chromatin (at Interphase various uncoiled thin chromosomes appear as a mass-like structure called chromatin) or the chromosomes (individuality of various chromosomes start appearing during the cell division due to increase in thickness arising because of coiling of chromatin fibres) are made up of DNA (33–43%), RNA (0.3–10%), histones (30–52%) and non-histones (4–25%). Some regions of the chromatin are lightly stained called **euchromatin** whereas other regions appear as darkly stained regions known as **heterochromatin**.

The **non-histones** show more diversity and their number varies from 12–20 in different organisms and tissues within the same organism. It is expected that they play a role in gene regulation.

The **histones** are highly conserved and they have more structural role than the gene regulatory role (generally they repress the gene activity whenever bound to DNA). There are five types of histones found in eukaryotic chromatin and chromosomes. **R.D. Kornberg** and **J.O. Thomas** (1974) gave the **nucleosome model** for the association of DNA and histones. According to this model two molecules each of the histones **H_2A, H_2B, H_3** and **H_4** form the octamer core and about 200 base pairs (bp) of DNA is found wrapped around this core to form 1.75 turns. The histone **H_1** helps in the stabilization of this DNA coiling to form the **nucleosome** having 11 nm (110 Å) diameter (Fig. 1.15). This 11 nm fibre is again and again coiled to give rise to 30 nm (300 Å) (solenoids) (Fig. 1.16) and thicker fibres that are stabilized by certain proteins, finally to give the Metaphase chromosomes having maximum thickness (therefore, chromosome counting for a cell is performed at Metaphase). Thus, in humans the largest chromosome containing about 85,000 μm long DNA is packed into a 0.5 μm thick and 10 μm long Metaphase chromosome (Fig. 1.17).

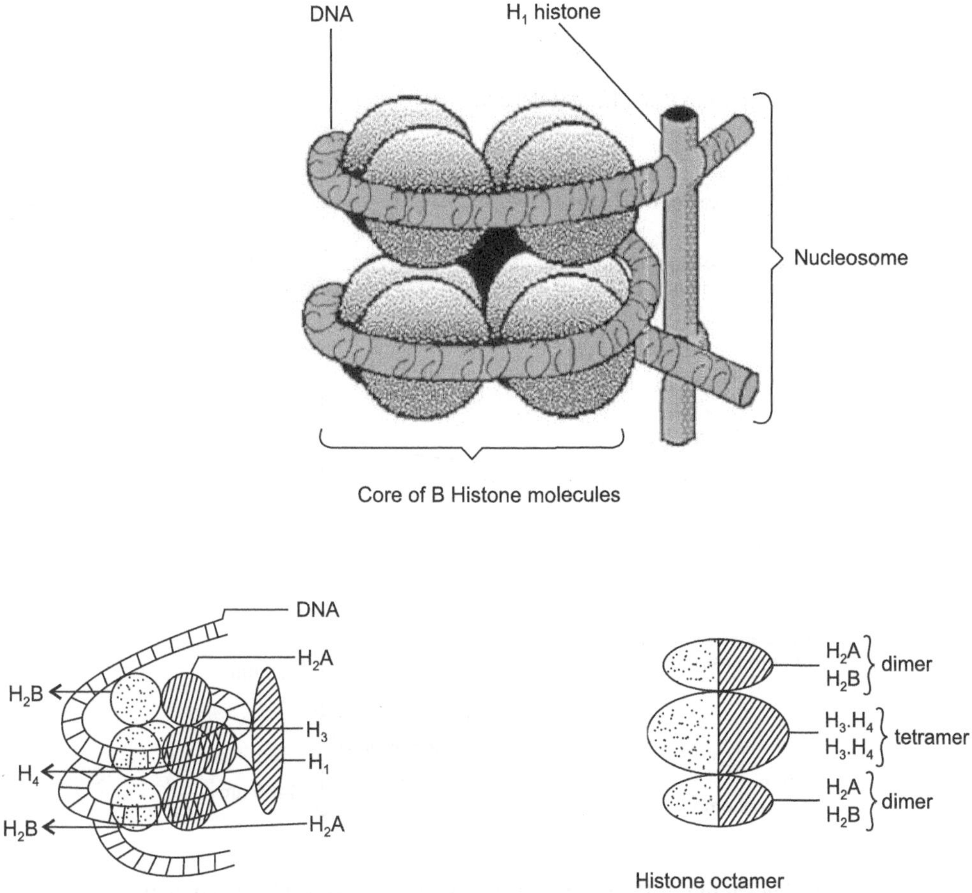

Fig. 1.15 Assembly of histones and model of nucleosome.

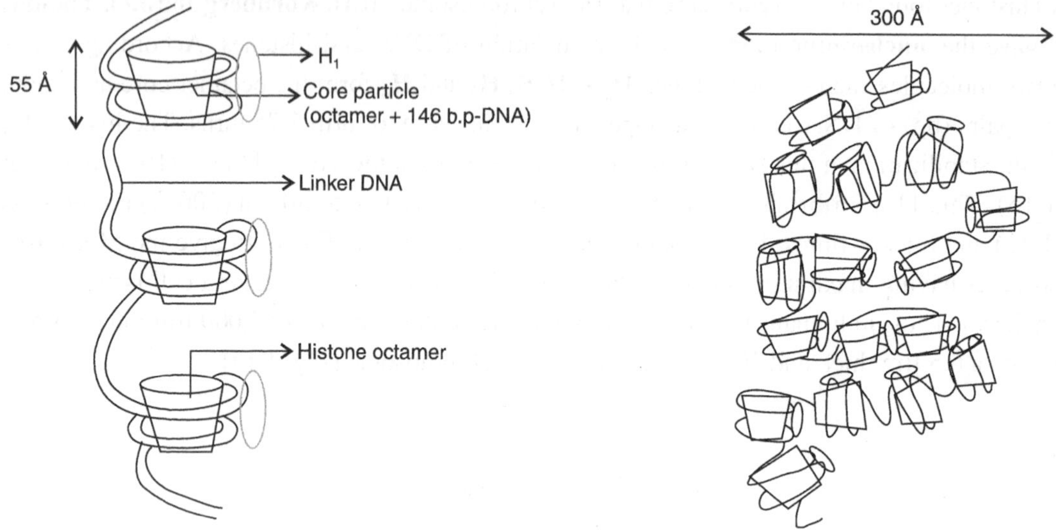

Fig. 1.16 A 11 nm thick fibre (left) and a 30 nm thick solenoid (right).

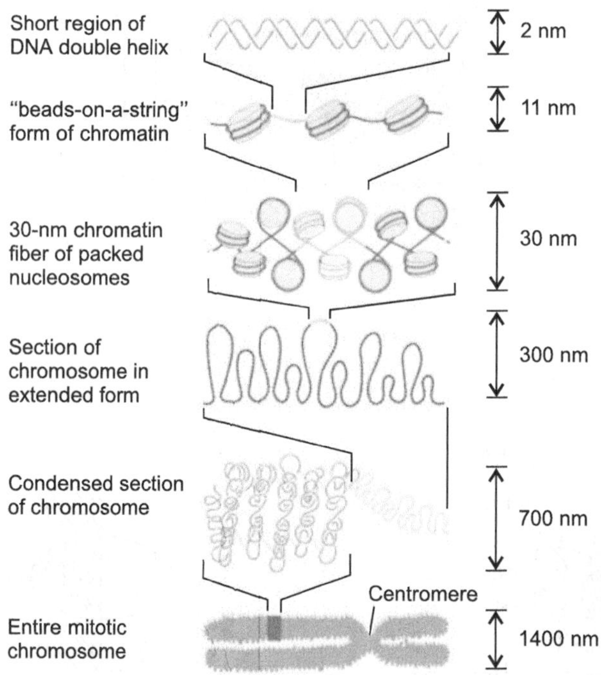

Fig. 1.17 Packaging of DNA into Metaphase chromosome (10,000 times).

(*iv*) ***Chromosomes:*** These are visible only during cell division and each is made up of two chromatids (half chromosome). Each species has a characteristic number, size, shape, etc., of the chromosomes that is represented by a diagram called ideogram (Fig. 1.18).

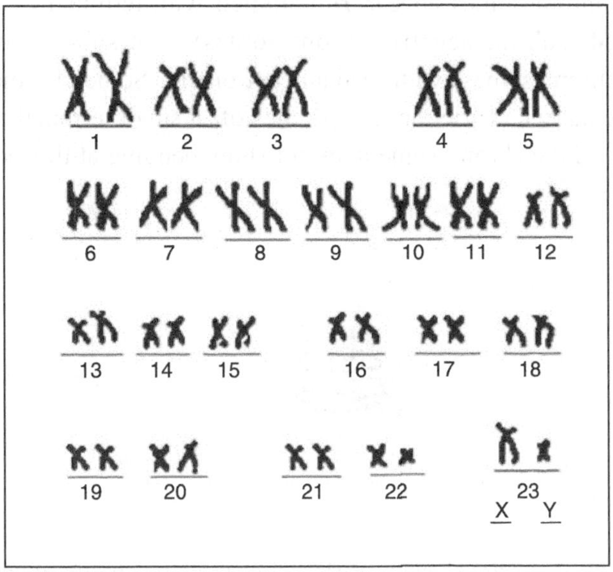

Fig. 1.18 An ideogram of chromosomes of a male human cell.

At metaphase and anaphase the chromosomes are made up of coiled gene-bearing filament, chromonema, in the matrix that is surrounded by the pellicle (probably made up of nucleolar material as it is formed when the nucleolus disappears). The terminal regions of chromosomes are called **telomeres** (Fig. 1.19). **Strasburger** (1875) called these thread-like structures as chromosomes (*chroma* means colour) because they have affinity for basic dyes. The diploid organisms have two sets of chromosomes, *Haplopappus gracilis* has **lowest number** of diploid chromosomes ($2n = 4$), whereas some pteridophytes contain more than 1,200 chromosomes in their diploid set.

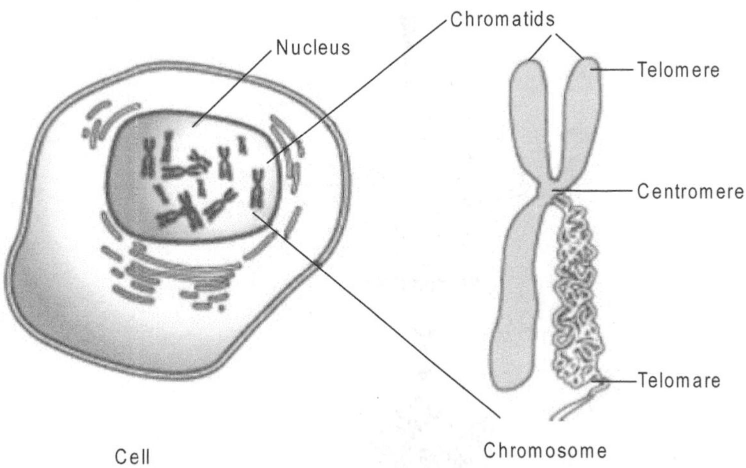

Fig. 1.19 A cell (left) showing chromosomes and a metaphase chromosome (right).

The **shape** of the chromosomes is observed at anaphase where the location of the centromere (primary constriction) determines the shape because chromosome bending occurs only at the location of centromeres.The chromosomes appear **rod-shaped** with terminal centromere (telocentric chromosomes), **J-shaped** (sub-metacentric chromosomes) with sub-terminal centromere and **V-shaped** (metacentric chromosomes) with median centromere. Some chromosomes, called **SAT chromosomes**, have secondary constriction that pinches off a small terminal fragment of chromosome known as satellite and the chromosome does not show bending at this point during anaphase (Fig. 1.20).

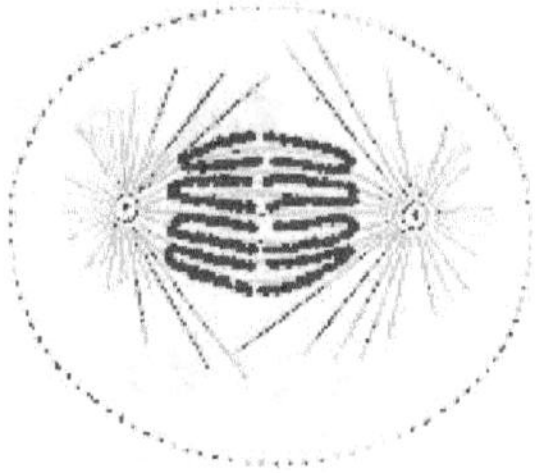

Fig. 1.20 Anaphase showing V-shaped metacentric chromosomes.

The **prokaryotic** cells lack a true nucleus and the naked (not associated with the proteins, histones and non-histones) genetic material, that is a single, circular molecule of double-stranded DNA, is not enclosed within any nuclear membrane. The folded and supercoiled circular DNA, called nucleoid (Fig. 1.21), with irregular outer boundary is found to be attached with the mesosome.

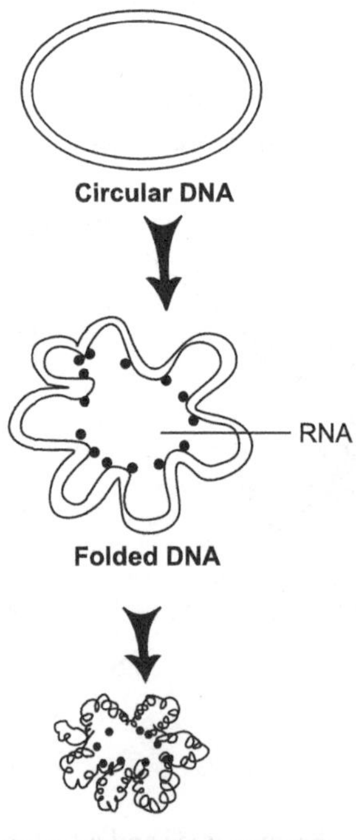

Fig. 1.21 Coiling of bacterial DNA.

Mitochondria

Mitochondria are found in aerobic **eukaryotic** cells and help in the respiration to produce energy, therefore, are also known as powerhouse of the cell. In aerobic respiration, the respiratory substrates are completely oxidized to CO_2 and H_2O to generate ATP that is required for the performance of various cellular activities, *e.g.*, protein synthesis. They multiply by binary fission method, like the bacteria. Their **number** varies from 1 to several thousands in different kinds of cells. Their **shape** may be spherical, oval, cylindrical, tubular, etc. Their **size** varies from 5–10 μm in length and 0.5–1.0 μm in diameter. The mitochondrion is **semiautonomous** as most of its protein requirements are fulfilled by its own mtDNA (mitochondrial DNA) and only some of the proteins are synthesized through nuclear DNA.

The mitochondrion is bounded two membranes (each about 7.5 nm thick), where the **outer membrane** is smooth and the **inner membrane** is infolded variously to form cristae that increase the physiologically active area. Various components of the respiratory chain are located in the inner membrane (Fig. 1.22). The cristae contains the enzyme ATPase responsible for the ATP synthesis during aerobic respiration. The inner membrane encloses the semi-fluid **matrix** that contains proteins, ribosomes, RNA, mtDNA, enzymes, components of citric acid cycle (Kreb's cycle) responsible for evolution of CO_2 during respiration, etc.

In **prokaryotes** the mitochondria are absent and the electron transport chain of respiration is located in **plasma membrane**.

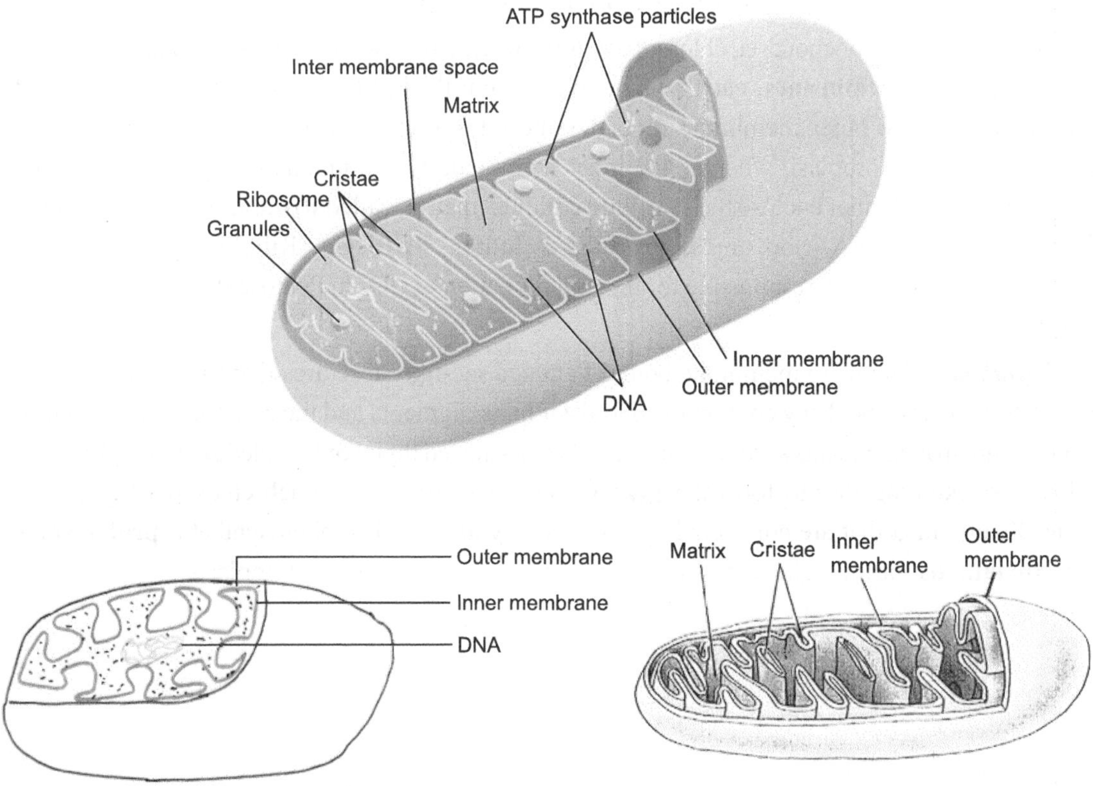

Fig. 1.22 Mitochondria.

Chloroplast

Chloroplasts are found only in photosynthetic organisms like **eukaryotic** algae, Bryophytes, Pteridophytes, Gymnosperms and Angiosperms, in those parts that are exposed to sunlight. Three types of plastids are found in plant cells:

1. **Leucoplasts** that are colourless found in non-green plants,
2. **Chromoplasts** are variously coloured ranging from yellow to red and impart the colour to flowers and fruits.
3. **Chloroplasts** are green plastids that also multiply by the **binary fission** method. They contain the **photosynthetic pigments**, like chlorophyll *a* and *b* (green in colour), carotenoids (colour ranges from yellow to orange) and xanthophylls (yellowish coloured), that capture the sunlight for the synthesis of organic carbohydrates during **photosynthesis**.

The **number** of chloroplasts varies from 1 (*e.g.*, the green alga *Ulothrix*) to several hundred (*e.g.*, green alga *Chara*) in a cell in different plants. Generally, each leaf cell contains 20–100 chloroplasts.

Shape: In higher plants the chloroplasts are generally disk shaped, but they may be rounded or club-shaped also. Their shape varies in different algae, *e.g.*, ribbon-like in the green alga *Spirogyra*, cup-shaped in the green alga *Chlorella*, reticulate in the green alga *Oedogonium*, etc.

Size: The **size** of the chloroplast varies from 4–10 μm in length and 2– 4 μm in breadth.

Structure: Like mitochondria, chloroplasts are also semiautonomous bodies. They are enveloped within **two smooth membranes**, each being about 7.5 nm thick, that are separated by a space of 10–20 nm width. The outer membrane regulates the transport of various materials between the cytoplasm and the chloroplast, whereas the inner membrane is continuous to thylakoids at many places. The two membranes enclose the semifluid colloidal matrix, called **stroma**, containing **chDNA** (chloroplast DNA), RNA, ribosomes, enzymes including **RuBisCo** (Ribulose Bis-phosphate Carboxylase enzyme) that is responsible for the initial fixation of CO_2 in Pentose Phosphate Pathway (Fig. 1.23)

Thylakoids: These are membrane bound flattened sac-like structures that run throughout the stroma to form a network. They contain the photosynthetic pigments and the components of electron transport chain. In higher plants at some places 2–100 disc-shaped thylakoids, called **granal thylakoids**, are closely stacked together to form the **grana** (like stacks of coins). Each chloroplast generally contains 40–60 grana that are connected by **stromal thylakoids**. The photosynthetic **prokaryotes** contain **thylakoids** for the light reaction of photosynthesis as they lack chloroplasts.

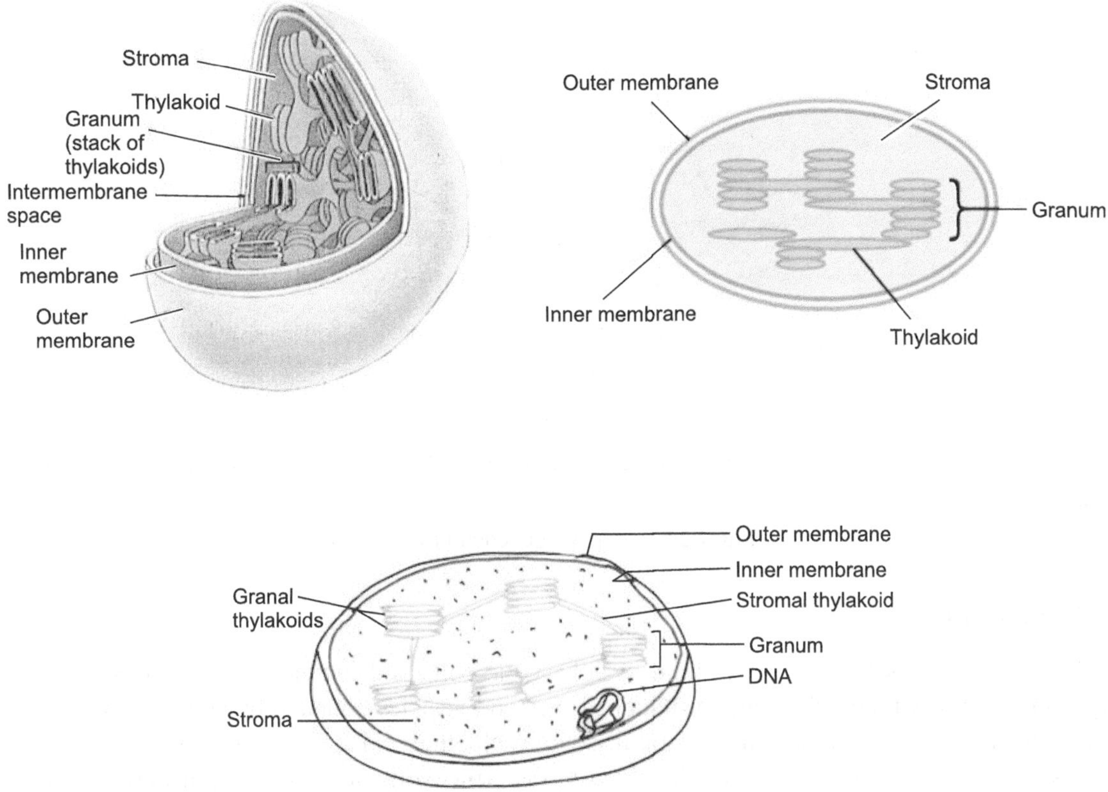

Fig. 1.23 Chloroplasts.

Endoplasmic Reticulum

The term endoplasmic reticulum was given by **K. Porter** (1953), which means fine network. It is the network of membranous tubules and flattened sacs in the cytoplasm of **eukaryotic** cells that are continuous with the outer nuclear membrane of nucleus and the plasma membrane. In some regions, the outer surface of the endoplasmic reticulum (ER) is covered with ribosomes to form the **rough endoplasmic reticulum** (RER) where the ribosomes are engaged in protein synthesis that are delivered into the lumen (cavity) of the RER. The other parts of ER that are not covered by any ribosomes are known as **smooth endoplasmic reticulum** (SER), which synthesizes lipids and sterols and it is continuous with the RER. Some regions of SER are continuous with Golgi apparatus. The RER and SER are involved (Fig. 1.24) in the intracellular transport of materials, like proteins and lipids.

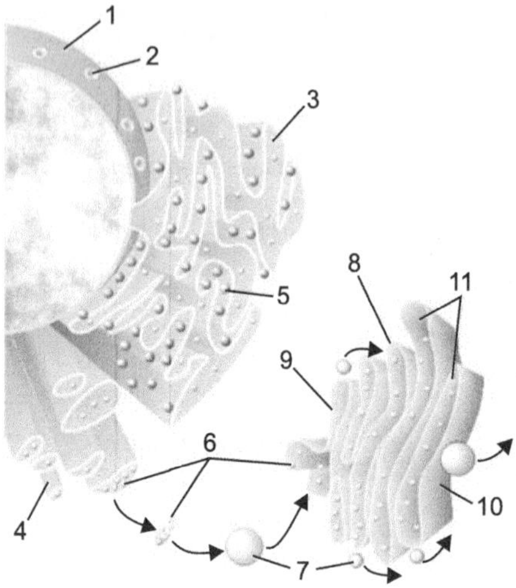

1 = Nuclear membrane, 2 = Pore, 3 = Endoplasmic reticulum, 5 = Ribosome, 6 = Primary vesicle, 8 = Golgi complex.

Fig. 1.24 Endoplasmic reticulum.

Golgi Apparatus

Golgi apparatus (complex) is found in eukaryotic cells and is made up of stacks, called **dictyosomes**, of 4–8 fluid-filled parallel-flattened sacs and vesicles bounded by single membrane, called **cisternae**. These cisternae have continuity with each other and are separated by thin layer of cytoplasm (Fig. 1.25). The apparatus is involved in the **processing** of products transported either directly through ER or by fusion of the vesicles budded off from the ER with the Golgi apparatus. The vesicles budded off from the ER and Golgi apparatus also give rise to the membranous organelles **lysosomes** and **peroxysomes**. The lysosomes (secretary vesicles) ultimately release their contents outside the cell by **exocytosis** and, thus, they also participate in recycling of the plasma membrane by adding new streches of the membranes during this process. Stretches of plasma membrane pinched off by pinocytosis (liquid substances outside the cell get enclosed within membranous vesicles formed by the invagination of the cell membrane) to form pinosomes or by phagocytosis (formed in a similar manner as pinosomes except for containing solid substances) to form phagosomes fuse with the lysosomes for digestion.

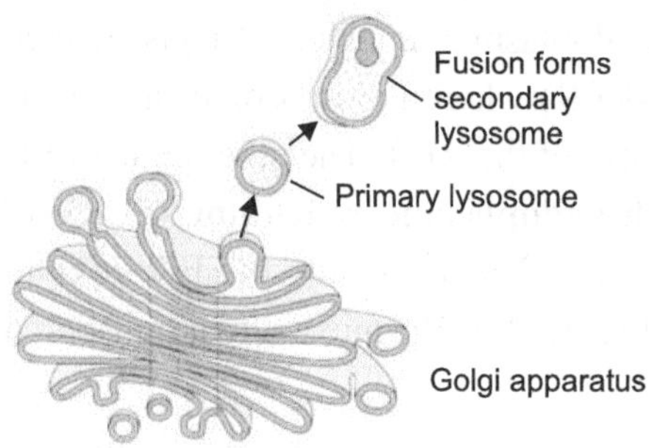

Fig. 1.25 Golgi apparatus.

Ribosomes

They are made up of rRNAs (ribosomal RNAs) and proteins and have about 23 nm average diameter. They have two subunits, one larger and the other smaller and at the time of translation process of protein synthesis, the two subunits bind with the mRNA. The mRNA attached with a number of ribosomes that are involved in the synthesis of the same protein constitutes a **polysome** (polyribosome).

In **prokaryotes** the ribosomes have 66% rRNAs and 34% proteins and are of **70S** (S is the Swedberg unit indicating the sedimentation coefficient) type with 30S smaller (16S rRNA and about 21 proteins) and 50S larger (23S rRNA, 5S rRNA and about 31 proteins) subunits (Fig. 1.26).

In **eukaryotes**, the ribosomes in the cytoplasm contain about 60% rRNA and 40% proteins and are of **80S** type with smaller 40S (18S rRNA and about 33 proteins) and larger 60S (28S rRNA, 5.8S rRNA, 5S rRNA and about 45 proteins) subunits.

These ribosomes are found either freely in the cytoplasm or attached with the endoplasmic reticulum (rough endoplasmic reticulum). In the organelles— mitochondria and chloroplasts (in photosynthetic organisms), the ribosomes are generally of 70S type, as in prokaryotes.

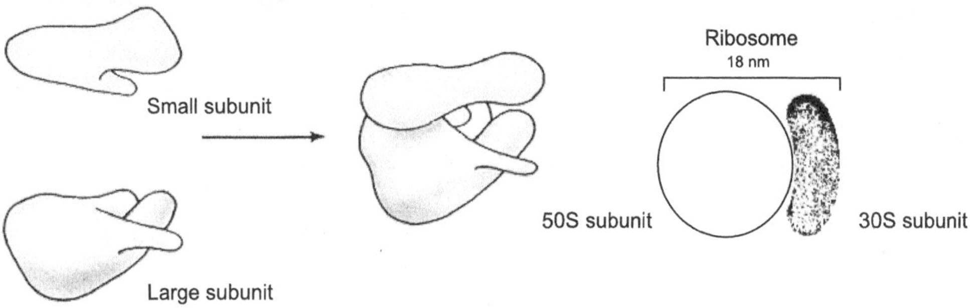

Fig. 1.26 Structure of ribosomes.

Lysosomes

Lysosomes are formed by budding off of the vesicles from Golgi apparatus and SER in **eukaryotes** and their diameter ranges from 0.2–0.8 μm. Their lumen is bounded by a single membrane and contains the **hydrolytic enzymes**, like nucleases, proteases, lipases, glycosidases, etc. that help in the digestion of the food material in the lysosome. Therefore, they are also known as **suicide bags** as their accidental lysis (breakage) may lead to the break down of various cellular constituents. Lysosomes of WBC (white blood cells) help in the defense mechanism. Acrosome of sperms in animals contains lysosomes that help in the penetration of egg during fertilization. Lysosomes are of four types:

(*i*) ***Primary Lysosomes.*** They are formed by pinching off of the membranous vesicles from Golgi apparatus and contain hydrolytic enzymes.

(*ii*) ***Secondary Lysosomes.*** Fusion of a primary lysosome with a phagosome leads to the formation of a secondary lysosome, where the solid food is acted upon by the hydrolytic enzymes and thus, the digested food diffuses into the cytoplasm.

(*iii*) ***Residual Bodies.*** After digestion of the food in secondary lysosome, the vesicle with leftover undigested food is called residual body that passes outward to fuse with the plasma membrane to throw out the waste products outside the cell by exocytosis.

(*iv*) ***Autophagosomes.*** They are formed by the fusion of primary lysosomes with degenerated intracellular organelles and help in the digestion of old and useless organelles. They also help in the digestion of stored food to provide energy during starvation.

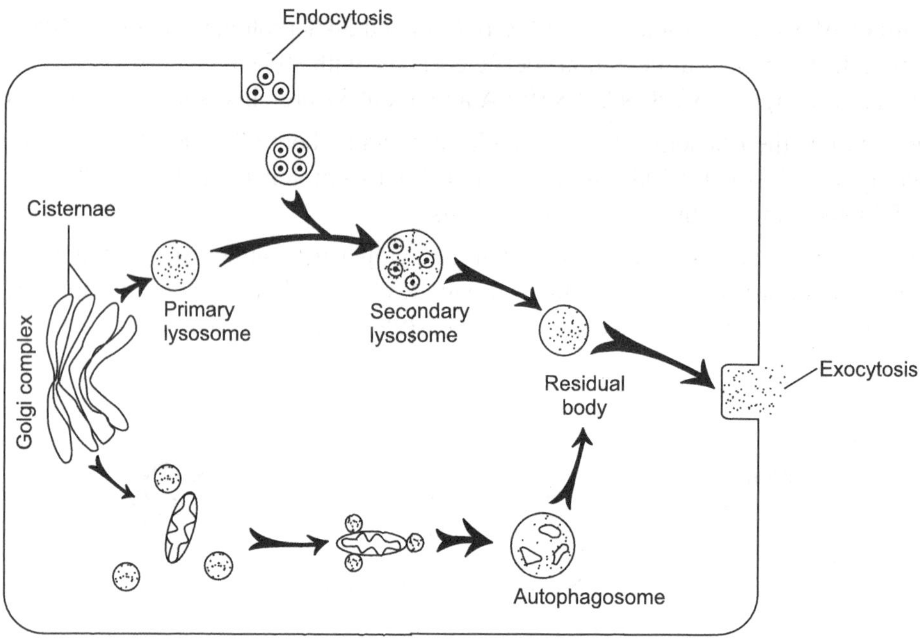

Fig. 1.27 Different types of lysosomes.

Vacuoles

The **eukaryotic** cell contains one or more **central vacuoles** (sap vacuoles) for the storage of nutrients and wastes, rapid exchange of solutes and gases between the cytoplasm and the adjoining fluid. They are more common and larger to restrict the cytoplasm into a thin layer in plants than in animals where they are fewer and smaller in size. It is covered by a semipermeable membrane, called **tonoplast**, to enclose the cell sap that contains minerals, sugars, amino acids, waste products, etc.

Some protozoa contain **contractile vacuoles** that help in osmoregulation and excretion.

Many **prokaryotes** (*e.g.*, many cyanobacteria) contain **air vacuoles** that gives buoyancy and stores metabolic gases.

Cytoskeleton

In eukaryotic cells the cytoskeleton facilitates changes of cell **shape**, regulates distribution and orientation of organelles through **protoplasmic streaming**, produces structures related to cell division like **spindles** (colchicines disrupts the microtubules of spindles to inhibit the movement of chromosomes during Anaphase and, thus, induces polyploidy) and organelles like **centrioles, basal bodies, cilia** and **flagella** and has role in **amoeboidal movement**. It is made up of three types of filaments consisting of proteins:

1. **Microtubules.** They are elongated , unbranched, cylindrical, hollow tubes of about 25 nm diameter occurring singly or in bundles in the cytoplasm. They also form the skeleton of cilia and flagella in motile organisms and of spindle in the dividing cells. Each microtubule encloses a **hollow core** of about 15 nm diameter and the **wall** is made up of 13 longitudinal filaments consisting of helically arranged two types of subunits, α and β, of a globular protein **tubulin.** Fig 1.28 see Appendix.

   They produce spindles and organelles like **centrioles, basal bodies, cilia** and **flagella** and facilitate changes in cell shape, protoplasmic streaming and cell movements along with microfilaments.

2. **Microfilaments.** They are 4–6 nm thin, cylindrical, rod-like filaments forming an extensive network in the cytoplasm and are associated with protoplasmic streaming, cell motion like amoeboid movements, change of cell shape, etc. They are mainly composed of the protein **actin** that may be associated with some filaments of the protein **myosin** and other proteins related to the process of contraction of microfilaments. Motility of non-muscle cells involves the interaction of actin and myosin in the microfilaments.

3. **Intermediate Filaments.** They are tough and durable, 8–10 nm thick, rope-like protein fibres (Fig. 1.29). One type of intermediate filament forms a network below the inner nuclear membrane, whereas, other types extends across the cytoplasm and gives mechanical strength to the cell.

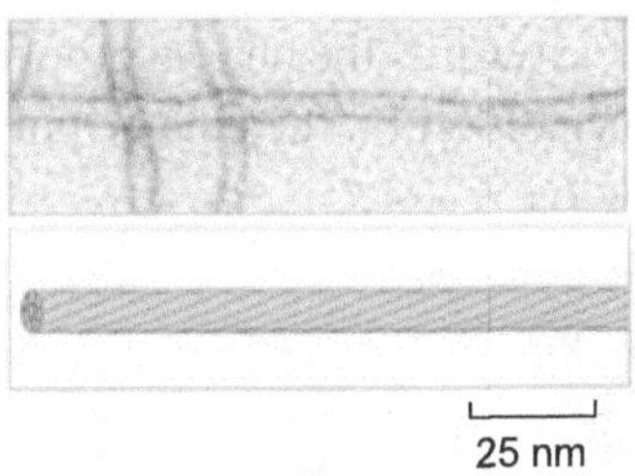

Fig. 1.29 Intermediate filaments.

Centrioles

Centrioles are generally found in **pairs**, arranged at **right angles** to each other near one pole of the nucleus only in (eukaryotes) animal cells and flagellated plant cells (*e.g.*, motile algae, motile sex cells of some primitive plants) (Fig. 1.30). During cell division, the two centrioles migrate to two opposite poles to form spindles. Each centriole is 0.5 μm long and 0.2 μm wide cylindrical, non-membranous structure open at both the ends. Their wall is made up of 9 groups (tilted at an angle of 40°) of triplets of **microtubules** that are arranged in a circle (**9 + 0** arrangement). Each microtubule is made up of the **tubulin** protein. The triplets of microtubules are connected to a proteinaceous hub in the centre through radial proteinaceous spokes. They are responsible for the formation of **spindles** at the time of cell division (plant cells lack centrioles and the spindles are formed without their aid) and, **cilia** (the centrioles divide many times and orient themselves along the surface of the cell to form numerous basal bodies of cilia) and **flagella** (centrioles move to the periphery of the cell to become basal body of the flagella).

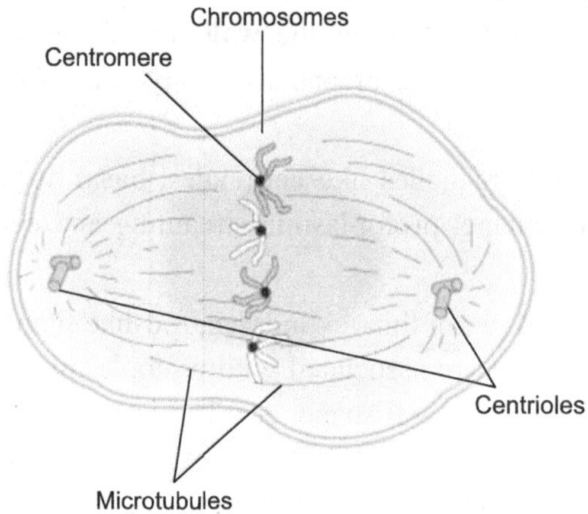

Fig. 1.30 Two pairs of centrioles in a dividing cell (Metaphase).

Cilia and Flagella

Cilia. In **eukaryotes, cilia** are 3-10 μm long and 0.5 μm thick, hair-like, numerous outgrowths that extend into the surrounding medium and move in a rowing manner for locomotion of some isolated cells, *e.g.*, some protozoa like *Paramoecium*. Cilia of tracheal and bronchial cells of respiratory tract move the inhaled dust particles towards the pharynx (outwards) to minimize their entry into the lungs (cigarette smoking may reduce or inhibit the ciliary movements of these cells) (Fig. 1.31).

Fig. 1.31 Cilia at the cell surface.

Flagella. It have the same structure as that of cilia except for having larger length of about 100–200 μm and they are one or few in number in a cell. These are also used for the locomotion of cells by the wave-like motion of the flagella (the organism is driven in the opposite direction to that of the location of flagellum), *e.g.*, flagellar movement of spermatozoa propel them within the fluid medium of the female reproductive tract (Fig. 1.32).

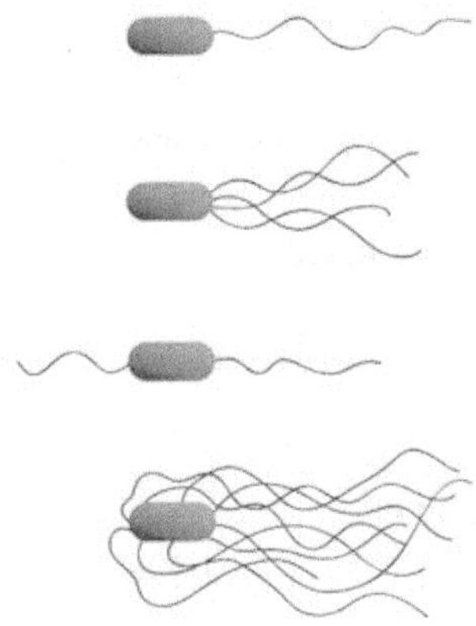

Fig. 1.32 Differential distribution of flagella over the cell surface.

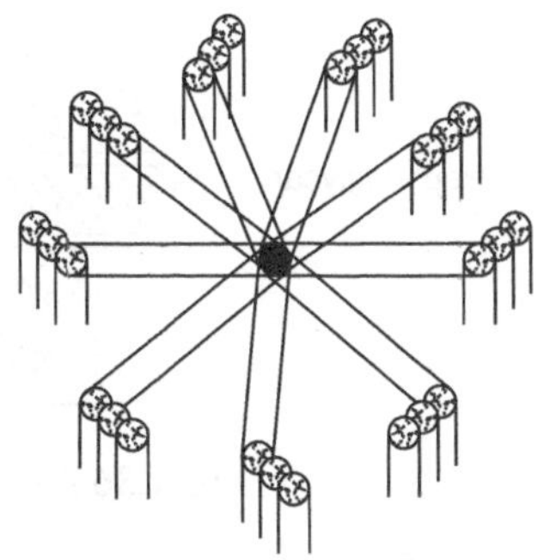

Fig. 1.33 Basal body of the eukaryotic flagellum (9 + 0 arrangement of microtubules).

The **basal bodies** (9+0 arrangement) are located at the base of cilia and flagella, and arise from the centrioles (in eukaryotes). Their structure is similar to that of centrioles (Fig. 1.33). The axis of cilia and flagella, called **axoneme**, is surrounded by a membrane (that is the extension of plasma membrane) and consists of microtubules that are arranged as 9 doublets at the periphery of a circle with 2 single microtubules in the centre (9 + 2 arrangement). The sliding movement of microtubules (that requires energy in the form of ATP) causes the movement of cilia and flagella (see Appendix Fig. 1.34).

In **prokaryotes**, the motile bacteria possess flagella emerging from the cell wall and are not covered within any membrane, as in eukaryotes. Their number varies from one (monotrichous) to two to many where the flagella may be arranged in a cluster at one pole (lophotrichous) or two clusters at two poles (amphitrichous) or all over the surface (peritrichous).

The prokaryotic flagellum has three basic parts:

(*i*) ***Basal Body:*** It is made up of a small rod with 2 (in G^+ bacteria) (Fig. 1.35) or 4 (in G^- bacteria) (Fig. 1.36) rings and helps in the attachment of flagellum to the cell wall and plasma membrane.

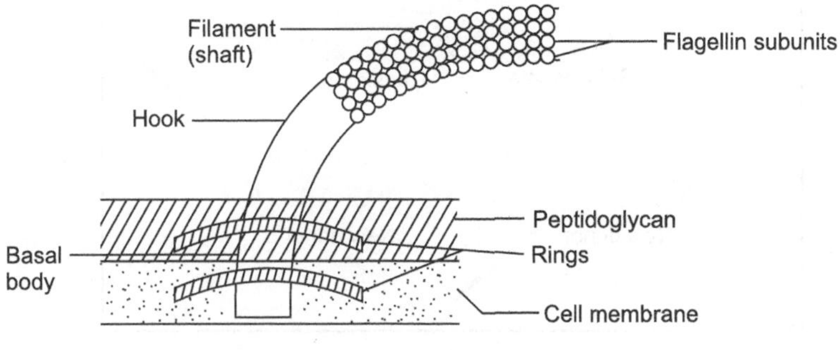

Fig. 1.35 A Gram positive bacterial flagellum.

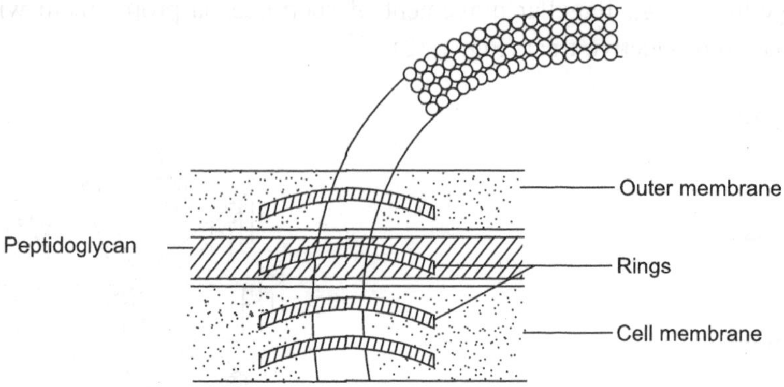

Fig. 1.36 A Gram negative bacterial flagellum.

(*ii*) ***Hook:*** It emerges outside the cell wall and is made up of different proteins.

(*iii*) ***Shaft:*** A long shaft (filament), extending from the hook is comprising of the globular protein flagellin, that are arranged helically around a hollow core.

Pili

In certain **G^- bacteria**, pili (fimbriae) are short (0.2-2.0 μm long), thin, straight, hair-like appendages found at poles or all over the surface of cell wall. They originate from the cytoplasm and

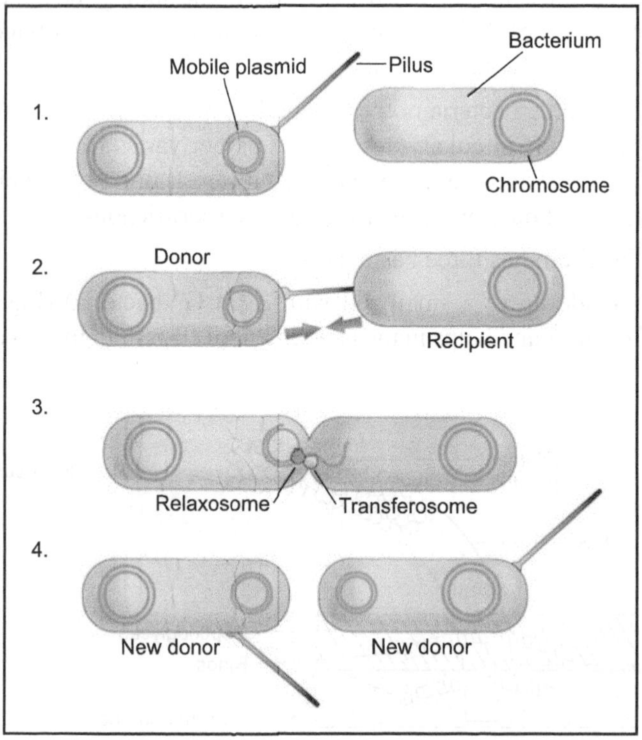

Fig. 1.37 Role of sex pilus in the process of conjugation in bacteria.

are not involved with motility of the cells. Their formation is generally governed by plasmids and is made up of a protein **pilin**. The sex pili help in the conjugation of two bacteria (one having the F factor or F^+ and the other lacking it or F^-) where they join together through their sex pili to form a conjugation tube. The fertility factor (F factor that is a plasmid) passes from F^+ to F^- bacterium through the conjugation tube (Fig. 1.37) and (Fig. 1.38 see Appendix). In other bacteria, the pili imparting adhesive properties help in the attachment of the bacterium to a substrate or with other cells to form aggregates.

Peroxysomes

They are spherical, single membrane bound 0.5–1.0 µm diameter structures found in eukaryotic cells and contain the enzymes for peroxide synthesis as well as catalase for metabolism of H_2O_2. In animals, they are involved in the metabolism of lipids, whereas in plants they perform photorespiration in association with mitochondria and chloroplast in which the carbohydrates and O_2 are used (in photorespiration) in the presence of sunlight to produce CO_2.

Glyoxysomes

They are found only in fat-rich parts of plants (*e.g.*, common in germinating oil seeds) and contain enzymes for the **β-oxidation of fatty acids** to produce acetyl coenzyme-A that is metabolized in glyoxylate pathway and for tryglyceride metabolism to produce carbohydrates.

Some important differences between prokaryotic and eukaryotic cells

S. No.	*Prokaryotic Cell*	*Eukaryotic Cell*
1.	**Nuclear Properties**	
	(*i*) An organized nucleus is absent, *i.e.*,	(*i*) A well organized **nucleus** with:
	(*ii*) Single, circular, coiled, naked DNA molecule, called **nucleoid,** that is not found to be associated with any histone and non-histone proteins, lies in the cytoplasm.	(*ii*) Many linear, coiled DNA molecules, called chromosomes, are found to be associated with **histones** and **non-histones.**
	(*iii*) Nuclear envelope, nucleoplasm and nucleolus are absent.	(*iii*) **Nuclear membrane, nucleoplasm** and **nucleolus** are present.
	(*iv*) DNA is attached with **mesosome** that is an invagination of the cell membrane.	(*iv*) Mesosome is absent.
2.	**Cytoplasmic Properties**	
	(*i*) Cell organelles like mitochondria, chloroplast, endoplasmic reticulum, Golgi apparatus, central vacuole, etc, are not found in the cells. In some prokaryotes	(*i*) Cell **organelles** mitochondria, like chloroplasts, **endoplasmic reticulum, Golgi apparatus, central vacuoles,** etc. are present.

	gas vacuoles are found that provide buoyancy and help in floating of the cells in aquatic environment.	
	(*ii*) Respiratory enzymes and other components are found in the cell membrane.	(*ii*) Double membrane bound **mitochondria** are meant for aerobic respiration.
	(*iii*) In photosynthetic prokaryotes, the photosynthetic components are found in the membranous sacs or tubes, called thylakoids.	(*iii*) Double membrane bound **chloroplasts** with granal and stromal thylakoids perform the photosynthesis in photosynthetic eukaryotes.
	(*iv*) Microtubules and microfilaments are absent.	(*iv*) The cytoplasm contains a network of **microtubules** and **microfilaments** that are made up of proteins and function as the cytoskeleton.
	(*v*) Centrioles are absent.	(*v*) In animal and motile cells a pair of **centrioles,** arranged right angles to each other and consisting of microtubules, are found at one side of the nucleus that produce spindle fibres during cell division.
	(*vi*) **Ribosomes** arc of **70S** type.	(*vi*) Ribosomes are of **80S** type in the cytoplasm, whereas, the cell organelles have generally **70S** type of ribosomes.
3.	**Cytoplasmic Movement** Cyclosis is absent.	**Cyclosis,** the cytoplasmic streaming, is present in eukaryotic cells that helps in the homogenous distribution of various components and movement of organelles.
4.	**Cell Wall** Cell wall if present, is generally made up of **peptidoglycan.**	Cell wall, if present, has different composition in different eukaryotes, e.g. **cellulose,** hemicellulose, pectin, **chitin** (in fungi), etc.
5.	**Flagella** In motile prokaryotes **flagella,** if present, are single-stranded consisting of flagellin protein. Here, the flagellum is not enveloped within any membrane.	In motile cells flagella, if present, are 11-stranded **(9+2 arrangement)** in the shaft portion and 9-stranded **(9+0 arrangement)** in the basal portion, where each strand is made up of α- and β-tubulin proteins. Moreover, the flagellum is **enveloped** in a membrane.

6.	**Cell Division** Mitosis and meiosis are absent and the cell divides by simple **binary fission** method in which the DNA replicates into two that are equally distributed into two daughter cells with the help of mesosome, as the spindles are absent.	Cell divides by **mitosis** and for sexual reproduction by **meiosis**, where the spindle fibres help in the separation of chromosomes into the daughter cells.
7.	**Protein Synthesis** (*i*) Transcription (DNA to mRNA rRNA or tRNA) and translation (mRNA to protein) occur simultaneously in the cytoplasm. (*ii*) During transcription, a gene in DNA leads to the formation of **mRNA** where the number of nucleotides is exactly same as in the corresponding gene in the DNA.	(*i*) Here, the transcription occurs in the nucleus, whereas, the translation in the cytoplasm. (*ii*) In eukaryotes many genes in DNA are split genes containing exons and introns, where the transcription leads to the production of **HnRNA** that is quite longer than the actual mRNA. Thus, the mRNA is produced after the processing of HnRNA where the intervening introns and other unwanted nucleotide sequences from the 5' and 3' ends are removed. Moreover, a cap of 5-methyl guonasine at the 5' end and a tail of poly A at the 3' end of mRNA is also added.
8.	**Sexual Reproduction** True sexual reproduction is lacking, as gametes are not formed in these organisms. Though, some kind of **parasexuality** through transformation, conjugation or transduction may occur that involves exchange of some genetic material.	Sexually reproducing eukaryotes produce gametes that unite to form the zygote (**true sexual reproduction**).

Important differences between G^+ and G^- bacteria:

S. No.	*G^+ Bacteria*	*G^- Bacteria*
1.	During Gram's staining* they retain **crystal violet** colour after the treatment with alcohol.	They do not retain crystal violet after alcohol treatment and get decolourised (as the outer membrane containing phospholipids gets degraded and the violet dye passes out of the cells). They are counter stained with reddish safranin to observe under the microscope.

2.	Outer membrane in the cell wall is absent.	**Outer membrane**, consisting of phospholipids and proteins, is found above the peptidoglycan.
3.	**Peptidoglycan** in the cell wall is **thicker** (20–80 nm), therefore, susceptibility to antibiotics (*e.g.*, penicillin) that act upon the peptidoglycan is high. Pentaglycine links the two adjacent tetrapeptides in peptidoglycan.	Peptidoglycan is comparatively quite **thinner** (8-12 nm) and thus, the susceptibility to penicillin-like antibiotics is low.Pentaglycine in peptidoglycan is absent.
4.	Cell wall has **low lipid** content due to absence of outer membrane.	Cell wall has **high lipid** because of the presence of outer membrane.
5.	Periplasmic space is absent.	A **periplasmic space** is present between outer membrane and plasma membrane.
6.	**Flagella**, if present, contain **two rings** in the basal body.	Flagella, if present, have **four rings** in basal body.
7.	Cell wall usually contains **teichoic acid**.	Teichoic acid is absent in the cell wall.
8.	These bacteria are mostly **spore-forming**.	They are generally non-sporous.
9.	Pili over the cell surface are absent.	Some G^- bacteria have **pili** (made up of protein pilin) over the cell wall that help in conjugation with other cell and attachment with some substratum or with other cells in aggregates.

***Gram's staining:** Different steps of Gram's staining are:

(*i*) A thin **smear** of the bacterial cells at the slide is made.

(*ii*) The smear is stained with **crystal violet** dye.

(*iii*) Now, it is treated with a mordant (**KI**), which fixes the crystal violet colour in the cells.

(*iv*) The smear is washed with 95% **alcohol**.

(*v*) After the alcohol treatment, which dissolves the lipids present in cell wall, the G^+ bacteria retain the violet colour (see Appendix Fig. 1.39), whereas, the G^- lose it and become colourless.

(*vi*) The decolourised cells of G^- bacteria are again stained with reddish coloured **safranin** dye (see Appendix Fig. 1.40).

Bacteria

Bacteria are pryokaryotic organisms devoid of any organized nucleus and membrane bound organelles like mitochondria, chloroplast, endoplasmic reticulum, central vacuole, etc. **Whittaker**

(1969) has grouped all autotrophic as well as heterotrophic prokaryotic organisms into the group Monera and according to *Bergey's Manual of Systematic Bacteriology*, the main groups of prokaryotic organisms are:

1. Bacteria (Eubacteria): They are unicellular microorganisms and have appeared on the Earth after the evolution of archaebacteria. They may be:

(*a*) Photosynthetic (purple or green coloured), where the main photosynthetic pigments are **bacteriochlorophylls** and they **do not evolve any O_2** during photosynthesis as the electron source are H_2S and other compounds (*e.g.*, *Rhodospirillum*, *Chlorobium*, etc.),

(*b*) Chemoautotrophic, deriving their energy from chemical reactions (*e.g.*, *Nitrsomonas*, *Nitrobacter* that oxidize NH_4^+ to NO_2^- and NO_2^- to NO_3^-, respectively).

(*c*) Heterotrophic (*e.g.*, *Escherichia coli*).

2. Archaebacteria: They are more primitive and ancient bacteria generally living in harsher environments, *e.g.*, *Methanobacterium* (strict anaerobic methanogenic), *Halobacterium* (extremely halophytic), *Thermococcus* (SO_4^{2-} metabolizing thermophilic)

3. Actinomycetes: They are filamentous bacteria, *e.g.*, *Streptomyces*, *Frankia*, etc.

4. Mycoplasma: These bacteria are devoid of a cell wall and are pleuromorphic in shape (variable shapes) and generally form colonies, *e.g.*, *Mycoplasma*.

5. Cyanobacteria: They are commonly known as blue-green algae and their structure ranges from unicellular, filamentous (branched or unbranched) to colonial organization. They are photosynthetic and have photosynthetic pigments **chlorophyll *a***, **carotenoids**, **xanthophylls** and blue coloured **phycocyanin** with or without red **phycoerythrin**. They **evolve O_2** during photosynthesis as the source of electron is H_2O.

Bacteria are **unicellular** microorganisms and were first discovered by Antony van Leeuwenhoek (1677) in pond water and the tarter scrapped from the teeth. They are universally distributed in water, soil as well as in air in moderate (**mesophilic**), very cold (**psychrophilic,** upto –22°C), hot (**thermophilic,** upto 90°C, *e.g.*, *Thermus aquaticus*), acidic (**acidophilic** upto pH 1), alkaline (**alkalophilic,** upto pH 11) or saline (**halophilic**, *e.g.*, *Halobacterium*) environments. They may be **free-living,** (*Azotobacter*, *E. coli*, etc.) or **symbiotic** (*e.g.*, *Rhizobium*, *Bradyrhizobium*, etc.), **photoautotrophic,** (*e.g.*, *Rhodospirillum*, *Chlorobium*, etc.), **chemoautotrophic,** (*Nitrosomonas*, *Nitrobacter*, *Thiobacillus*, etc.), heterotrophic **saprophytic,** (*e.g.*, *E.coli*, *Bacillus*, etc.), heterotrophic **parasitic,** (*e.g.*, *Vibrio cholerae*, *Diplococcus pneumonae*, etc.) in nature. Many bacteria can form spores that help in their survival under unfavourable conditions of temperature, pH, water availability, etc.

According to the requirements and non-requirements of O_2, bacteria are also divided into following:

(*i*) **Aerobic,** (*e.g.*, *Azotobacter*, *Pseudomonas*, etc.) requiring O_2 for respiration,

(*ii*) **Anaerobic,** (*e.g.*, *Clostridium*) where the O_2 becomes toxic and they grow in the absence of O_2.

(*iii*) **Facultative,** aerobic or anaerobic, that do not require O_2 for the growth but can use it whenever available in the environment, (*e.g.*, *Shigella*, *Salmonella*, etc.) or microaerophilic requiring low levels of O_2 for growth.

The bacteria may have different shapes (Fig. 1.41):

(*i*) **Cocci,** that are spherical or ovoid cells occurring singly or in groups (*e.g.*, *Staphylococcus*) or chains (*e.g.*, *Streptococcus*), groups of 2 (*e.g.*, *Diplococcus pneumoniae*), 4 or many.

(*ii*) **Bacilli,** are rod-like cells found singly or in groups, *e.g.*, *Bacillus*.

(*iii*) **Spirillium** are spirally coiled cells, *e.g.*, *Azospirillium*.

(*iv*) **Vibrio** are comma-shaped bacteria, *e.g.*, *Vibrio cholerae* causing cholera.

(*v*) **Mycelial** bacteria are filamentous and aseptate, *e.g.*, *Actinomycetes*, *Streptomyces*.

(*vi*) **Pleuromorphic** bacteria can occur in many shapes at different stages, e.g. *Rhizobium*.

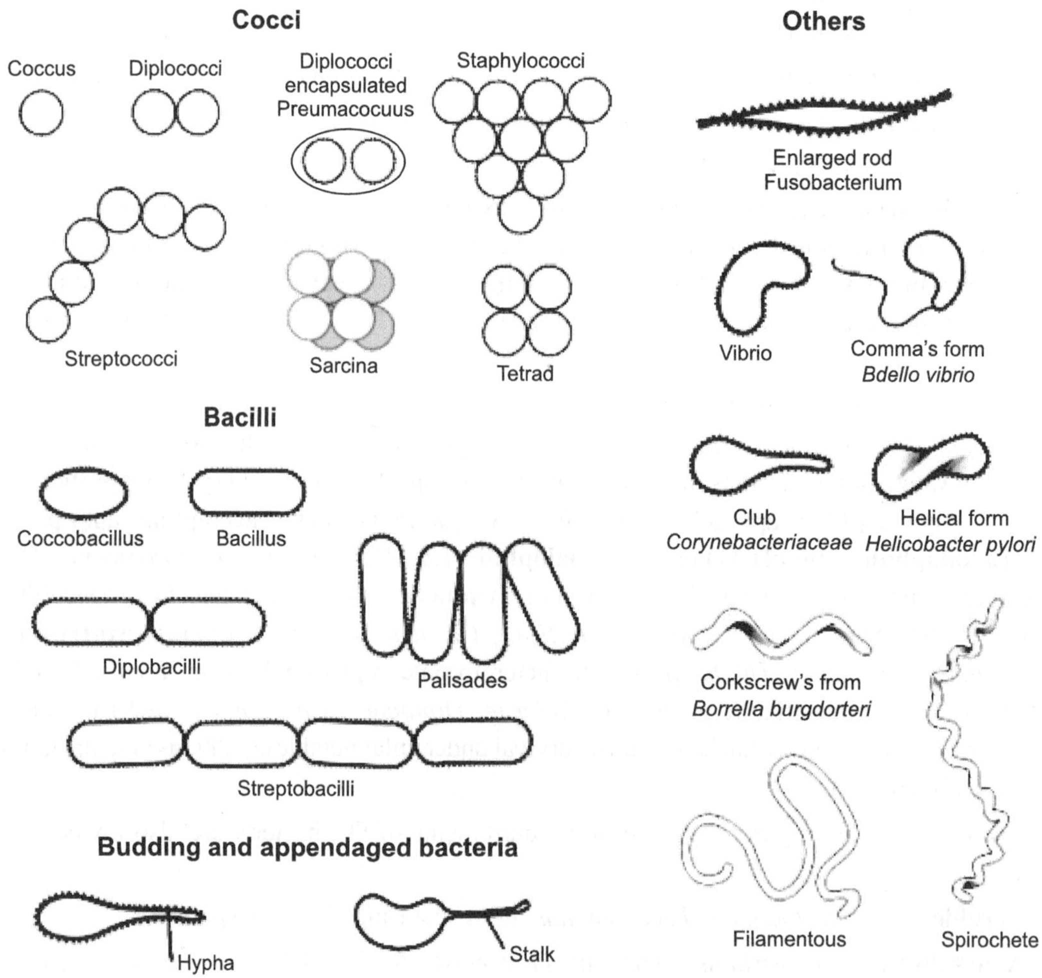

Fig. 1.41 Different shapes of bacterial cells.

The bacteria may or may not have flagella that help in the motility of cells (Fig. 1.32).

The cell wall of most of the bacteria is made up of **peptidoglycan** and on the basis of Gram's staining they have been classified into G^+ and G^- bacteria. In Gram's staining the bacteria are first stained violet with the dye, crystal violet. Then the dye is fixed in the bacterial cells by treatment with 0.5% KI solution. When the cells are washed with absolute alcohol, some bacteria having low lipid content in their cell wall retain the crystal violet dye to remain violet and are known as **G^+ bacteria** (*e.g.*, *Bacillus subtilis*, *Lactobacillus*, etc); whereas other bacteria are classified as **G^- bacteria** (*e.g.*, *Escherichia coli*, *Rhizobium*, etc.) as they have high lipid content in their cell wall due to presence of outer cell membrane and they lose the dye to become colourless and are counter stained with safranin to attain reddish colour.

They reproduce asexually by **binary fission** (Fig. 1.42) method and true sexual reproduction,

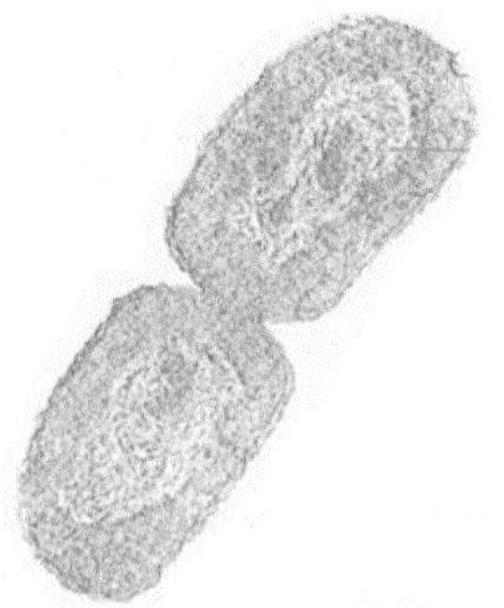

Fig. 1.42 Division of cell by binary fission method in *E. coli.*

by the union of male and female gametes is lacking. Some kind of genetic exchange may takes place by:

(*i*) *Conjugation,* where the free plasmid or plasmid inserted in bacterial chromosome is transferred from donor to the recipient cell through a conjugation tube (Fig. 1.37).

(*ii*) *Transduction,* where during infection by bacteriophages a small segment of the bacterial chromosome may get incorporated into the progeny phage genetic material and transferred to another host bacterial cell during the fresh infection by these progeny phages (see Appendix Fig. 1.43).

(*iii*) *Transformation,* in which the free DNA segments present in the external medium may be taken up by the bacteria and get integrated into their chromosome. The first experiment on transformation was conducted by **Avery, Mc Cleod** and **Mc Carthy** (1944) in the bacterium *Diplococcus pneumoniae.*

Though, most of the bacteria are harmful for human beings as they spoil the food (food poisining by *Clostridium botulinum*, *Staphylococcus aureus*, etc.) and drinking water, cause various diseases to humans (*e.g.*, *Staphylococcus aureus*, *Salmonella typhi*, *Vibrio cholerae*, *Shigella dysenteriae*, *Mycobacterium tuberculorisis*, etc.), animals (*e.g.*, *Bacillus anthracis*) and plants (*e.g.*, *Xanthomonas citrii*, *Agrobacterium tumefaciens*, etc.). Many bacteria (*e.g.*, denitrifying bacteria) may lead to loss of fertility of soils by converting fixed NH_4^+ back into the inert N_2 gas that finally escapes into the atmosphere (*e.g.*, *Thiobacillus denitrificans*). Many bacteria are **economically important** and have been used successfully in various fields like industries, agriculture, **medicines** (vaccines, antibiotics, interferons, insulin, growth hormone, etc.), **sewage treatment**, **bio-mining** (*e.g.*, Cu, Au, Fe, U, etc. by *Thiobacillus ferroxidans*), **fuels** (*e.g.*, bio gas that mainly contains CH_4 gas by the activity of the anaerobic methanogens like *Methanobacterium*), etc.

In **industry,** they are used in food industry (*e.g.*, dairy products, pickles, vinegar, lactic and citric acids, idli, etc.), anaerobic retting of fibres (like jute, flax, hemp, etc. by the bacteria *Clostridium*, *Pseudomomas fluorescence*, etc.), removal of hairs from the hides in leather industry, commercial production of various useful chemicals (like enzymes, amino acids, organic acids, etc.), etc.

In **agriculture,** many bacteria have played important role in reclamation of alkaline and saline soils, composting, nutrient cycling and, thus, maintaining the fertility of soil by making available many nutrients, like NO_3^- (*e.g.*, *Rhizobium*, *Azotobacter*, etc. convert the non-utilizable atmospheric N_2 into NH_4^+, *Nitromonas*, etc, oxidize NH_4^+ into NO_2^-, *Nitrobacter*, etc. further oxidize the NO_2^- to NO_3^-, etc.), HPO_4^{2-} (rock phosphate solubilising bacteria like *Pseudomonas*, *Bacillus*, etc.), SO_4^{2-}, etc. for the growth of crop plants. Many bacteria are commercially produced as biofertilizer, like *Rhizobium*, *Azotobacter*, *Azospirillum*, *Pseudomonas*, etc. that replace the harmful chemical fertilizers to some extent.

VIRUSES

Viruses are 20–300 nm, **acellular** living organisms as they are generally made up of a core of genetic material (nucleoid) of only one type of **nucleic acid**, either DNA (mostly double stranded but in some cases single stranded) or RNA (mostly single stranded but in some viruses double stranded), that is surrounded by a protective **protein** coat (capsid) and lack the cellular structures like cytoplasm, various organelles, etc. The capsid is made up of capsomeres the number of which varies from 12 (bacteriophage ϕX174) to 2130 (plant virus TMV). They may contain a few enmzymes, *e.g.*, **lysozyme** that helps in the penetration of host cell wall, **reverse transcriptase** in some RNA viruses (*e.g.*, retroviruses with single stranded RNA) to produce temporary viral DNA that integrates into the host DNA and is required for the formation of genomic RNA as well as the mRNA. In some viruses additional envelope may be present around the protein coat.

They are obligate (strictly) **parasites** and require a host cell for their **multiplication** where they neither divide nor grow in size and their various components are formed independently that assemble together to produce the daughter viruses. The viruses have very little water and they neither respire nor have any independent metabolism (like synthesis of proteins, nucleic acids, etc.) and the antibiotics are not effective against them. Outside the host cell (in cell free environment) they are found in the inactive state and become active only after entering their host cell, where they utilize the synthetic machinery of the host cell. Therefore, they cannot be cultured in cell-free medium (can be cultured only in the presence of their host cells).

The viruses have been classified in many ways depending upon their specificity to hosts, their structure, nature of nucleic acid, etc. On the basis of the specificity to their hosts the viruses have been classified into different groups:

Bacteriophages

They infect **bacteria**, like *E. coli*, and contain generally double stranded DNA (*e.g.*, T2, T4 bacteriophages) or rarely single stranded DNA (bacteriophage φX174 and M13 with circular DNA) or single stranded RNA (bacteriophages R-17, MS-2) as their genome..

Their body generally has a polyhedral (commonly icosahedral with 20 faces, but may be tetrahedral with 4 faces or dodecahedral with 12 faces) **head** (bacteriophages φX174, λ, T1, T5) that may be associated with a helical **tail** (*e.g.*, bacteriphages T2, T4) or it may be **filamentous** (bacteriophage M13) in shape.

The T2 **bacteriophage** infects *E. coli* and has an **icosahedral** (20-sided) **head** (with 95 nm length and 65 nm breadth) and a helical **tail** (115 nm long and 17 nm wide). Around 2000 capsomeres of head enclose a **double stranded DNA** molecule as the nucleoid. The tail has a contractile sheath consisting of 144 capsids that are arranged helically and enclose a core tube. This tail contains the enzyme lysozyme that helps in the penetration of the host cell wall through which the phage DNA is passed into the host cell. A proteinaceous **neck** with circular collar is found between the head and the tail. The basal part of the tail is occupied with a **base plate** that is made up of protein that bears 6 **spikes** (tail pins) and 6 long tail fibres at the corners. The **tail fibres** remain folded in the free form but spread out at the time of infection to help the attachment of the bacteriophage at the cell wall of the bacterium (Fig. 1.44).

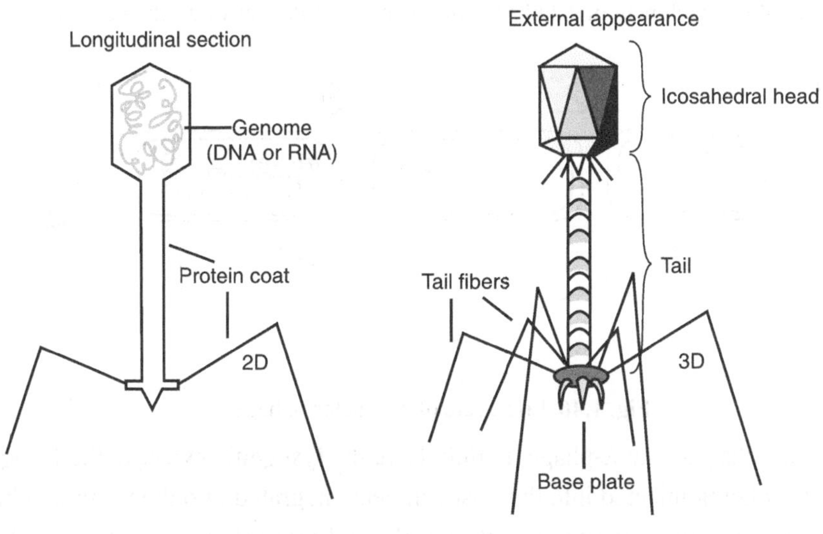

Fig. 1.44 Structure of T2 bacteriophage.

For **multiplication**, the T2 bacteriophage **lands** on the host cell wall with the help of their tail fibres and then the tail pins make contact with the cell wall. Ultimately, when the tail comes in contact with the bacterial cell, the tail sheath contracts and the tail core tube passes through a central hole in the base plate into the cell wall and the **DNA is injected** into the host cell (Fig. 1.45). The **bacterial DNA is then broken down** by the phage coded nucleases and the phage **utilizes the**

host's protein synthesis machinery (ribosomes, RNA polymerase, etc.) to synthesise its own **proteins**. The injected phage **DNA replicates** to produce many copies of the phage genome that gets incorporated into the newly synthesized protein capsids to produce **progeny phages**. The progeny phages are **released** by sudden lysis (bursting) of the host cell that occurs within 20–60 min of infection. The individual progeny phages may again infect a fresh host cell and this multiplication cycle is known as **lytic cycle** (Fig. 1.46).

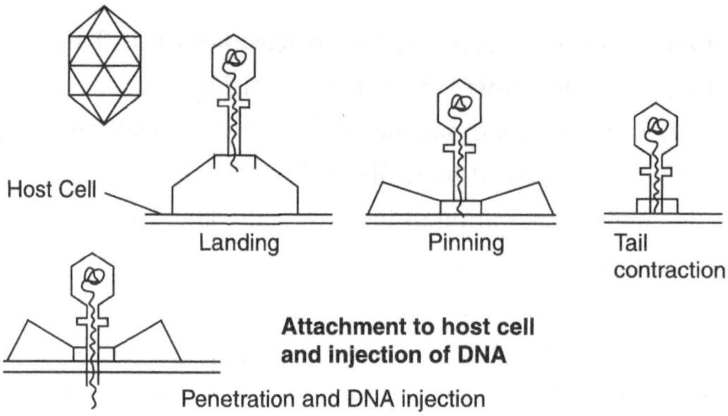

Fig. 1.45 Attachment of T2 phage and injection of its DNA into the host cell.

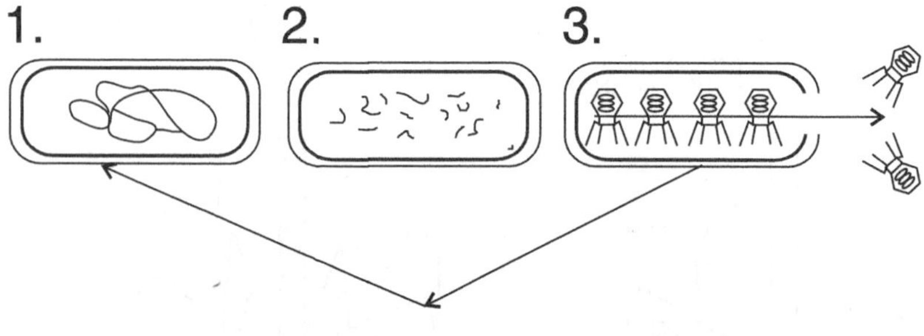

Fig. 1.46 Life cycle of T2 bacteriophage.

Some bacteriophages, like λ-phage, mainly have the **lysogenic cycle.** In the **lysogenic cycle,** the phage DNA after being injected into the host cell gets integrated into the bacterial chromosomal DNA (and in this form the phage is known as the prophage that is noninfectious) that replicates along with the host chromosome. In this form the phage is called **temperate phage** and the host bacterium is known as **lysogenic bacterium**. Sometimes, the prophage enters the **lytic cycle** when the prophage accidentally gets detached from the host chromosome to produce and release progeny phages by bursting the host cell (Fig. 1.47).

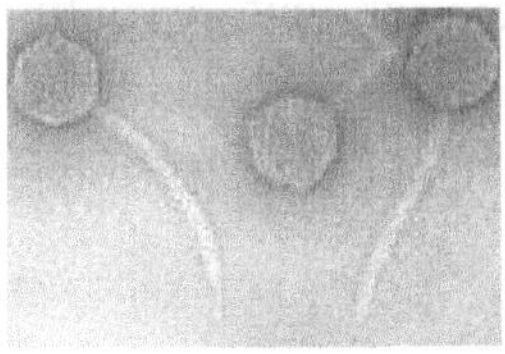

λ phage

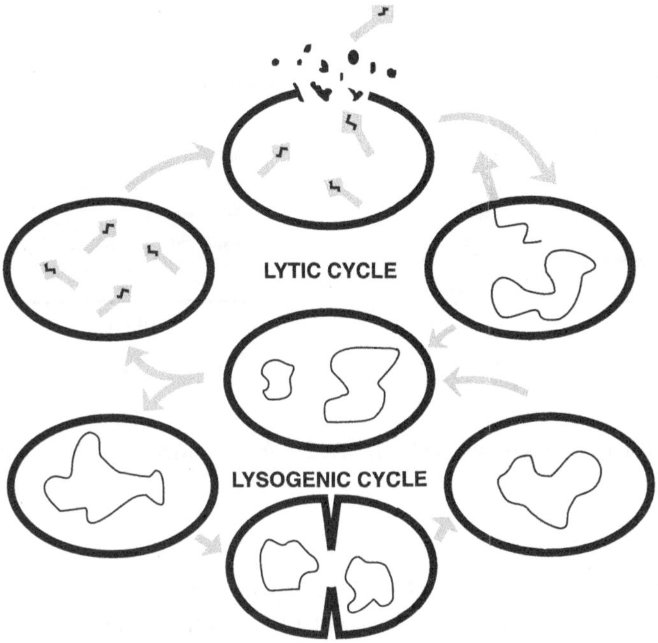

Fig. 1.47 Structure (top) and life cycle of λ-phage.

Cyanophages

They infect cyanobacteria (blue-green algae) and most of them have the structure similar to the bacteriophages with icsahedral head, with or without a tail. Examples of some of the cyanophages are LPP-1 (infecting *Lynbya*, *Plectonema* and *Phormidium*), SM-1 (hosts *Synechococcus* and *Microcystis*), N-1 (infects *Nostoc muscorum*), etc.

Plant Viruses

Plant viruses usually have single stranded RNA as their nucleoid (*e.g.*, TMV), but some have double stranded DNA as the genetic material, *e.g.*, CaMV (Cauliflower Mosaic Virus), DMV (Dahlia Mosaic Virus).

Tobacco Mosaic Virus (TMV) is rod shaped (300 nm long and 17.5 nm diameter) with spirally arranged capsomeres (2,130 protein sub-units, each turn contains 16 capsomeres) enclosing a tubular core (4 nm wide) that contains single stranded helically coiled RNA (6,400 nucleotides, each turn having 49 nucleotides) as the nucleoid. It infects tobacco and other members of Solanaceae family (Fig. 1.48).

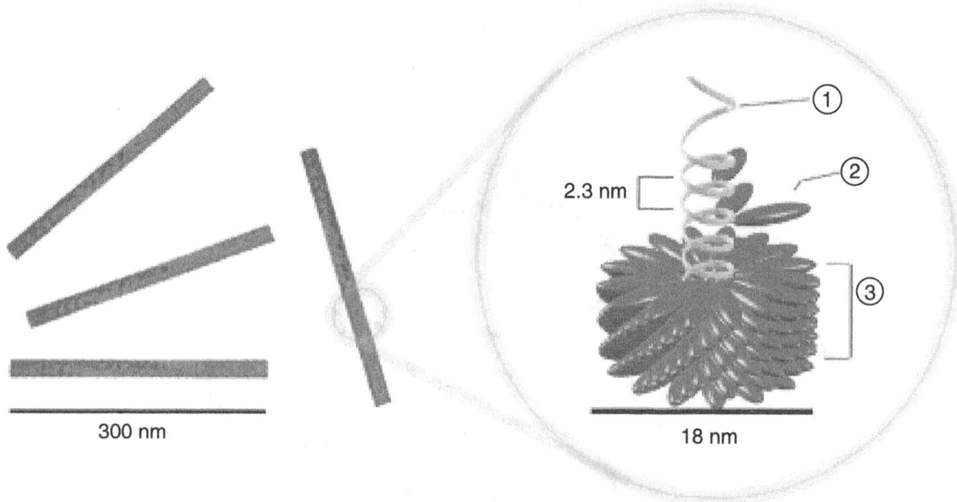

Fig. 1.48 Tobacco mosaic virus (1 = RNA, 2 = A capsomere, 3 = Protein coat).

Animal Viruses

Some viruses, in addition to nucleic acid and proteins, also contain a membranous envelope (*e.g.*, Herpes virus, mumps virus, influenza virus, etc.), consisting of proteins, lipids and carbohydrates, which may bear spikes (projections). It is generally derived from their host cells. They commonly contain **single stranded RNA** (*e.g.,* foot and mouth disease virus that is icosahedral in shape) as the nucleoid. Retroviruses (*e.g.,* Rous sarcome virus that is spherical in shape with an envelope, mouse leukemia virus) contain single stranded RNA as their genetic material and the enzyme **reverse transcriptase** that helps in the formation of DNA from the genetic RNA, that in turn gives rise to genetic RNA as well as the mRNA (Fig. 1.49 and 1.50)

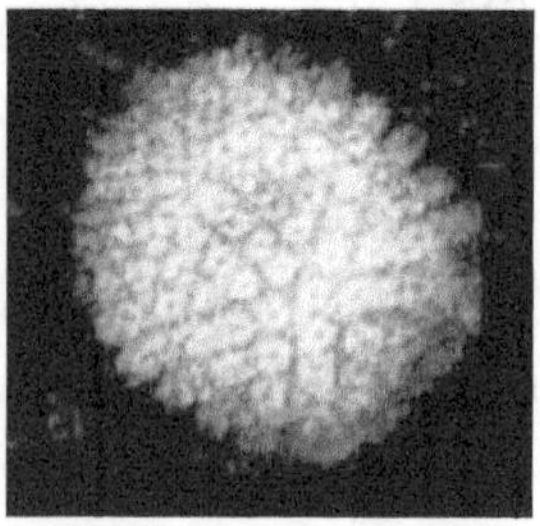

(*a*)

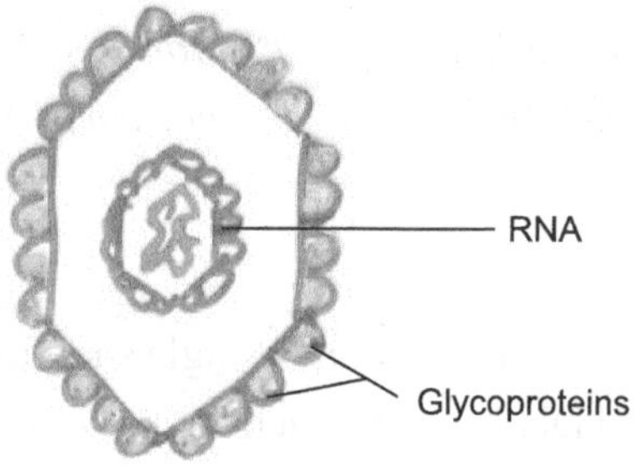

(*b*)

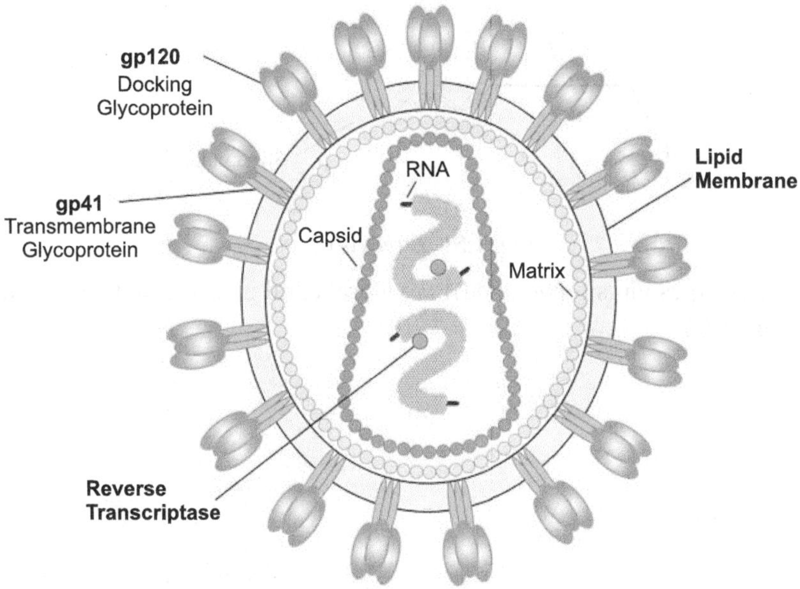

(*c*)

Fig. 1.49 Retroviruses

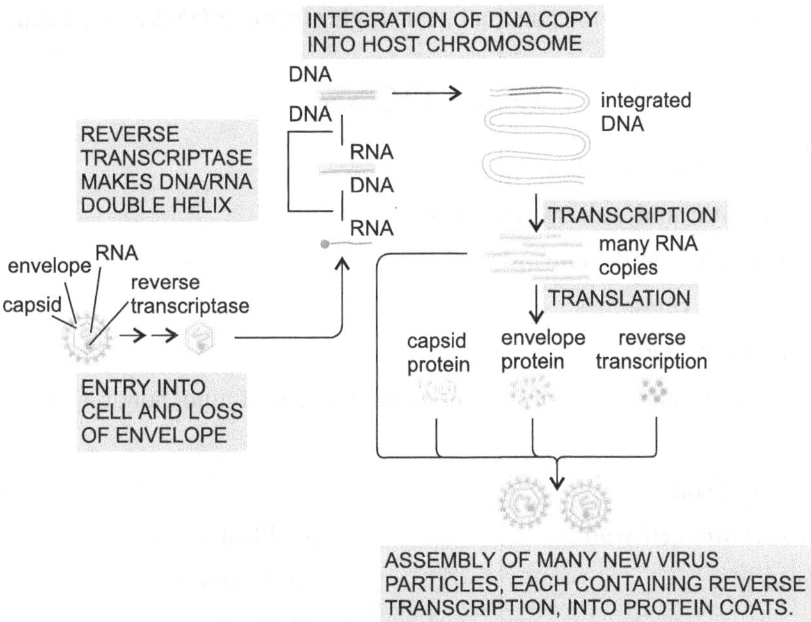

Fig 1.50 Life cycle of a retrovirus.

Some viruses contain **double stranded RNA** (*e.g.*, Reo viruses, etc.) as their genetic material (Fig. 1.51).

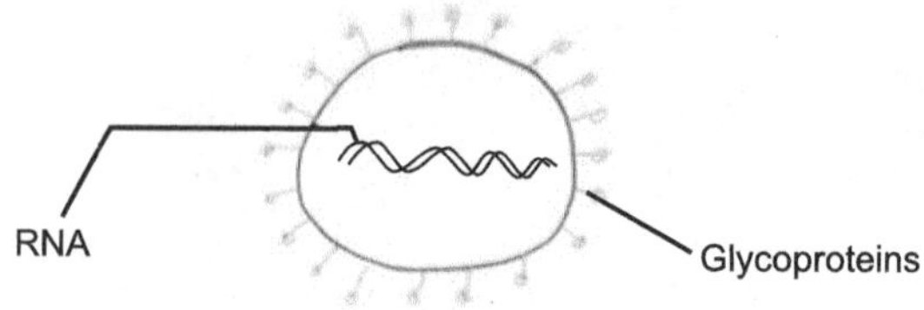

Fig. 1.51 A reovirus.

Adenoviruses [Herpes virus (enveloped)] are icosahedral in shape and contain linear **double stranded DNA** (Fig. 1.52).

Fig. 1.52 Adenoviruses.

Parvoviruses (*e.g.*, Adenovirus) contain linear **single stranded DNA** as their genetic material.

QUESTIONS

1. Differentiate between the following:
   - **(*i*) Prokaryotic and eukaryotic cells**
   - **(*ii*) Mitochondria and chloroplast**
   - (*iii*) 80S and 70S ribosomes
   - (*iv*) G^+ and G^- bacteria
2. Explain in brief the composition, structure and functions of mitochondria and chloroplast.
3. Write short notes on:
   - **(*i*) Prokaryotic cell wall**
   - (*ii*) Peptidoglycan
   - (*iii*) Vacuoles
   - (*iv*) Bacteriophages
   - (*v*) Plant viruses
   - (*vi*) Animal viruses
   - (*vii*) Nucleus
   - (*viii*) Chromosomes

Chapter 2

CARBOHYDRATES

For various living organisms, 16 macronutrients and 11 micronutrients (total 27) are essential. In living organisms, H, O, C of N are most abundant elements, constituting about 99% of mass, where C = 50–60%, O = 25–30 and N = 8–10%.

Carbohydrates are organic substances mainly consisting of C, H and O; some carbohydrates also contain S, N and P. The word carbohydrates is derived from the French word '*hydrate de carbon*', which means **hydrate of carbon**. Its general formula is $\mathbf{C_nH_2O_n}$ (n = generally 3–7) or $\mathbf{C_x(H_2O)_y}$, where H and O are in 2 : 1 ratio, as in H_2O. But, many substances are not carbohydrates, but have the similar general formula of carbohydrates,

e.g., (*i*) **formaldehyde** — (CH_2O) – (where, n = 1)

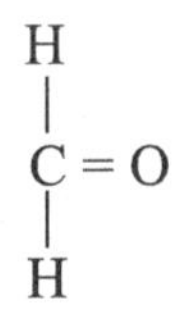

(*ii*) **acetic acid** — (CH_3COOH) or ($C_2H_4O_2$) – (where, n = 2)

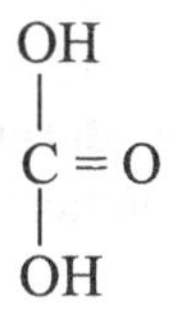

Similarly, some carbohydrates do not fulfill the general formula of carbohydrates,

e.g., (*i*) **deoxyribose** — ($C_5H_{10}O_4$) (*ii*) **rhamnose** — ($C_6H_{12}O_5$)

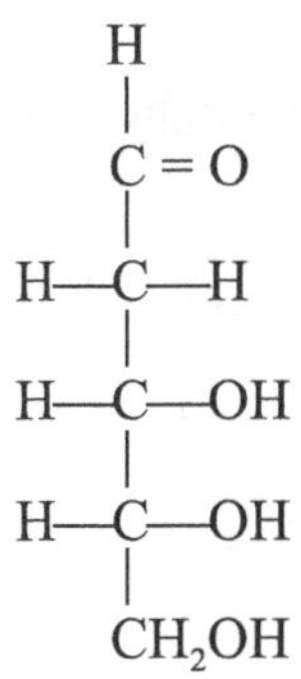

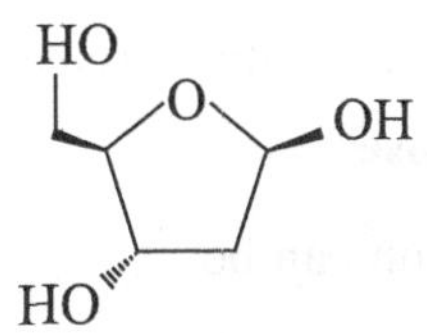

Thus, more precise definition of carbohydrates is that they are **polyhydroxy aldoses** or **ketoses** or **their condensation products** (chains or rings of C with 2 or more –OH groups).

(*i*) **aldoses**

(– CHO in terminal position)

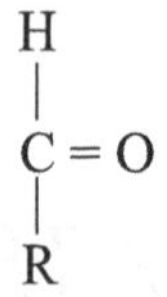

(*ii*) **ketoses**

(>C=O in sub-terminal position)

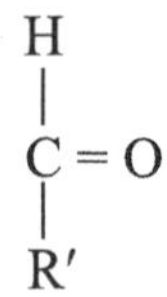

Depending upon the number of monomeric units, carbohydrates are of three types:

I. MONOSACCHARIDES

The word *saccharide* means, sugar. These are made up of only one monomeric unit and cannot be hydrolysed. They are commonly known as sugars and are sweet in taste and soluble in water. All monosaccharides and nearly all disaccharides are **reducing sugars,** as they can reduce a number of reagents, *e.g.*,

(*a*) Fehling's solution ($Cu^{++} \longrightarrow Cu_2O$ red ppt.),

(*b*) Benedict's reagent ($CuSO_4 \longrightarrow Cu_2O$ red ppt.),

(*c*) Tollen's reagent ($AgNO_3 \longrightarrow Ag$)

They have 3–7 C in backbone and are also found as building blocks of oligosaccharides and polysaccharides. Many are intermediates of respiratory and photosynthetic pathways.

All monosaccharides, except some, *e.g.*, dihydroxyacetone that does not contain any C* (asmymmetric C atom, which is attached to 4 different groups), are optically active (bend the plane of polarised light) and exhibit **stereoisomerism**. Stereoisomers have similar chemical properties but, different physical and biological properties. The number of possible isomers for a sugar is $\mathbf{2^n}$ where, *n* is the number of C*.

Thus, due to the presence of asymmetric **C*** atoms, the sugar can occur in two absolute configuration, D and L, that are mirror images of each other:

1. D-configuration, that can be:

(*a*) Dextrorotatory (*d*) or (+), which rotate the plane of polarized light to right (clockwise) *e.g.*, **D-(+)-glucose** or

(*b*) Levorotatory (*l*) or (–), which rotate the plane of polarized light to left (anticlockwise) *e.g.*, **D-(–)-fructose**

2. L-configuration, that again can be:

(*a*) Dextrorotatory or

(*b*) Levorotatory

Most of the naturally occurring sugars are usually found in D-configuration.

Table 2.1

Sugar	*C**	2^n		(*d*)	(*l*)
Aldoses					
1. Trisoses	$n = 1$	2^1	(2×1)	1	1
	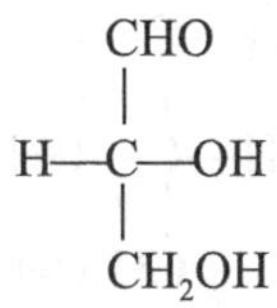 D-glyceraldehyde		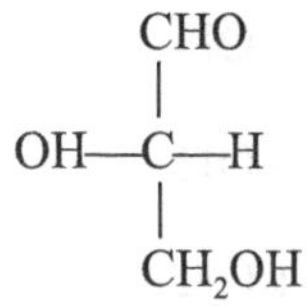 L-glyceraldehyde		
(Glyceraldehyde is considered as standard and all the optically active compounds are compared with its D- or L-structure.)					
2. Tetroses	$n = 2$	2^2	$(2 \times 2 = 4)$	2	2
3. Pentoses	$n = 3$	2^3	$(2 \times 2 \times 2 = 8)$	4	4
4. Hexoses	$n = 4$	2^4	$(2 \times 2 \times 2 \times 2 = 16)$	8	8
				(4 in living system)	
Ketoses:					
1. Tetroses	$n = 1$	2^1	$(2 \times 1 = 2)$	1	1
2. Pentoses	$n = 2$	2^2	$(2 \times 2 = 4)$	2	2
3. Hexoses	$n = 3$	2^3	$(2 \times 2 \times 2 = 8)$	4	4

Classification of Monosaccarides

On the basis of number of C atoms in the backbone, the monosaccharides have been classified as:

1. Trioses

They have 3 C in backbone, *e.g.*, $C_3H_6O_3$.

e.g., (*i*) **D-glyceraldehyde** (*ii*) **Dihydroxy acetone**

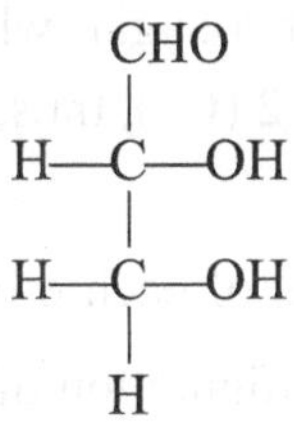

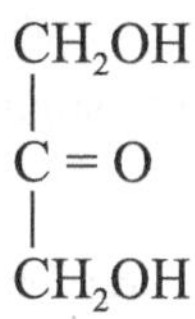

Since, it does not contain any asymmetric C*, it is **not optically active.**

2. Tetroses

They have 4 C in backbone, *e.g.*, $C_4H_8O_4$.

e.g., (*i*) **D-erythrose** (*ii*) **D-threose**

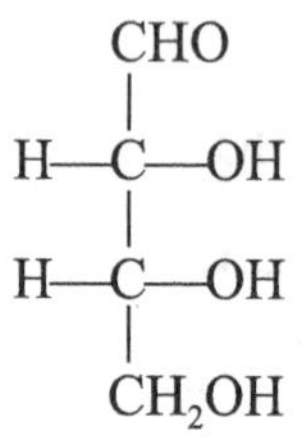

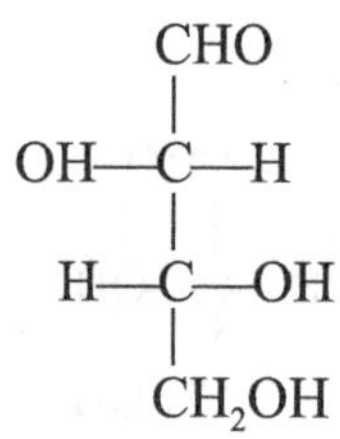

(*i*) **D-erythrose** (*ii*) **L-erythrose** (*iii*) **D-threose** (*iv*) **l-threose**

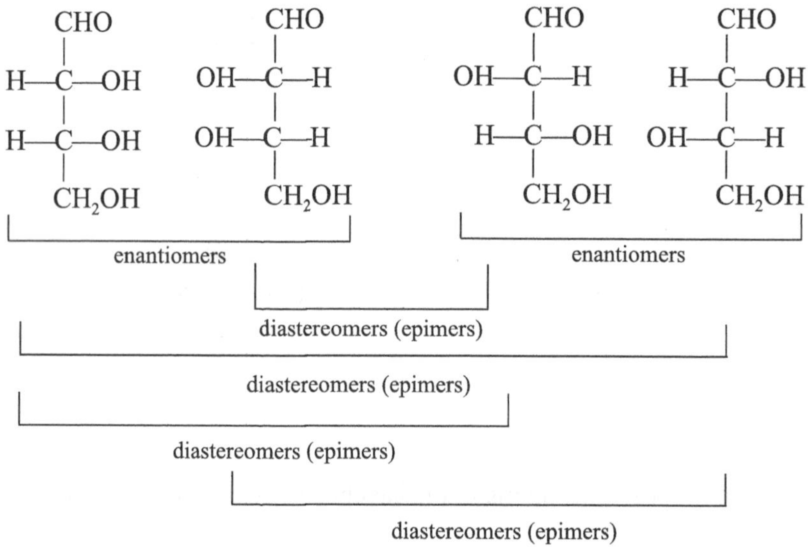

Epimers (*diasteriomers*) are pairs of stereoisomers that are not mirror images of each other and they differ in the configuration of one of the several asymmetric C atoms (C*) present in their molecules. There are 2^n possible diastereomers for a sugar, where, n is the number of asymmetric C atoms (C*) in that sugar. 4C aldoses have $n = 2$ (C*), thus, they can have 4 isomers ($2^2 = 2 \times 2 = 4$).

Enantiomers are stereoisomers that are mirror images of each other.

Anomers are stereoisomeric pairs that differ in the configuration of first asymmetric C atom (1C*) that is involved in the ring formation in ring forms. (C* = asymmetric C) *e.g.*, α-D-erythrose and β-D-Erythrose.

3. Pentoses

They have 5 C in backbone **(5C)**, *e.g.*, $C_5H_{10}O_5$

e.g., (*i*) **D-ribose** (*ii*) **D-xylose**

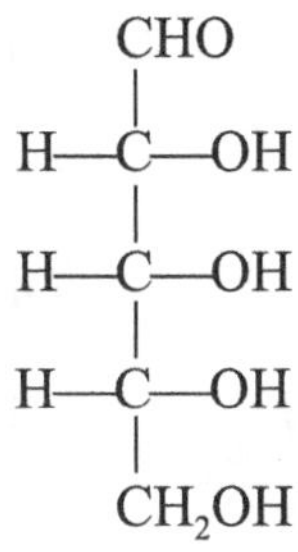

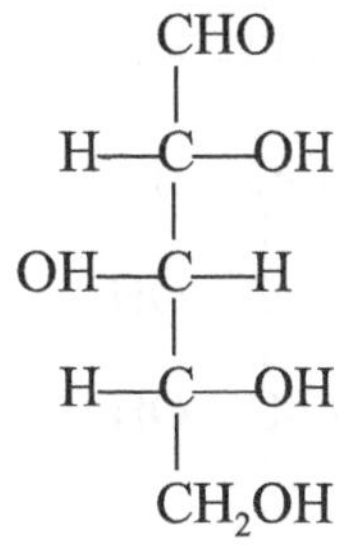

(Found in ATP, RNA, FAD, NAD)
(Found in wood)

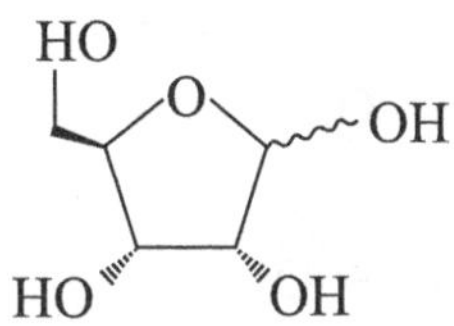

Ring form of D-ribose

(*iii*) **D-ribulose**

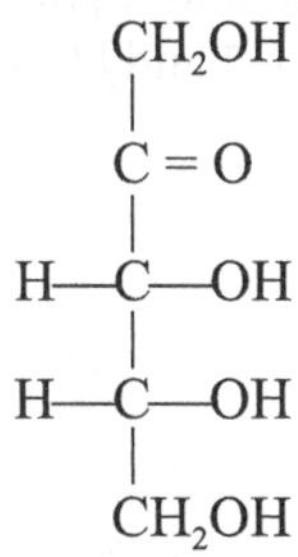

(*iv*) **D-xylulose**

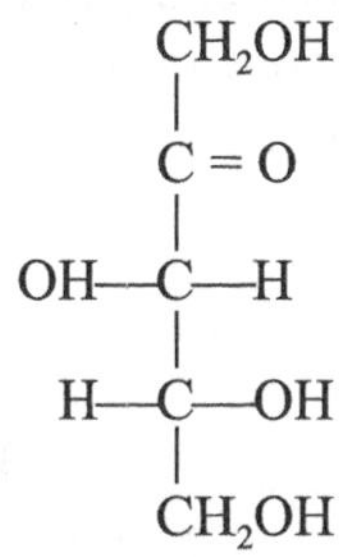

[Some sugars are found in L-form, *e.g.*, **L-arabinose** that is a 5C aldose].

4. Hexoses

They have 6 C in backbone **(6C)**, *e.g.*, $C_6H_{12}O_6$.

(*i*) **D-glucose** (*ii*) **D-mannose** (*iii*) **D-galactose**

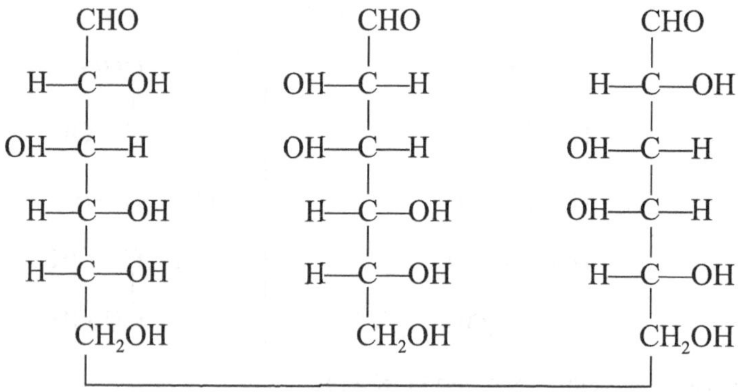

Epimers (diasteriomers)

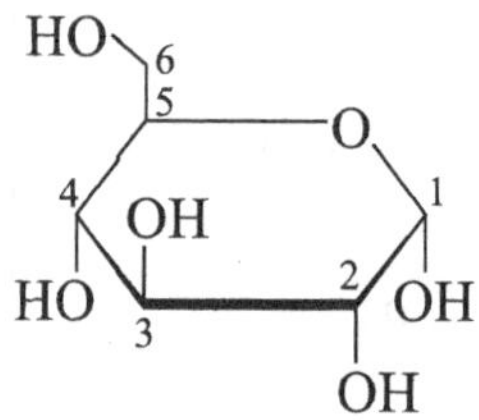

Ring form of D-glucose

e.g., in 6 C aldoses with 4 C* , where 16 optically active forms are possible, (of which, 8 pairs of enantiomers are mirror images) some of the commonly found sugars are:

a pair of D- and L-glucose, a pair of D- and L-mannose and a pair of D- and L-galactose, where, D-glucose and D-mannose differ in configuration of 2nd C* and D-glucose and D-galactoset differ in configuration of 3rd C*.

(*iv*) **D-fructose**

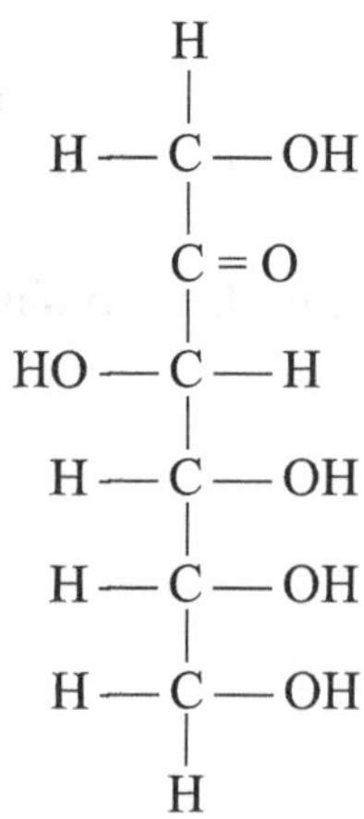

Some common hexoses are:

(*i*) **Glucose:** It is commonly known as grape sugar or corn sugar or blood sugar, found in honey, urine of diabetics, etc. Since, it is dextrorotatory, it is also known as dextrose. It is most abundant in all hexoses, is a respiratory substrate and its oxidation provides immediate energy. It is 3/4th as sweet as sucrose (sweetening power of various sugars: **Fructose** > **Invert sugar** > **Sucrose** > **Glucose** > **Maltose** > **Lactose**). Monomeric unit of starch, glycogen, and cellulose is glucose. Commercially it is obtained from starch by treating it with hot mineral acids. Glucose has 4 secondary and 1 primary – OH groups.

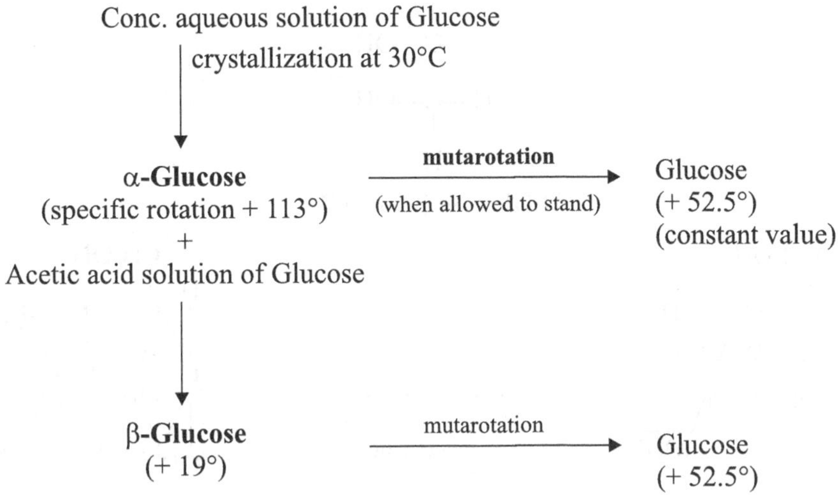

α-D-Glucose and β-D-Glucose are **anomers**.

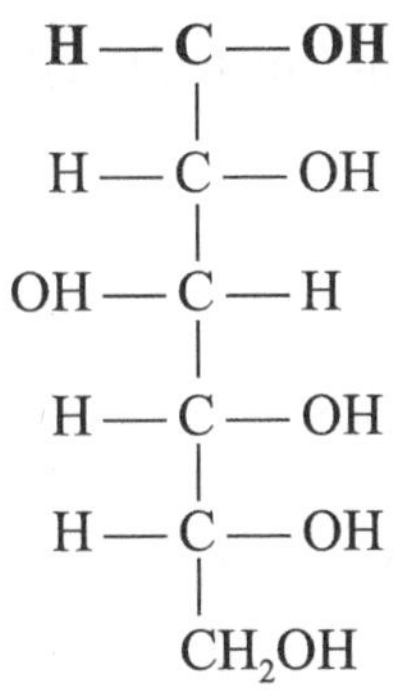

α–D–Glucose

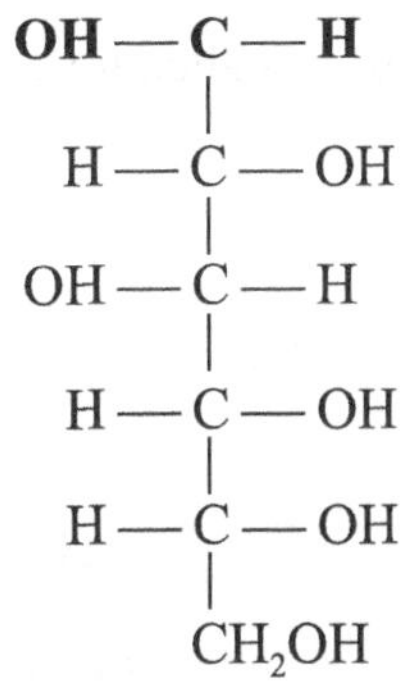

β–D–Glucose

C1 becomes asymmetric in cyclic forms:

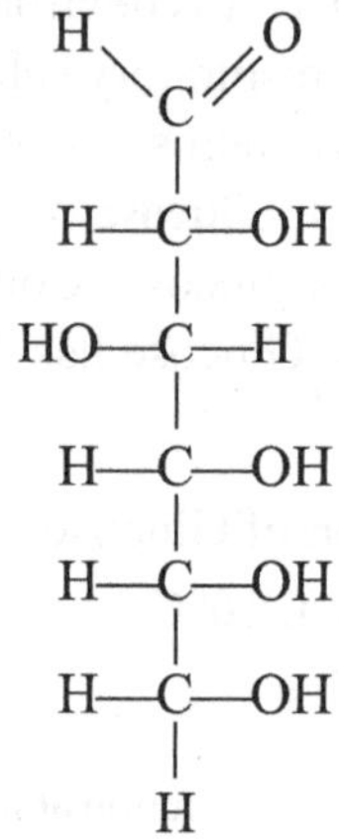

D-Glucose (with 4C*)

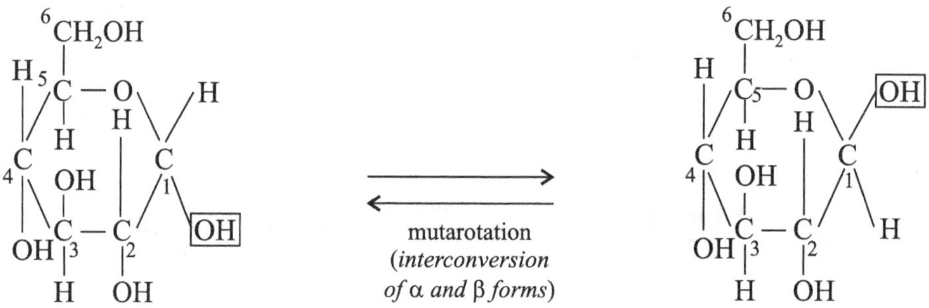

α-D-Glucose (glucopyranose)
(1/3)

β-D-Glucose
(2/3)

Pyranose form of glucose is called glucopyranose in which C1 and C5 are linked by an atom of O:

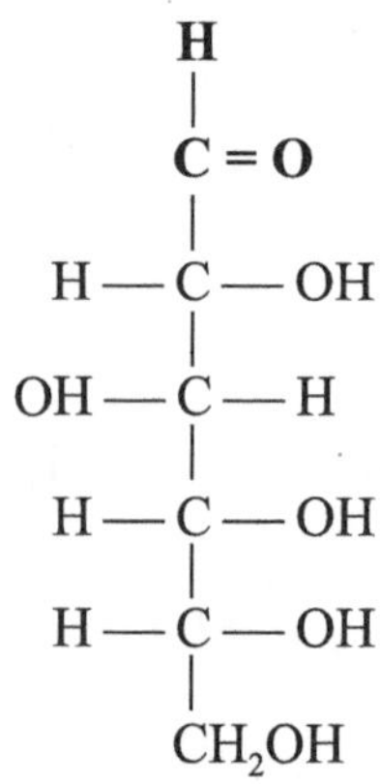

D-Glucose

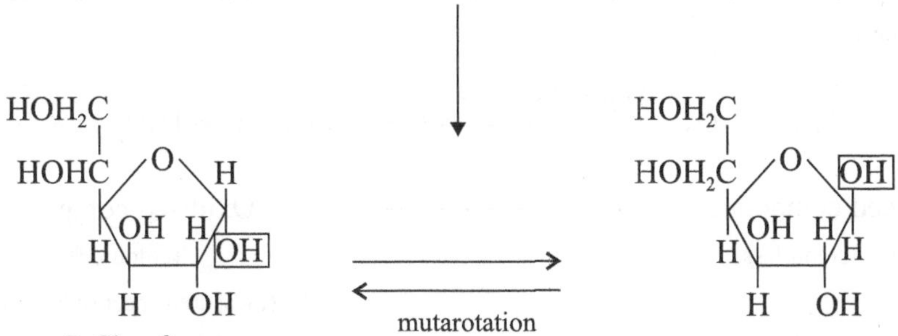

α-D-Glucofuranose   mutarotation   β-D-Glucofuranose

Fermentation of glucose in yeast yields alcohol:

$$\text{Glucose} \xrightarrow{\text{glycolysis}} \text{Pyruvate} \longrightarrow \text{Acetaldehyde} \longrightarrow \mathbf{C_2H_5OH} + CO_2$$

Oxidation of glucose yields:

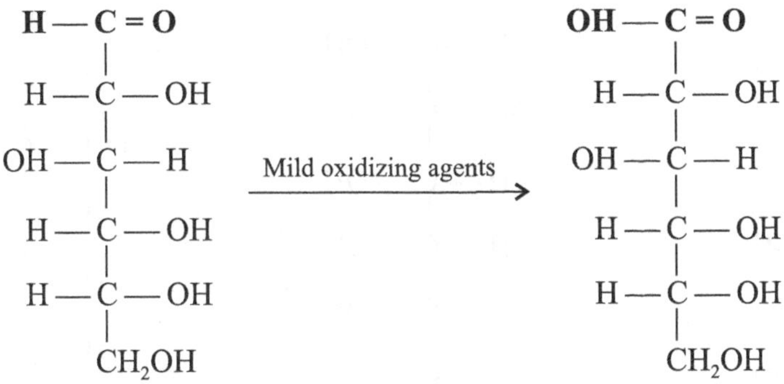

Glucose   **Gluconic acid**

Reaction with Fehling's reagent: (or Benedict's reagent)

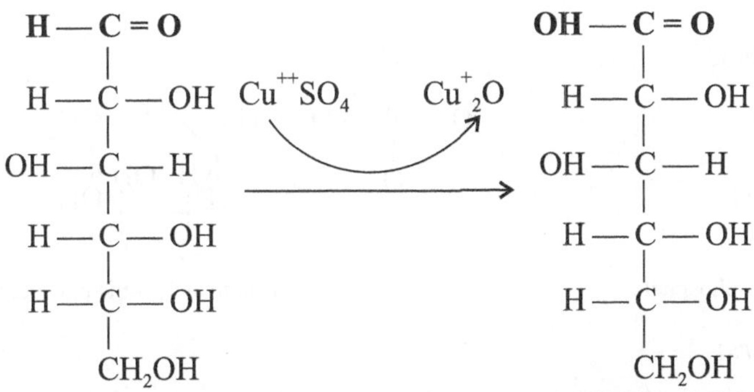

Glucose   **Gluconic acid**

Earlier, this reaction was used for the diagnosis of diabetes, but nowadays, **diabetes** is diagnosed using biosensor:

$$\text{Glucose} + O_2 \xrightarrow{\textbf{Glucose oxidase}} \text{Gluconolactone} + H_2O_2$$

$$\underset{\text{(colourless)}}{\text{Reduced compound}} + H_2O_2 \xrightarrow{\textbf{Peroxidase}} \underset{\substack{\text{(coloured)}\\ \text{(measured spectrophotometrically)}}}{\text{Oxidised compound}}$$

Some important **glucose derivatives** are:

(*a*) ***Glucose phosphates*** that are trapped sugar (cannot pass through plasma membrane), *e.g.*,

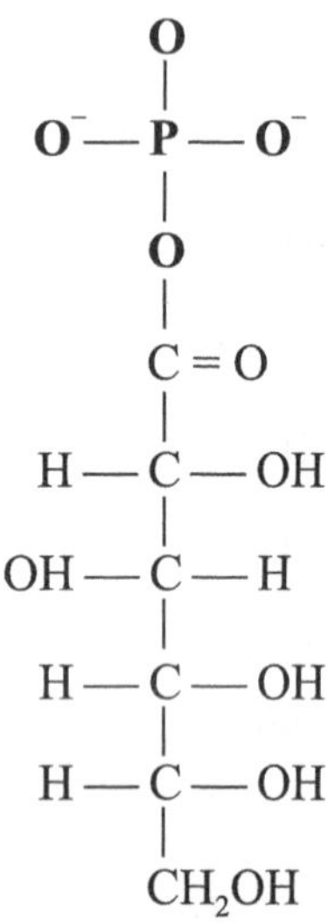

Glucose-1-phosphate

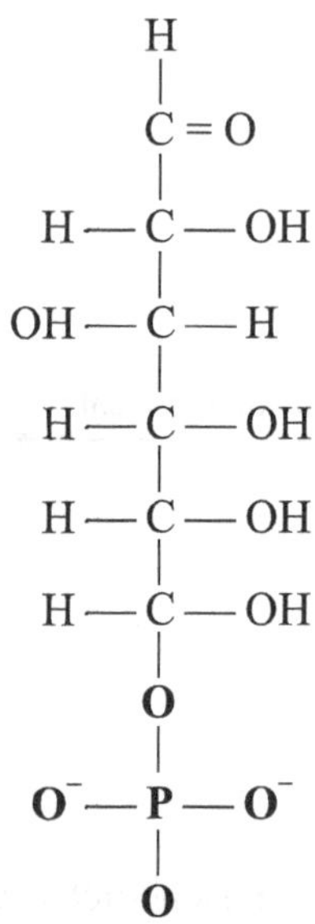

Glucose-6-phosphate

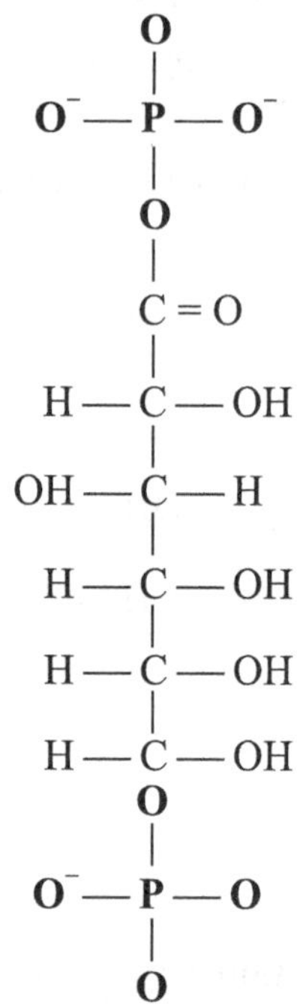

Glucose-1, 6-phosphate

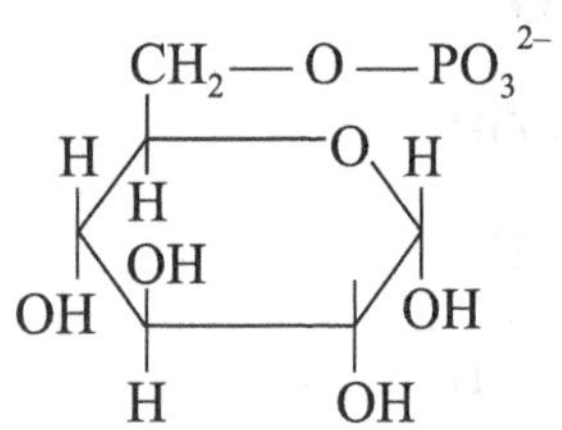

Ring Form of β-D-glucose-6-phosphate

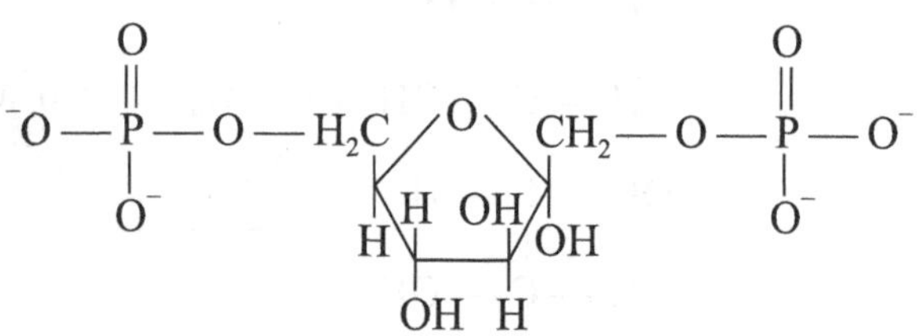

Fructose-1, 6-bis phosphate

(*b*) ***Deoxy sugars:***

e.g., 2-deoxy-D-ribose (furanose) (derivative of ribose)

(*c*) ***Amino glucose derivatives (sugars):***

e.g., found in bacterial cell wall (peptidoglycan) (Fig. 2.1)

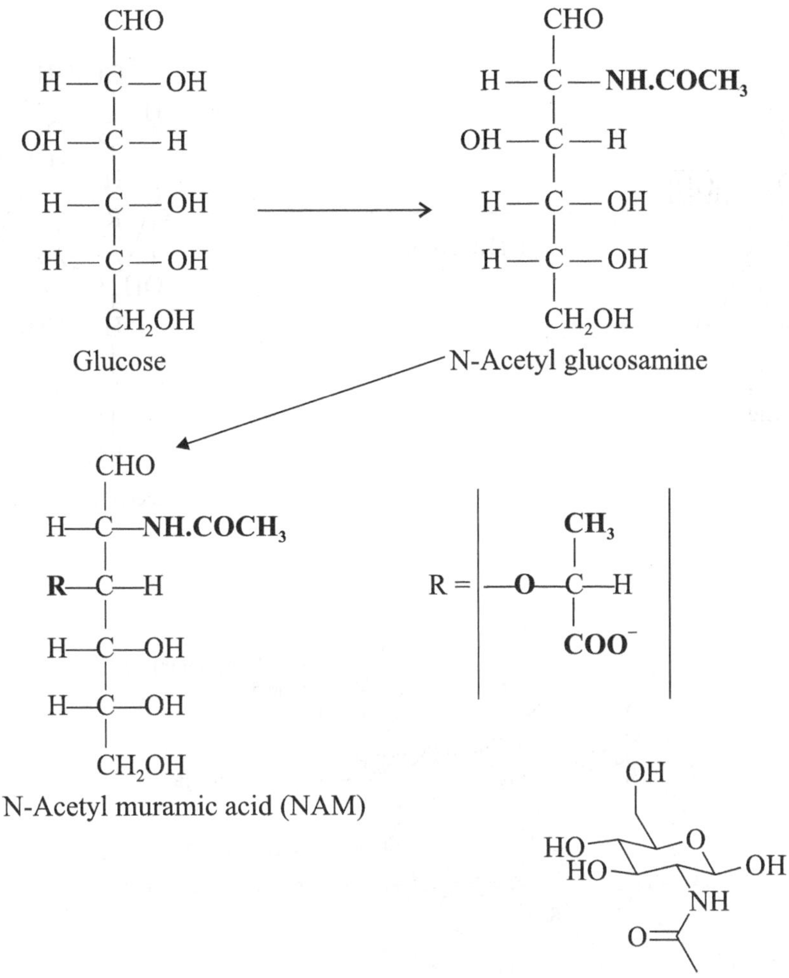

Ring form of NAM

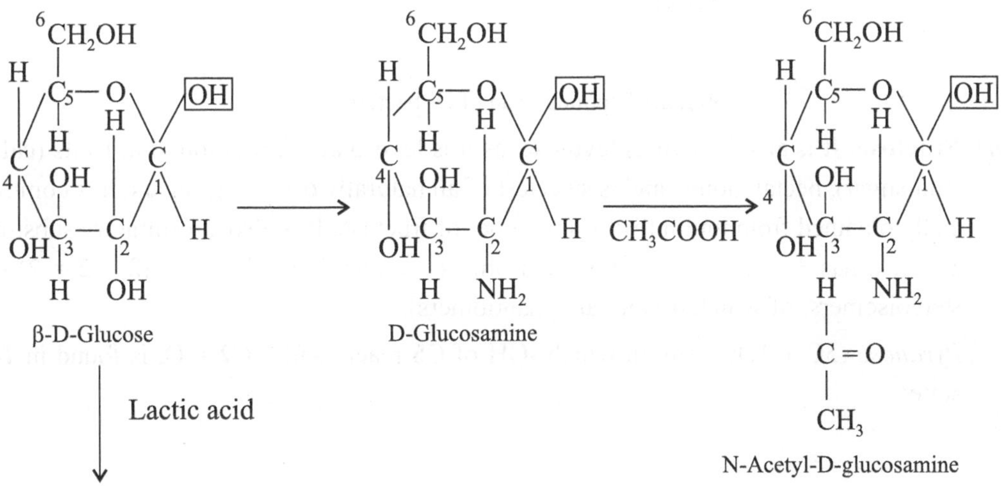

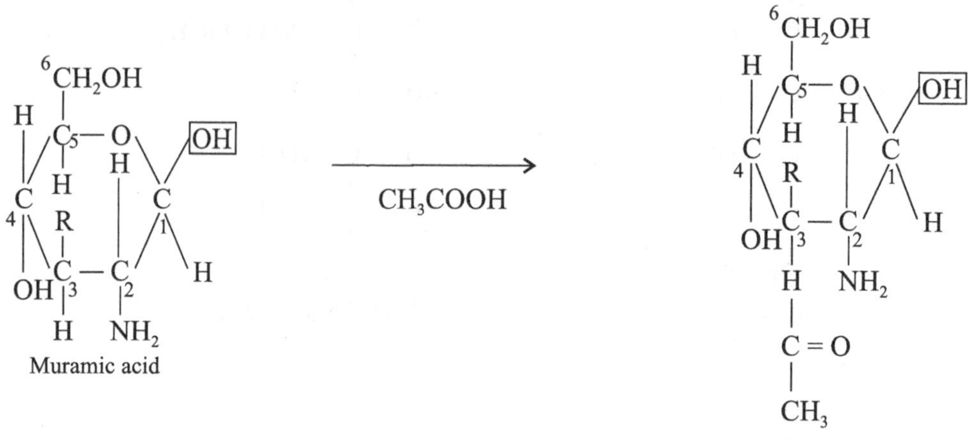

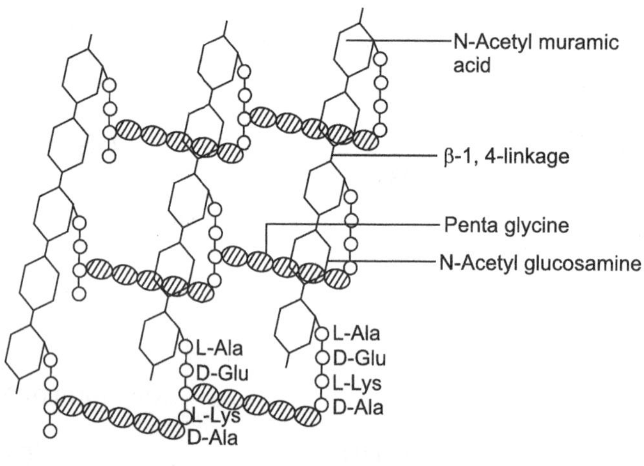

Fig. 2.1 Peptidoglycan of G^+ bacteria.

(*ii*) **Fructose:** It is also known as levulose, as it is levorotatory. It is found in fruits (called fruit sugar), nectar, honey and is sweetest of all naturally occurring sugars. It is commercially obtained from inulin that is a polymer of fructose. It is also a respiratory substrate. Since, it has 3 C*, a total of 8 stereoisomers are possible [$2^n = 2^3 = (2 \times 2 \times 2) = 8$ stereoisomers, of which 4 pairs are enantiomers].

Pyranose (5C + 1O) form, in which -OH of C**6** reacts with >C**2** = O, is found in **free** state.

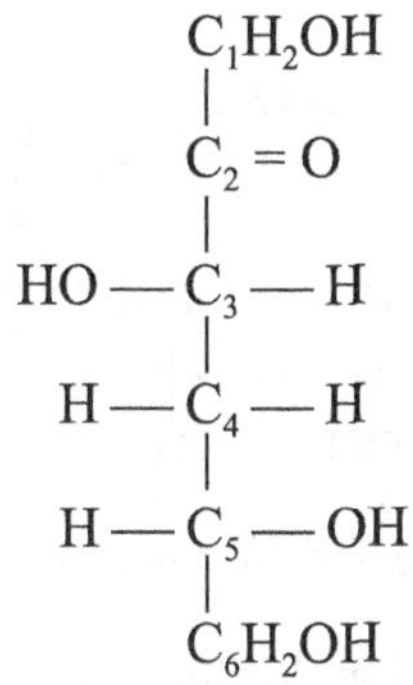

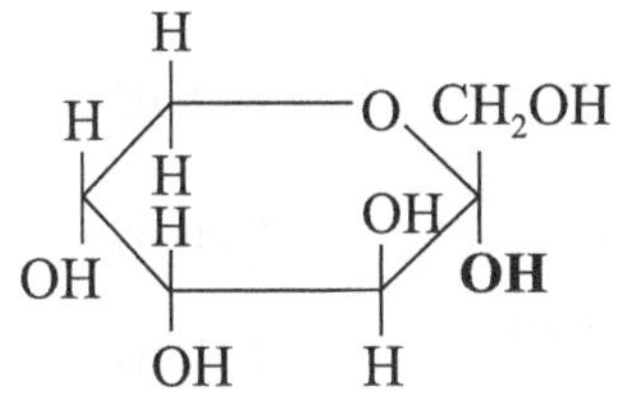

α-D-fructose (fructopyranose)
(– 21°)

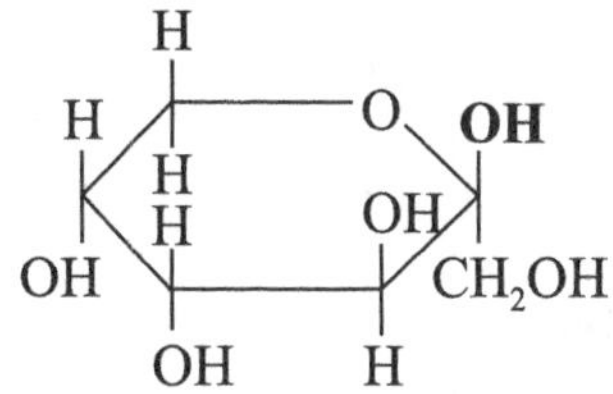

β-D-fructose
(– 133°)

At equilibrium optical rotation = **– 92.3°**

Whereas, *furanose* (4C+1O) form, in which -OH attached to C_5 reacts with $>C_2 = O$, and is found in **combined** state.

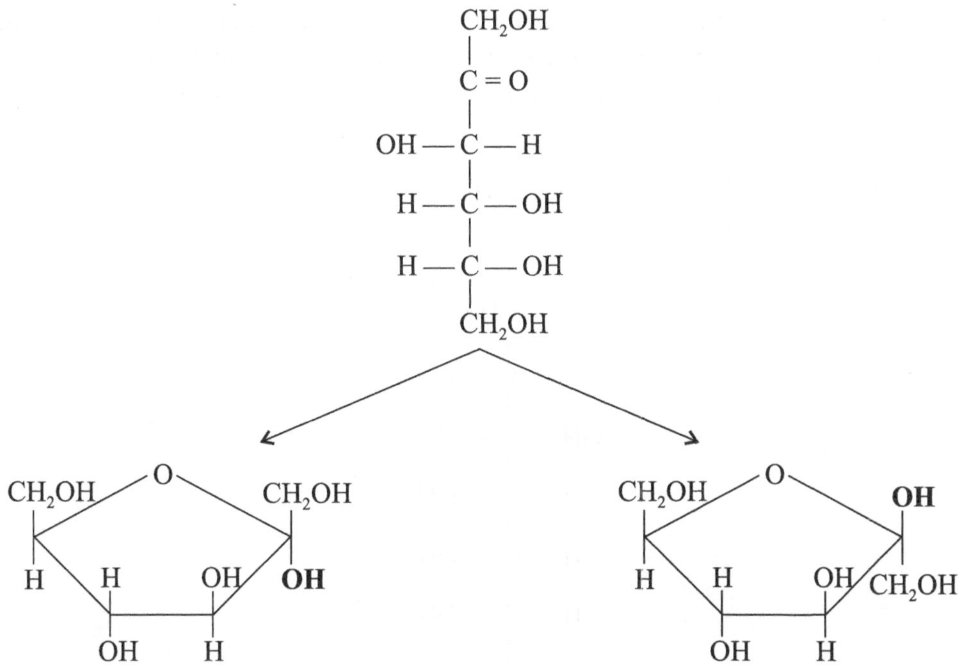

α-D-Fructose (fructofuranose)

β-D-Fructose (More abundant)

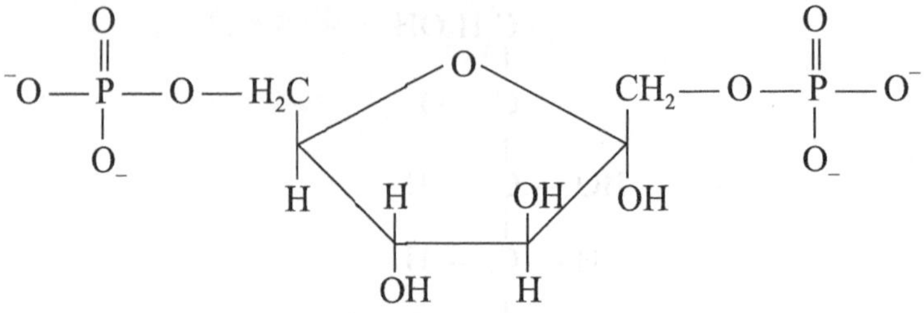

Fructose-1, 6-bis phosphate

Oxidation: It reduces Fehling's solution, Tollen's or Benedict's reagent due to the presence of a-hydroxy ketonic group (it itself is oxidised by these reagents).

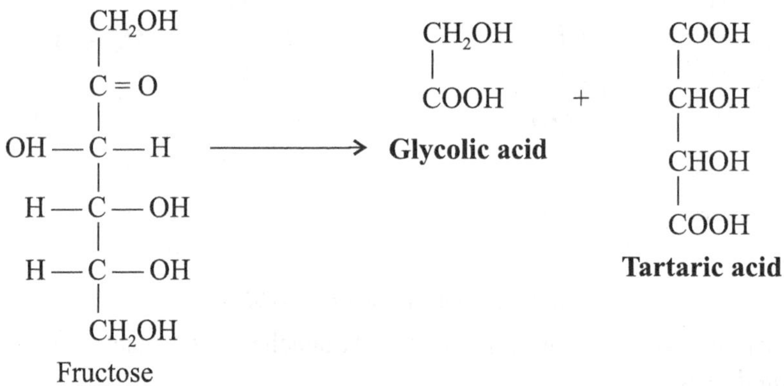

(*iii*) **D-Galactose and D-mannose:** Less commonly, they are found in free state and generally, found in various polymers, as constituent of cell wall, etc., and their main function is structural. Galactose is a component of lactose sugar (a disaccharide).

5. Heptoses

They have 7 C in backbone **(7C),** *e.g.*, $C_7H_{14}O_7$

e.g., (*i*) D-sedoheptulose.

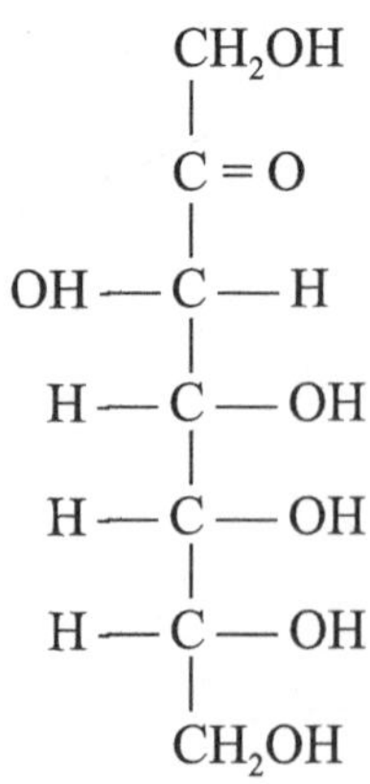

Ring Form Monosaccharides

Those **aldehydes** (monosaccharide) with **4C** and more than 4C in their backbone and; those ketones (monosaccharide) with **5C** and more than 5C in their backbone, predominantly are found in cyclic or ring forms in aqueous solution that generates an additional asymmetric C (C*). For example, C1 in Glucose becomes asymmetric in aqueous solution, which otherwise in chain form is found as symmetric C, thus, two more stereoisomers, α-glucose and β-glucose (anomers) becomes possible. Two common types of ring forms are:

(*i*) 6-membered rings (5C + 1O), called pyranoses as they resemble **pyran**. They are more stable and found predominantly in 6C aldoses.

(*ii*) 5-membered rings (4C + 1O), called furanoses as they resemble **furan**.

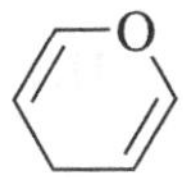

Pyran

Furan

II. OLIGOSACCHARIDES

They contain **2–9** (<10) monomers (or their derivatives) and their names end in suffix, –ose. Upon acid hydrolysis, they yield the constituent monosaccharides.

The monosaccharides are linked by **glycosidic bonds** that is formed between a –OH of C1 of one monomer and other –OH of second monomer by the condensation reaction by elimination of a molecule of H_2O (dehydration synthesis). There are many types of glycosidic bonds:

(*i*) **1,4-linkage,** where C_1 of one monomer and C_4 of other monomer are linked together (between two –OH groups in α-position), found in unbranched chains of polysaccharides, **maltose**, etc.

(*ii*) **β-1,4-linkage** formed between C_1 of one monomer and C_4 of other monomer (two –OH groups in β-position), *e.g.*, lactose.

(*iii*) **1,6-linkage** formed between C_1 and C_6 at the points of branching of a chain, *e.g.*, **isomaltose.**

[Polymerization reactions in proteins, carbohydrates, nucleic acids are also condensation reactions and their hydrolysis needs incorporation of H_2O molecules]

Like monosaccharides, they can also be reducing or non-reducing in nature:

(*i*) **Reducing sugars:** They have a free –CHO or >C=O group (*e.g.,* maltose, lactose).

(*ii*) **Non-reducing sugars:** In these sugars, the reducing groups are involved in glycosidic linkage, *e.g.,* **sucrose**.

On the basis of number of monomeric units, oligosaccharides are classified into the following:

1. Disaccharides

They are the most important oligosaccharides, which are soluble in water and, like monosaccharides, are sweet in taste, *e.g.,* (*i*) **Sucrose:** Having formula $C_{12}H_{22}O_{11}$, and commonly known as table sugar, cane sugar or beet sugar. It is also found in honey, pineapple, banana and other fruits. Commercially, it is obtained from sugar cane and sugar beet. It is **optically active**, **dextrorotatory**, with specific rotation +66.5, but does not show mutarotation [no α- and β-sucrose is found, as it does not contain any free –CHO or >C=O group. It is non-reducing (no free –CHO or >C=O group). **α-1, β-2 glycosidic bond** is found between the monomeric units, glucose and fructose.

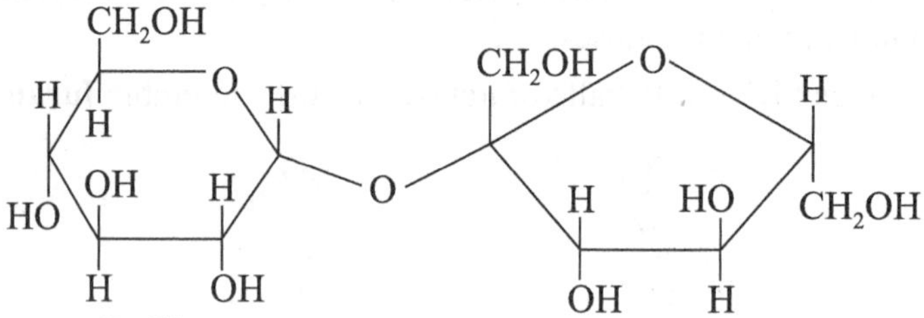

α–D–Glucose

β–D–Fructose (furanose)

Sucrose

Its **hydrolysis yields:**

$$\underset{\substack{\text{Sucrose} \\ (+\,66.5^\circ)}}{C_{12}H_{22}O_{11}} + H_2O \xrightarrow[\text{Mineral acid}]{\text{Invertase or}} C_6H_{12}O_6 + \underset{\substack{\text{Fructose} \\ (-\,92^\circ)}}{C_6H_{12}O_6}$$

Invert sugar (with specific rotation of – **20°** at equilibrium)

Invert sugar [named due to **optical inversion** from dextro Sucrose (*d*) to levo invert sugar (*l*) due to more (*l*) nature of fructose is used as a substitute for honey.

Fermentation of sucrose yields:

$$\underset{\text{Sucrose}}{C_{12}H_{22}O_{11}} + H_2O \xrightarrow{\text{Invertase}} \underset{\text{Glucose}}{C_6H_{12}O_6} + \underset{\text{Fructose}}{C_6H_{12}O_6}$$

$$\xrightarrow{\text{Zymase (yeast)}} C_2H_5OH + CO_2$$

(*ii*) **Maltose:** Its molecular formula is $C_{12}H_{22}O_{11}$ and is found in germinating seeds. Commercially, it is obtained from starch:

$$\text{Starch} \xrightarrow{\text{Diastase } (\alpha\text{-amylase})} \text{Maltose}$$

It is a **reducing sugar** having a free –CHO group. Moreover, because of the presence of a free –CHO group and due to the relative positions of –OH group, two anomers, **α- and β-maltose**, are possible.

Its **hydrolysis** yields:

$$\text{Maltose} \xrightarrow{\text{Maltase}} 2\ \text{Glucose}$$

It has **α-1,4 glycosidic bond** between two glucose units.

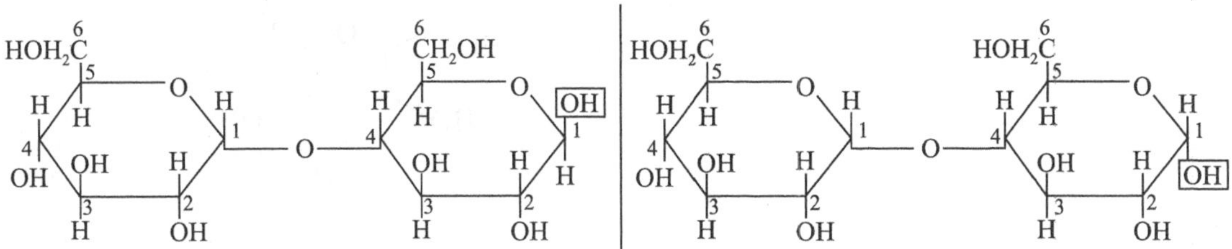

(α-D-Glucose) (β-D-Glucose) (α-D-Glucose) (β-D-Glucose)

β-Maltose | α-Maltose

(*iii*) **Lactose:** Its molecular formula is $C_{12}H_{22}O_{11}$ and is commonly known as milk sugar, as it is present in milk and synthesized in mammary glands. It may be present in urine during pregnancy. Souring of milk is due to the activity of certain bacteria that drop the pH of milk because of conversion of lactose into lactic acid:

$$\text{Lactose} \xrightarrow{\text{lactobacilli}} \text{Lactic acid}$$

It is a **reducing sugar** (having free –CHO) and exhibits **mutarotation** to give α- and β-lactose in equimolar quantities. It has **β-1,4-glycosidic linkage** between galactose and glucose.

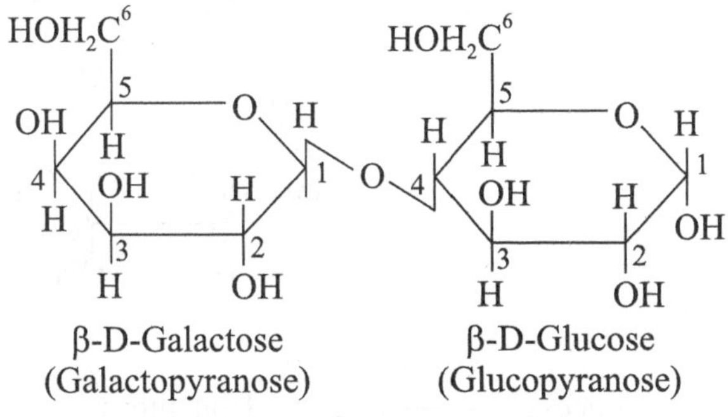

β-D-Galactose (Galactopyranose) β-D-Glucose (Glucopyranose)

α-Lactose

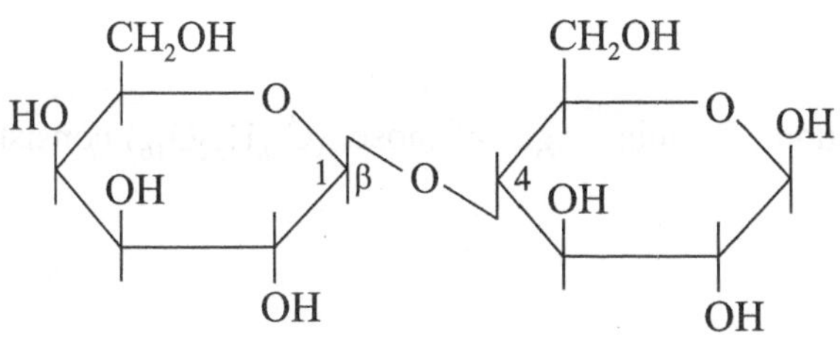

β-Lactose

(*iv*) **Cellobiose:** Commercially, it is obtained from cellulose and is a **reducing sugar**. The two glucose molecules are linked by **β-1,4-glycosidic bond.**

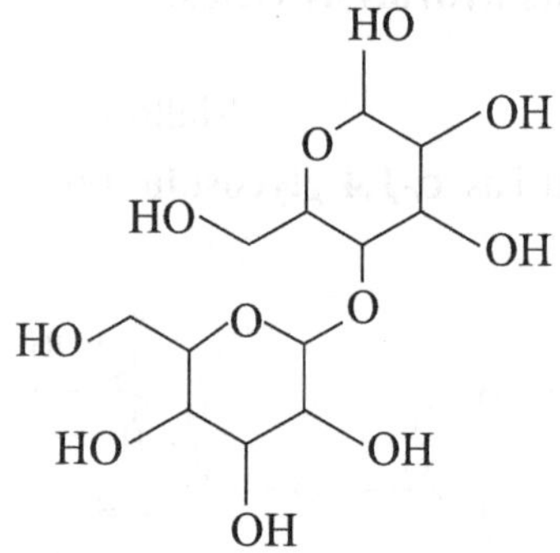

Glucose + Glucose ⟶ Cellobiose

(*v*) **Isomaltose:** It is a **reducing sugar** where two glucose molecules are linked by **α-1,6-glycosidic linkage**.

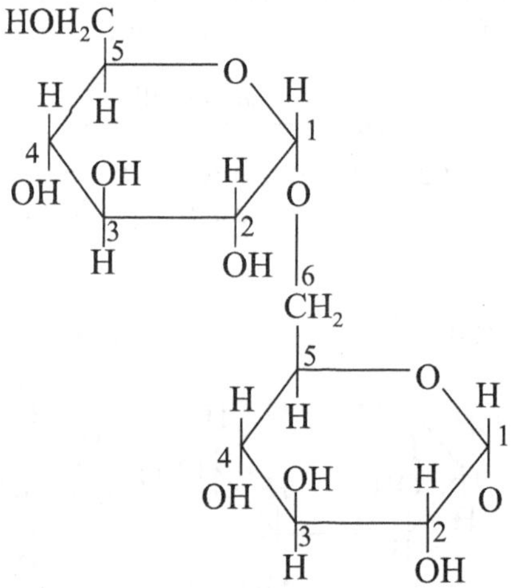

Glucose + Glucose ⟶ Isomaltose

2. Trisaccharides

They have three monomeric units, *e.g.*, raffinose, ($C_{18}H_{32}O_{16}$) consisting of D-Glucose, D-Fructose and D-Galactose.

III. POLYSACCHARIDES

They have **10 or more** monosaccharide units and are usually insoluble in water or they may form colloidal suspension. They are tasteless amorphous solids and are called **glycans**, *e.g.*, **glucans** are polymers of glucose, **fructans** are polymers of fructose, etc. Depending upon the types of monomers found in polysaccharides, they are of two types:

(*i*) **Homo-polysaccharides:** Consisting of one type of monomers (about 95%, rarely in pure form), *e.g.*, glucans, like starch, glycogen, cellulose that are consisting of glucose, chitin that is made up of N-acetyl glucosamine (amino derivative of glucose) that are linked by β-1.4-linkage (glucan). Some important homo-polymers are:

(*a*) ***Glycogen:*** It is found as reserve food in animals (liver and muscles), bacteria, fungi, *etc.* Structurally, it is profusely branched (branches after 8-10 glucose) with **α-1,4-glycosidic linkage** between α-D-glucose units). It is a non-reducing carbohydrate and gives red colour with iodine.

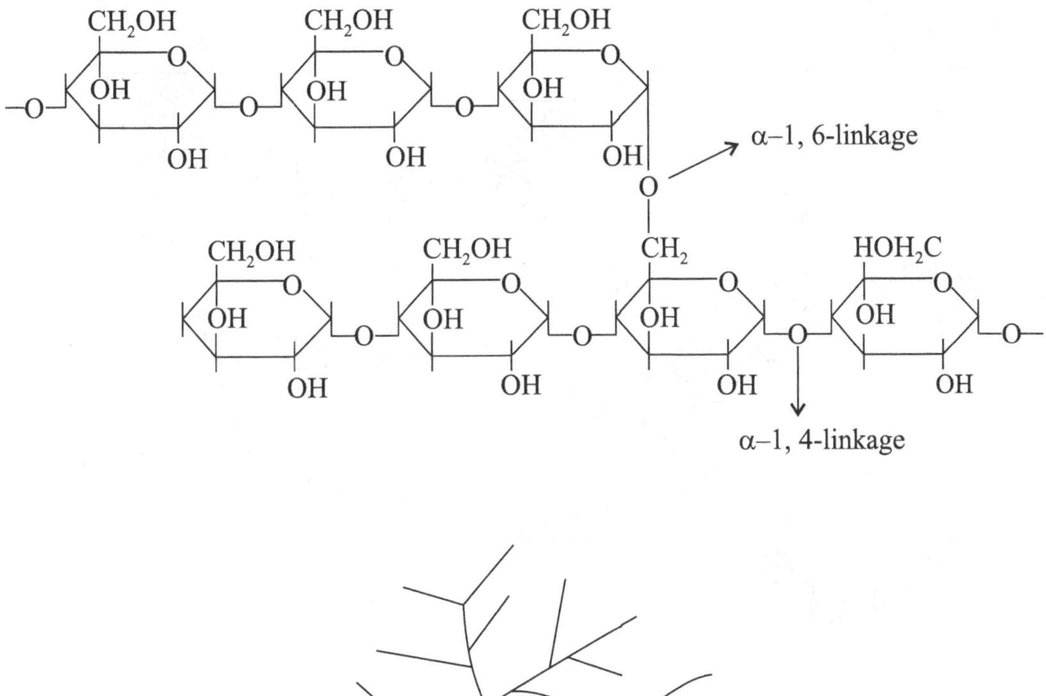

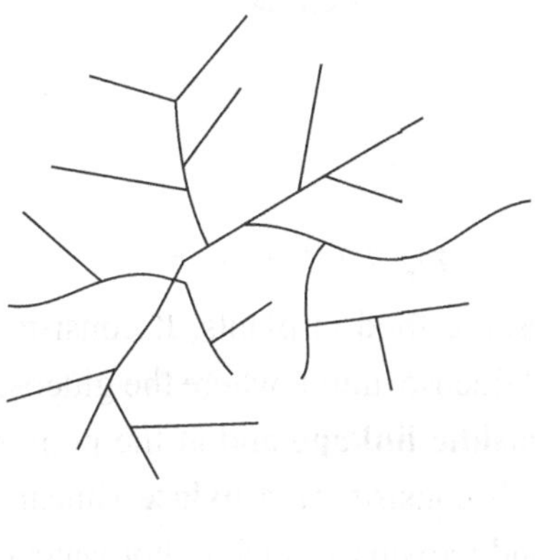

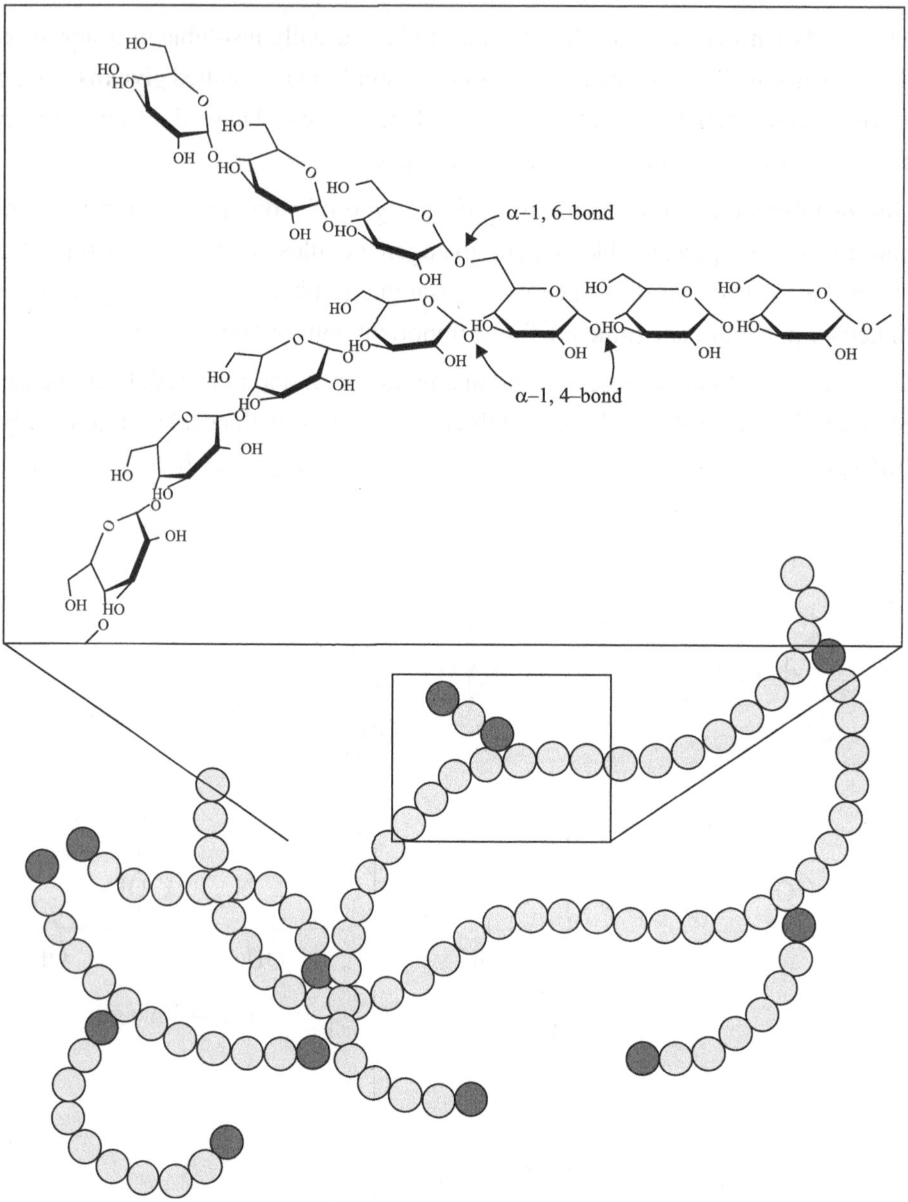

Fig. 2.2. Glycogen.

(*b*) ***Starch:*** It is found as reserve food in plants. It consists of branched chains of glucose (branches after about 25 Glucose units), where the glucose (α-D-Glucose) units in chains are linked by **α-1,4-glycosidic linkage** and at the point of branches the bond is **α-1,6-glycosidic linkage** type. It consists of **amylose** (linear and soluble in hot water) and amylopectin (branched and remains as such in hot water) present in 1 : 4 ratio (Fig. 2.3).

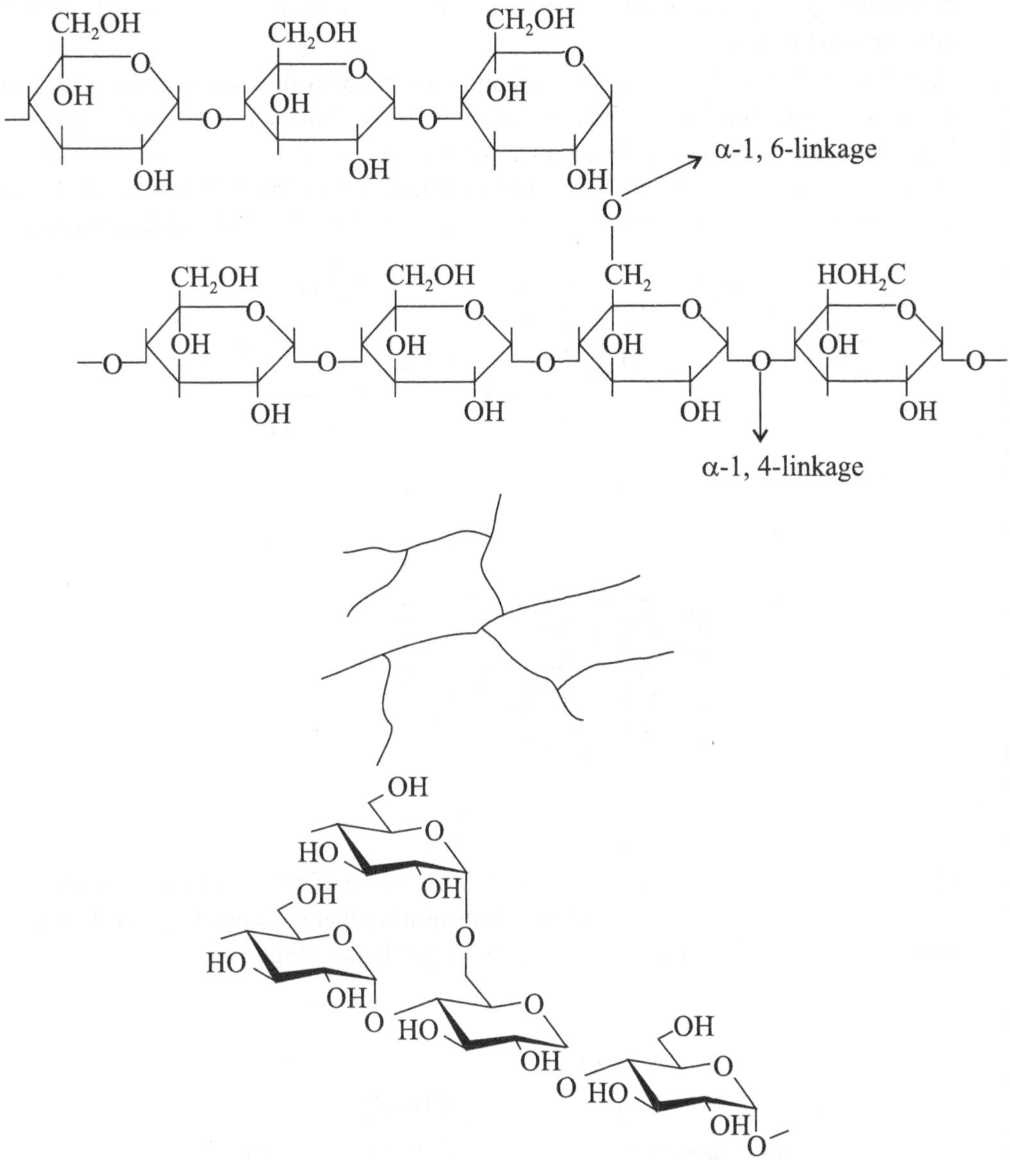

Amylopectin

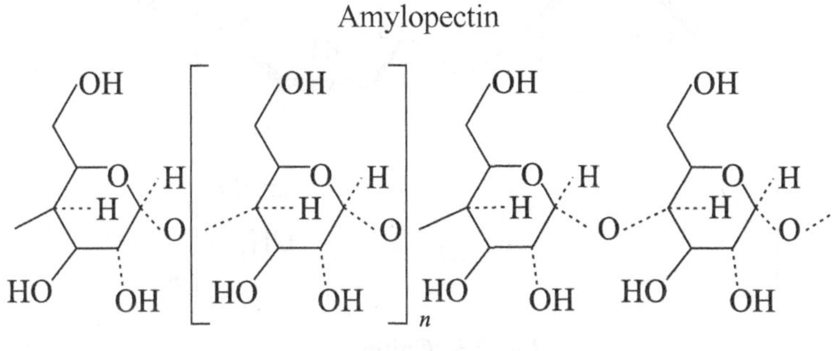

Amylose

Fig. 2.3. Starch.

Starch upon hydrolysis by α-amylase yields glucose and maltose. The partial hydrolysis of starch by α-amylase produces dextrins, which by the action of the enzyme dextrinase gives rise to glucose.

(*c*) ***Cellulose:*** It is found in plant cell wall and consists of **unbranched** chains of glucose. The glucose microfibrils are held together by hydrogen bonds. It is insoluble in water and highly resistant to hydrolysis. It can not be digested by human beings. The bacteria present in the rumen of cattle, termites, some fungi contain the enzyme cellulase that helps in the digestion of cellulose. Various β-D-glucose units are linked by β-1,4-linkages (Fig. 2.4).

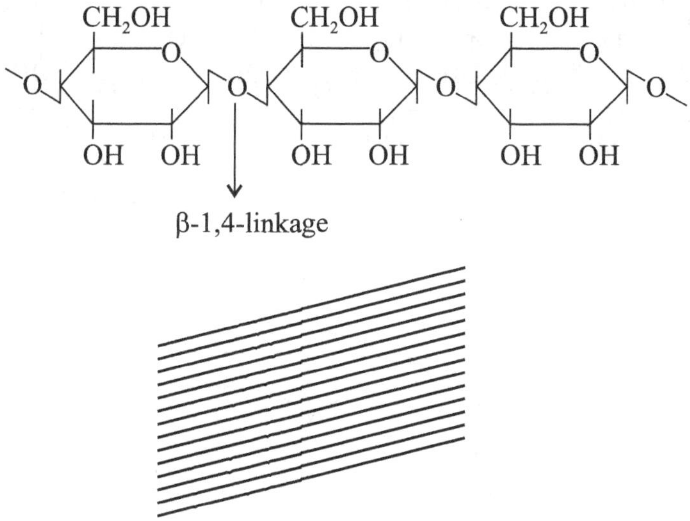

Fig. 2.4. Cellulose.

(*d*) ***Chitin:*** It is found in fungal cell wall, hard exoskeleton of arthropods (insects and crustaceans), etc. It is a polymer of **N-acetyl glucosamine** that are linked by **β-1,4 glycosidic bonds**. The chains are held together by hydrogen bonds (Fig. 2.5).

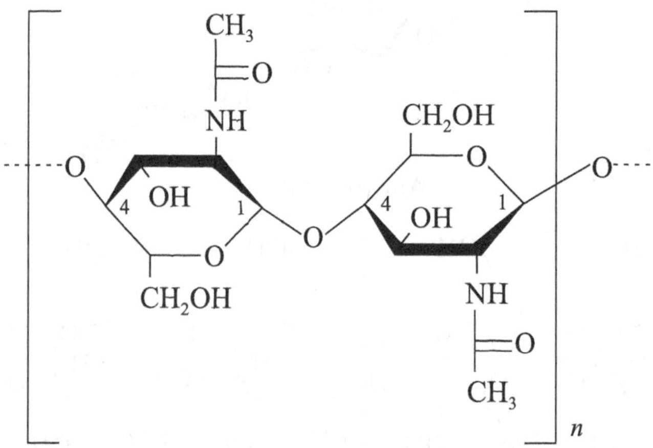

Fig. 2.5. Chitin

(*e*) ***Inulin:*** It is present as reserve carbohydrates in many plants and is the polymer of **fructose** (furanose) (fructan) that are linked by **β-1,2 glycosidic bonds** (Fig. 2.6).

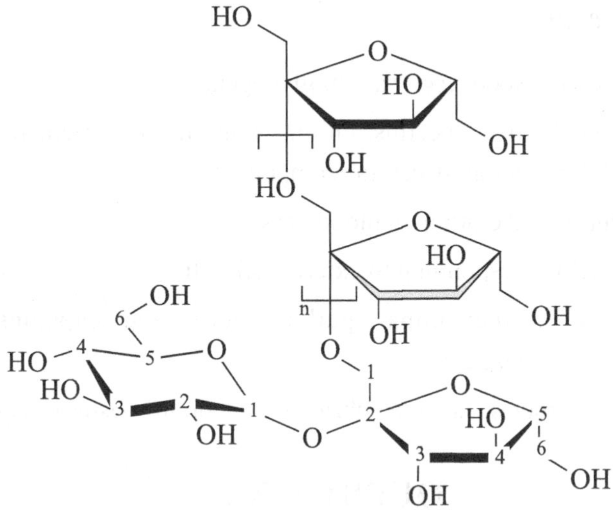

Fig. 2.6. Inulin

(*f*) ***Mannans:*** These are polymers of mannose and are found in yeast cell wall. The mannose units are linked by α-1,6 linkages and at branches the bonds are α-1,5 linkage or α-1,3 linkage type.

(*g*) ***Xylans:*** These are polymer of D-xylose that are linked by β-1,4 linkage. It is a type of hemicellulose.

(*ii*) **Heteropolysaccharides:** They contain more than one types of monomers (generally 2), *e.g.*, pectins, hemicellulose, some gums, mucilage, agar, etc. Some important heteropolysaccharides are:

(*a*) ***Agar:*** It is made up of D-galactose and L-galactose (galactan) and is used in bacteriological medium. It is obtained from certain sea weeds (red algae).

(*b*) ***Pectins:*** It is found in cell wall, intercellular spaces of plants and consists of arabinose, galactose and galacturonic acid (acid of galactose).

(*c*) ***Hemicellulose:*** It is found in the cell wall (in addition to cellulose) of plants and is made up of xylose, L-arabinose, D-galactose, L-rhamnose, D-mannose and D-glucuronic acid.

(*d*) ***Gums:*** It is extruded from plants on mechanical injury or after infection and seals the wound.

(*e*) ***Murein:*** It is found in cell wall of many bacteria and consists of a polysaccharisde (consisting of alternate units of N-Acetyl glucosamine or NAG and N-Acetyl muramic acid or NAM) and short peptides (tetrapeptide of L-*Ala*, D-*Glu*, L-*Lys*, D-*Ala* and pentapeptide of L-*Gly*) that is known as peptidoglycan. Lysozyme enzyme attacks on the peptidoglycan by hydrolysing NAG-NAM bonds.

Role of carbohydrates

(*i*) It is found as reserve food, *e.g.*, starch, glycogen, etc.

(*ii*) Cellulose, hemicellulose, pectins, chitin, murein are present in cell walls of many organisms and, thus, act as structural components.

(*iii*) Heparin prevents blood clotting in blood vessels.

(*iv*) Receptor sites help in responding to external stimuli.

(*v*) In many pathogens, they impart pathogenicity, *e.g.*, capsules in many bacteria (*Diplococcus pneumoniae*).

(*vi*) Antibodies, blood group substances, fibrinogen, albumin, plasma proteins are glycoproteins.

QUESTIONS

1. Explain what are **carbohydrates**?
2. Write short notes on:

   (*i*) Monosaccharides (*ii*) Disaccharides (*iii*) Polysaccharides.

Chapter 3

PROTEINS

Biomolecules are mostly organic compounds, *e.g.* nucleotides, nucleic acids, amino acids, proteins, lipids, carbohydrates, vitamins, hormones, etc.; a few are inorganic, *e.g.* H_2O, salts, mineral ions, etc. Some are small molecules (micromolecules), *e.g.* minerals, water, amino acids, sugars (mono and disaccharides), lipids, nucleotides, whereas, others are larger in size (macromolecules), *e.g.,* proteins, nucleic acids, polysaccharides.

INTRODUCTION

Protein word is derived from the **Greek** word, which means "*of primary importance*" or "*of the first rank*". These are linear polymers of amino acids and are macromolecules, which consist of mainly **C, H, O, N** and in some cases **S** that is present in the amino acids cysteine and methionine. Their molecular weight ranges from 12,000 daltons to several millions daltons. Molecular weight of proteins is determined by ultracentrifugation method. About 1,000–2,000 types of proteins are found in bacteria and 1,00,000 types in human cell. Most of the proteins have less than 2000 amino acids. The proteins can be degraded by **Sanger's reagent** (2,4-dinitrofluoro benzene). They are denatured by heat, UV, alcohol, acetone, acid, detergents, disinfectants, soap, etc, in which the 3-D structure of protein is lost and it becomes nonfunctional and, thus, gets precipitated.

Small proteins are soluble in water and are not coagulated by heat, *e.g.* histones; but as the size of globular proteins increases, solubility in water decreases and heat coagulability increases, *e.g.* egg albumin, serum globulins, etc. **Fibrous** proteins are insoluble in water. Glutelins are found in wheat and rice seeds.

Only plants and some microbes can synthesize proteins from inorganic nitrogenous materials. All proteins are **levorotatory** due to α-amino acids. Acid-base properties of proteins are due to terminal —NH_2, —COOH and R-groups.

Protein Tests: Some of the important methods to test the proteins are:

(*i*) **Biuret Test:** It is the test of peptide bonds:

Protein + 10% of NaOH

↓ Biuret

Heat

↓

1 drop $CuSO_4$

↓

Reddish-violet colour

(*ii*) **Millon's Test :** Protein + Million's reagent
[$HgNO_3$ + $Hg(NO_3)_2$ + HNO_3 + HNO_2]

↓

Red precipitate

(*iii*) **Ninhydrin Reaction :** α-Amino acid + Ninhydrin

↓

Blue colour

(*iv*) **Xanthoproteic Reaction :** Protein + conc. HNO_3
(like skin)

↓

Yellow colour

AMINO ACIDS

Amino acids are separated by paper chromatography, gel electrophoresis, etc. They are highly soluble in water and insoluble in organic solvents. They have high melting points.

General formula of amino acid is :

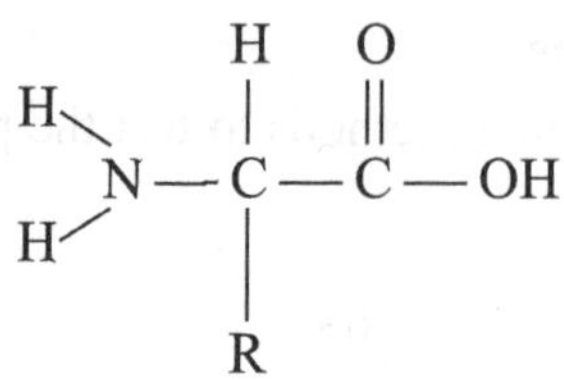

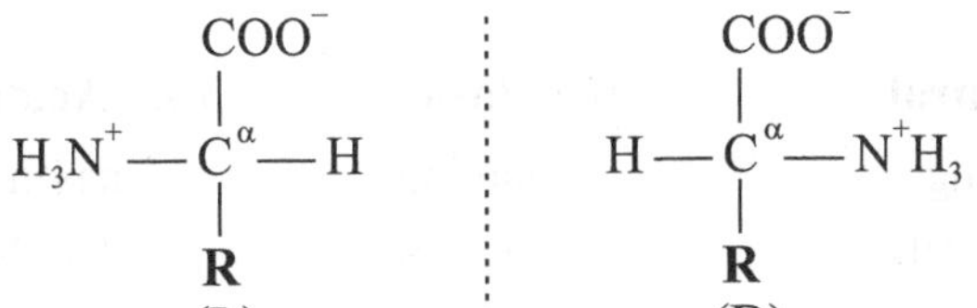

(L) (D)

(M)

Mirror images of the amino acid.

Proline amino acid (cyclic) is exception to this structure:

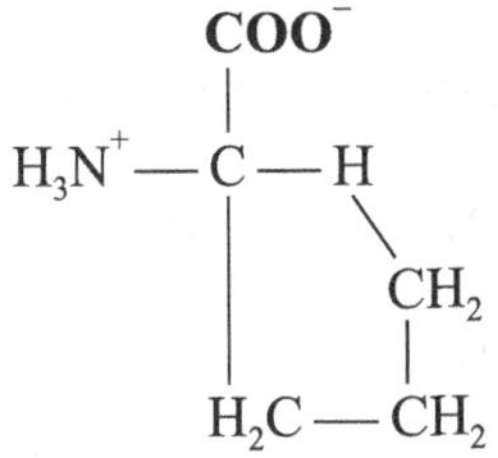

[**Racemic mixture** is the mixture of D- and L-forms of molecules.]

D- and L-forms of all the optically active compounds are compared with D- and L- forms of glyceraldehydes :

L (+) Amino acid L(–) Glyceraldehyde

[The —COO$^-$ groups of amino acid is comparable to the —CHO group of glyceraldehyde as after reaction with an oxidizing agent, —CHO group of amino acid is oxidized to —COOH group.]

The **α-C** of all amino acids is **asymmetric** (attached to 4 different groups) and, thus, are optically active, except **glycine,** in which α-C is symmetric and, thus, it is optically inactive:

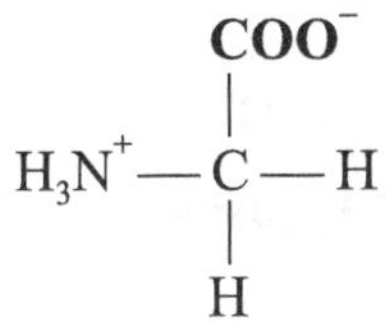

Optically active compounds rotate the plane of polarized light and their D and L (absolute configurations) stereoisomers, that are mirror images of each other, are possible. All naturally occuring amino acids are present in **L-form,** except in some cases they are found in D-form, *e.g.* peptidoglycan of bacterial cell wall, some antibiotics.

Amino acids can be:

(*i*) **Neutral**	(*ii*) **Basic**	(*iii*) **Acidic**
having	having	having
1 —NH_2	2 —NH_2	1 —NH_2
1 —COOH	1 —COOH	2 —COOH
e.g. Glycine, Alanine	Lysine, Arginine	Aspartic acid, Glutamic acid

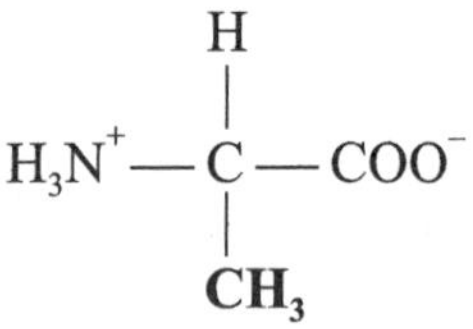
Alanine

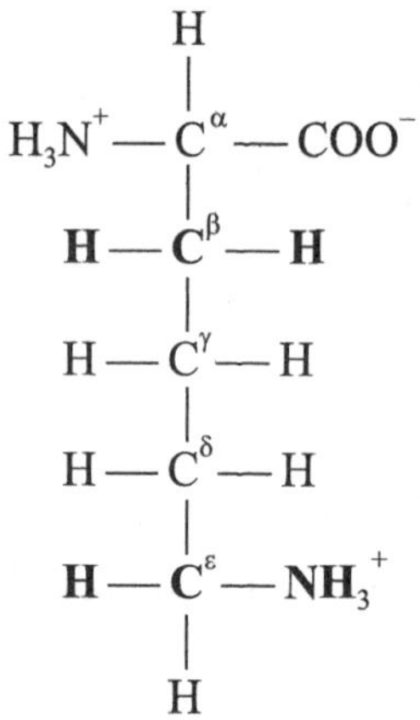
Lysine

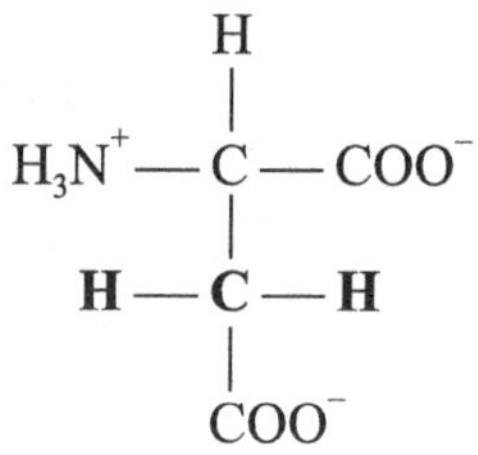
(*i*) Aspartate

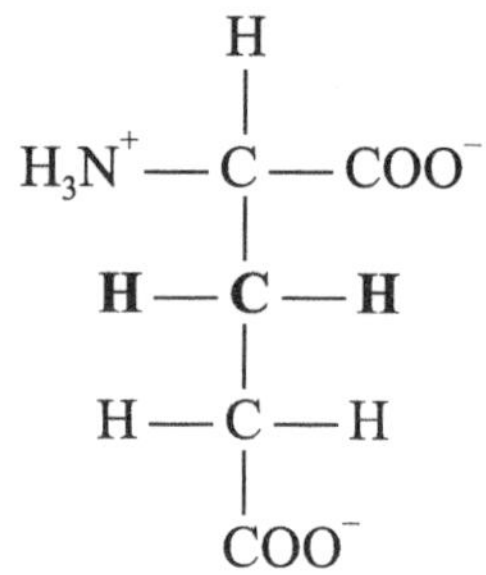
(*ii*) Glutamate

There are 20 **standard** amino acids which constitute the proteins. The **aromatic** amino acids (R → aromatic) absorb maximally in UV region.

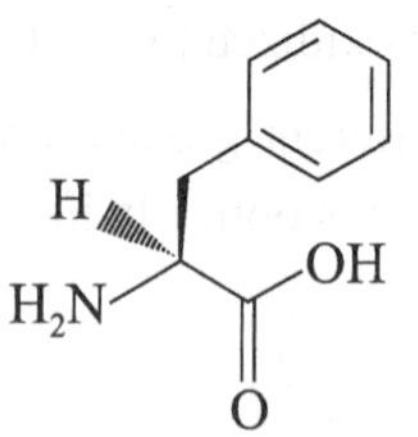
Phenylalanine

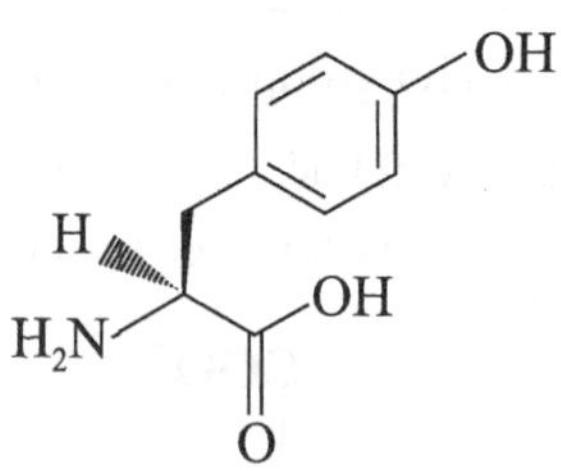
Tyrosine

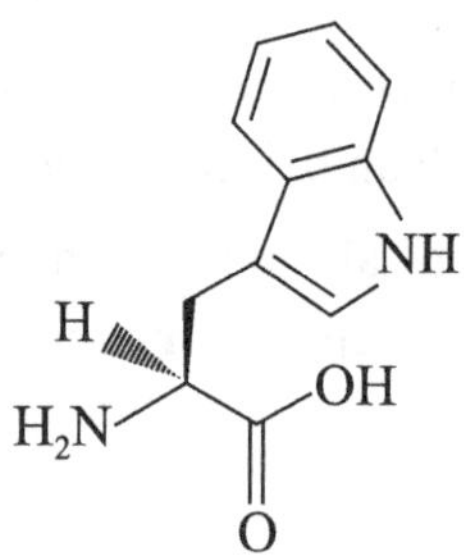
Tryptophan

Two amino acids are S-containing:

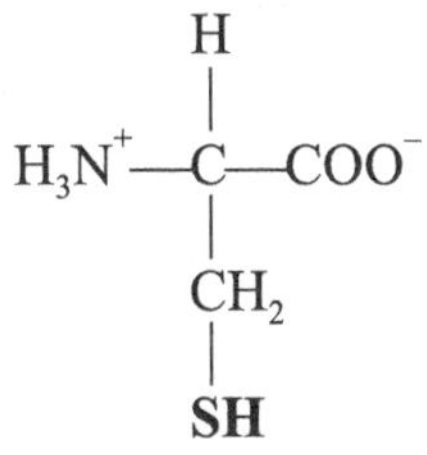

Cysteine

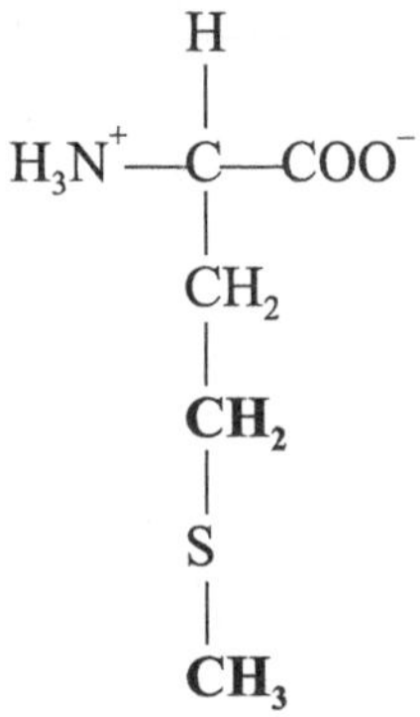

Methionine

In addition to these standard amino acids, there are many **non standard amino acids** that are either:

(*i*) Found in proteins and derived from standard amino acids. Here, modification occurs after incorporation into the proteins, *e.g.* hydroxyl-proline, hydroxyl-lysine, etc.

(*ii*) Not found in proteins and are intermediates of various biosynthetic pathways (about 300 types) *e.g.* ornithine, citrulline.

Ionization of Amino Acids:

(*i*) At **isoelectric point** (specific for each amino acid, about **pH ~ 6**) the —COOH group of amino acids becomes negatively charged and -NH_2 group positively charged and the molecules at such points are called **zwitter ions** (dipolar ions) that is a German word, which means hybrid ions. In this situation, the molecule can act as an acid (H^+ donor) or as a base (H^+ acceptor) and does not migrate in an electric field.

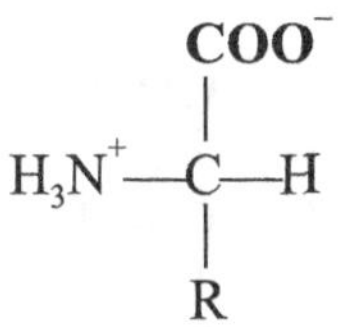

Zwitter ion

Thus, the pH at which a molecule has equal number of positive and negative charges is known as isoelectric point (no net charge) and on titration with a base, the —NH_3^+ group acts as the H^+ donor and gets neutralized to —NH_2 group. Whereas, on titration with an acid, the —COO^- group acts as H^+ acceptor and gets neutralized to —COOH.

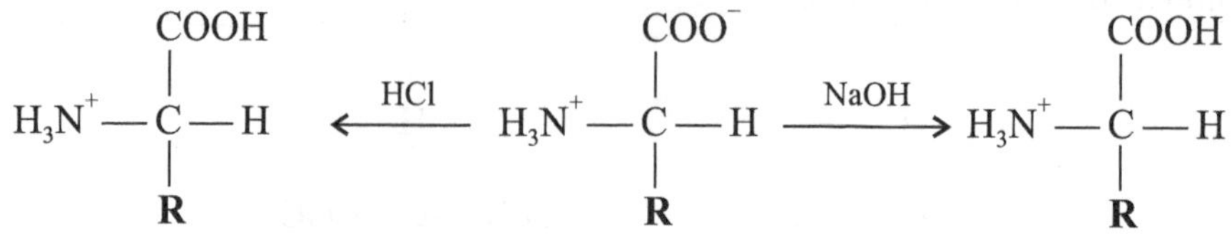

Cation Zwitter ion Anion

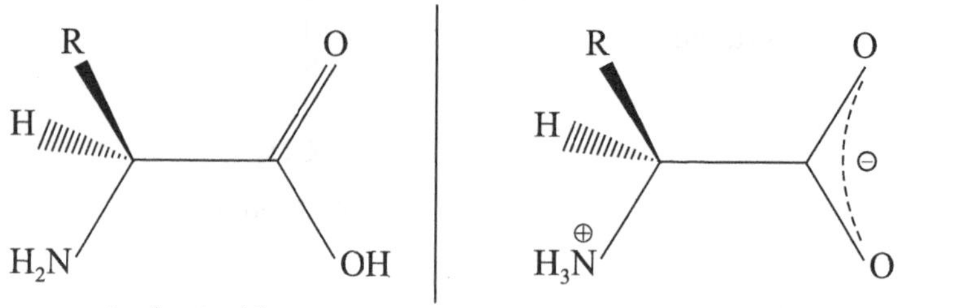

Amino acid

Zwitter ion form of the amino acid

(*ii*) At **lower pH** (acidic), the —COO^- group after accepting the H^+ from the acid becomes neutralized to —COOH, whereas the —NH_3^+ group remains as such. The amino acid, thus, becomes a cation and moves towards the cathode in an electric field.

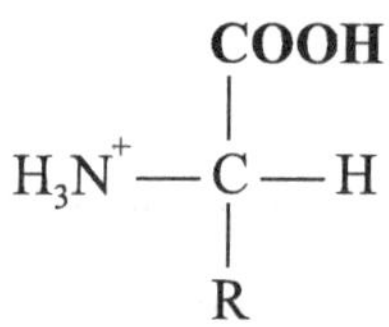

(*iii*) At **higher pH** (basic), the —NH_3^+ group after donating a H^+ becomes NH_2, whereas, the —COO^- group remains as such, and the amino acid becomes an anion and moves towards the anode in an electric field.

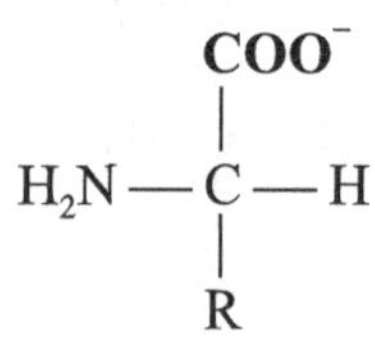

This form of amino acid cannot exist at any pH:

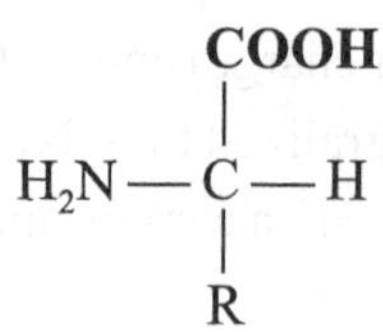

The adjacent amino acids in proteins are covalently linked by **peptide bonds,** that is formed by the condensation reaction through elimination of H_2O between —NH_2 group of one amino acid and —COOH group of other amino acid (Fig. 3.1).

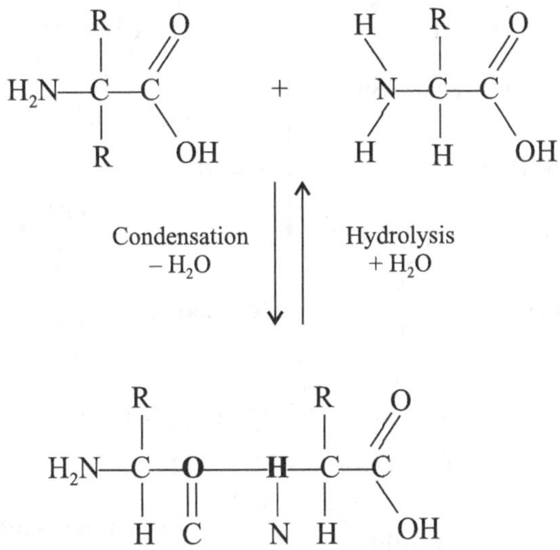

Peptide linkage

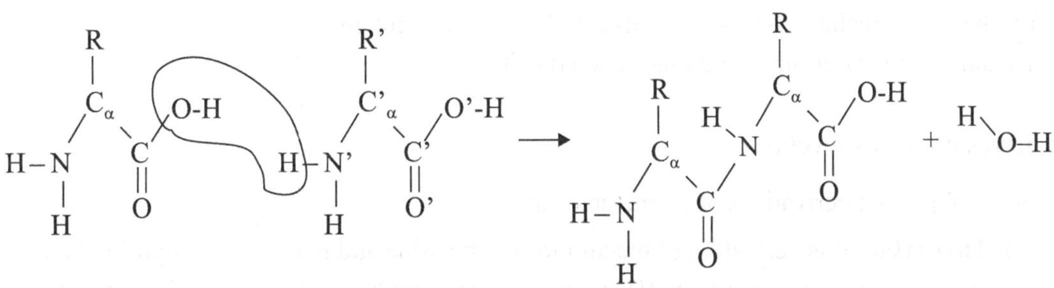

Amino acid 1 Amino acid 2 Two amino acids linked by peptide bond Water

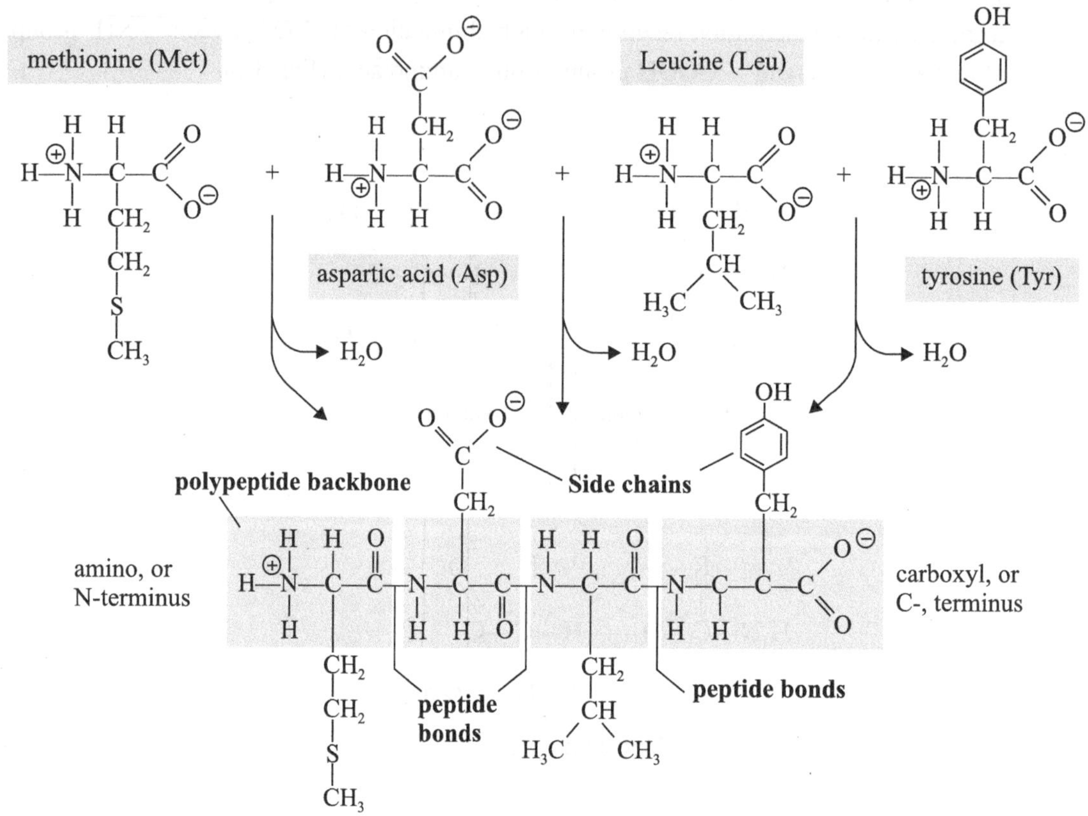

Fig. 3.1 Mechanism of peptide bond formation.

After elimination of a molecule of H_2O, the amino acid is known as residue and is shown by the suffix -lyl, *e.g.* Glycine + Alanine → Glycyl-alanine. The protein chain has —NH_2 group at one end (left) and —COOH group at the other end (right).

Derivatives of Amino Acids

Some important derivatives of amino acids are:

(*i*) **Thyroxine:** It is derivative of the amino acid tyrosine and is formed in thyroid glands. It contains four atoms of iodine (I) and helps in the regulation of growth and metabolism.

(*ii*) **Epinephrine** (*adrenaline*): It is also a derivative of tyrosine and is produced in adrenal medulla (above kidneys). It regulates heart beat and blood pressure.

(*iii*) **IAA** (*Indole-3-acetic acid*): It is a plant hormone and is a tryptophan derivative.

(*iv*) **Nicotinamide:** It is a vitamin and is tryptophan derivative.

(*v*) **Diamino-pimelic acid:** It is derivative of lysine and is found in G^- bacterial cell wall (peptidoglycan).

(*vi*) **Melanin:** It is a skin pigment and is the derivative of tyrosine.

Classification of Proteins

1. On the basis of **composition,** the proteins are of two types:

(*i*) **Simple Proteins:**

They are made up of only amino acids (or their derivatives), *e.g.* albumins (in blood), globulins, histones (basic proteins), ribonuclease, etc.

(*ii*) **Conjugated Proteins:**

In addition to amino acids, they also contain non-protein prosthetic groups, *e.g.* glycoproteins (immunoglobulins), phosphoproteins (casein), lipoproteins, chromoproteins (haemoglobin, where haeme is the prosthetic group, cytochromes having iron porphyrin), metalloproteins (nitrogenase has Mo, plastocyanin has Cu), etc.

2. On the basis of **size,** the poteins are classified as:

(*i*) **Peptides**

They contain 2–20 amino acid residues and may be present in free state (may be protein degradation products or in combination with carbohydrates (*e.g.* peptidoglycan or mucopeptides). Some examples are:

Cystine, where two cysteine residues are linked by disulphide linkage (—S—S—).

Glutathione, which is a tripeptide, γ-Glu-Cys-Gly.

Oxytocin consists of nine amino acids and is secreted by posterior pituitary (in brain). It helps in contraction of uterus in labour.

Vasopressin is secreted by posterior pituitary and helps in constriction of blood vessels, increases blood pressure, promotes re-absorption of water in kidney tubules.

Gramicidin, an antibiotic is produced by *Bacillus brevis* and is a cyclic decapeptide.

Tyrocidin, again an antibiotic, is a cyclic decapeptide.

(*ii*) **Polypeptides:** They contain 20–50 amino acids (molecular weight below 10,000)

(*iii*) **Proteins:** They have generally more than 50 amino acids. The proteins may contain more than one polypeptide chains. Some examples are:

Ribonuclease contains one polypeptide chain that is folded by disulphide bonds (–S–S–). It degrades RNA.

Lysozyme also contains one polypeptide chain, which is folded by disulphide bonds (–S–S–). It degrades peptidoglycan by cleaving glycosidic bonds between NAM-NAG units.

Papain has one polypeptide chain, which is folded by disulphide bonds (–S–S–). It is a proteolytic enzyme and is obtained from the latex of papaya.

Insulin is excreted from pancreas (*Islets of Langerhans*) and is a hormone that converts glucose to glycogen. It was the first protein of which amino acid sequence was determined by **Sanger** (1951). It contains two polypeptide chains that are covalently linked by –S–S– bonds. It has 51 amino acid residues, of which 21 amino acids make α-chain and rest of the 30 amino acids form β-chain (Fig. 3.2).

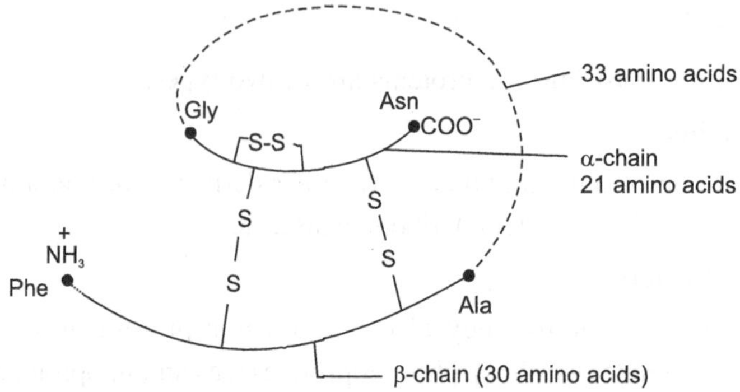

Fig. 3.2 Structure of insulin

Glucagon is a pancreatic hormone that converts glycogen into glucose and consists of 29 amino acids.

Haemoglobin contains four polypeptides, where, two are α-chains and two are β-chains. **Sickle cell anaemia** is caused due to difference of only one amino acid in β-chain, so that the 6th amino acid, glutamate, is replaced by valine. In this disease, the erythrocytes becomes sickle shaped and have reduced O_2 exchange capacity in comparison to the normal ones.

	1 2 3 4 5 6 7 8
Normal (HbA)	$^+NH_3$—Val—His—Leu—Thr—Pro—**Glu**—Glu—Lys----
Sickle cell anaemia (HbS)	$^+NH_3$—Val—His—Leu—Thr—Pro—**Val**—Glu—Lys----

A part of β-globin protein of normal and sickle cell haemoglobin.

Structural Levels of Proteins

Just after the protein synthesis, the protein gets folded into 3-dimensional form to attain secondary, tertiary and quaternary structures. Thus, four levels (Fig.3.9) of organizations in proteins are:

1. Primary Structure

They consist of amino acids linked by peptide bonds and are generally non-functional, *e.g.* few plant proteins (Fig. 3.3). Insulin, which regulates glucose metabolism in mammals, was the first protein whose primary structure was determined by **Sanger** (1951).

$^+NH_3$—Val—His—Leu—Thr—Pro—**Glu**—Glu—Lys-----COO^-

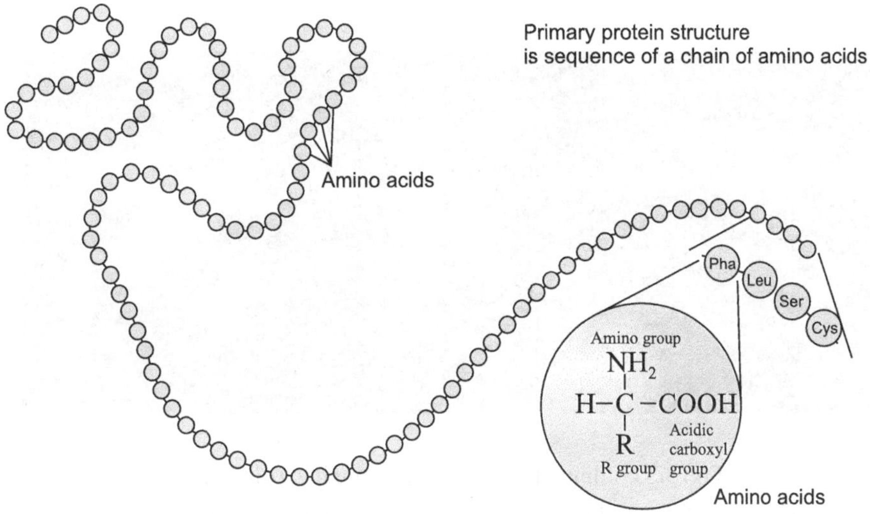

Fig. 3.3 Primary structure of a protein.

2. Secondary Structure

Here, polypeptide chains are folded or coiled to attain second level of organization through hydrogen bonding between various functional groups, *e.g.*

$$\begin{array}{ccc} —\overset{|}{C} = O & & H—\overset{}{N}— \\ | & & | \\ O & \cdots\cdots\cdots\cdots & H \end{array}$$

Examples of such proteins are fibrous proteins (thread-like), which is mainly structural, collagen, α-keratin, fibroin, myosin in muscle. These are insoluble in water. Two main secondary structures are:

(*i*) **α-Helix:** Here, the amino acid back bone is right-handed coiled (helical) and it is the most common coiling found in proteins (Fig. 3.4 and 3.5). It is more stable for L-amino acids than left handed coiling and has 3.6 amino acids / turn (18 a.a. / 5 turns). It is hard and their examples are α-keratins, found in fur, claws, hooves, feathers, hair, nails, **shell of tortoise** (hardness due to extensive –S–S– bonds).

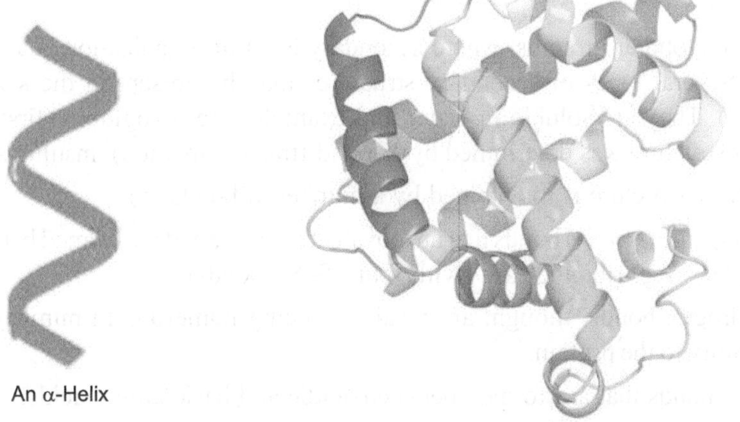

Fig. 3.4 α-Helices

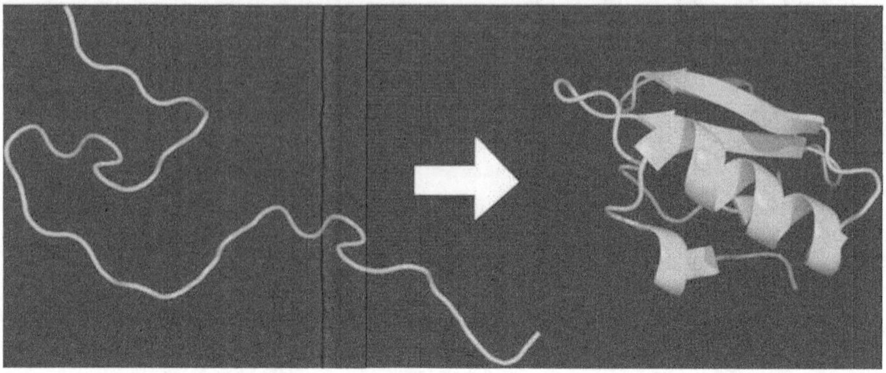

Fig. 3.5 Folding of a primary structure into α-helix.

(*ii*) **β-conformation:** It is 3-D (three dimensional) structure and is also known as β-pleated sheet as the structure is folded into flat sheet with zig-zag backbone. Adjacent fibres are hydrogen-bonded (Fig. 3.6). This type of structure is flexible and some examples are β-keratin of silk fibroin, spider web fibroin, etc.

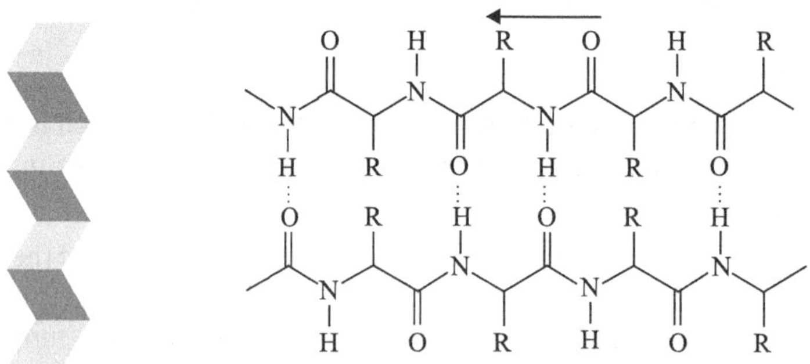

Fig. 3.6 β-Pleated sheet (β-conformation).

3. Tertiary Structure

These are globular proteins, where secondary level of organization is again twisted and folded. Several types of secondary structures may be present in the same polypeptide (Fig. 3.7). They are soluble in water and examples are myoglobin (first protein whose globular structure was determined by X-ray diffraction method), insulin, albumin, etc.

The tertiary structure is maintained by a number of bonds, *e.g.*

1. Disulphide (–S–S–) bonds that are covalent in nature and are formed between 2 cysteine residues, *e.g.* ribonuclease (with four –S–S– bonds).
2. Hydrogen bonds, though, are weak but being numerous in number provide quite stability to the protein.
3. Ionic bonds that are formed between acidic and basic amino acids.
4. Hydrophobic interactions occurring between hydrophobic amino acids that are organized in the interior of protein, between non-polar R-groups (in aqueous medium

maximum number of polar groups get organized towards the surrounding medium, whereas, the non-polar groups directed internally), etc.

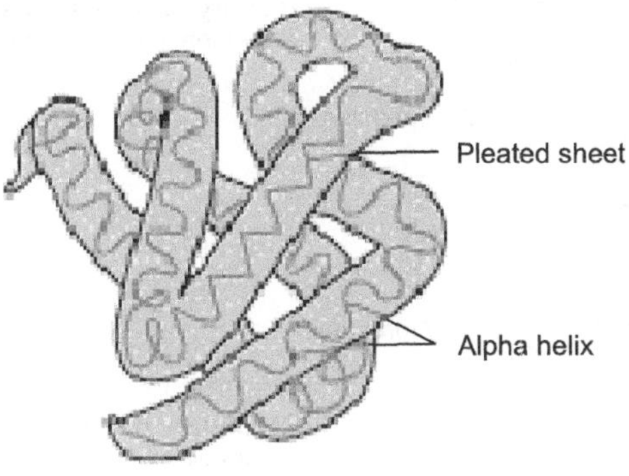

Fig. 3.7 Tertiary structure of a protein.

4. Quaternary Structure

It contains more than one polypeptide chains (subunits) that are associated together non-covalently. These subunits can be homogeneous (identical) or heterogeneous (non-identical, *e.g.*, haemoglobin consisting of two α- and two β-chains) (Fig. 3.8). Most common structures are dimers, trimers, tetramers, pentamers (*e.g.* RNA polymerase) and decamers. Glutamine synthetase enzyme has 12 sub units and proteins with a maximum of 32 subunits have been reported.

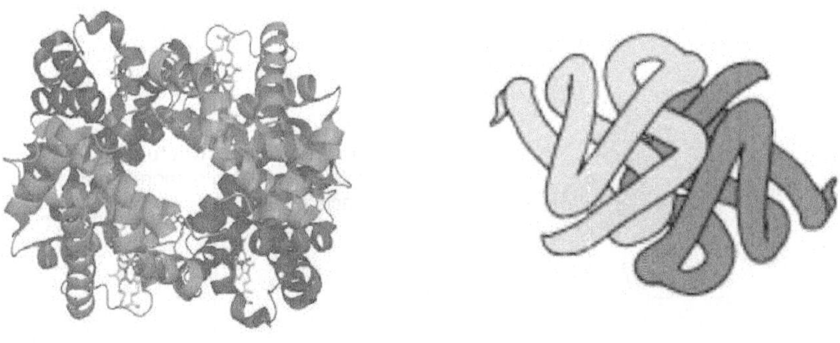

Fig. 3.8 Quaternary structure of haemoglobin having four polypeptides (left) and a protein with two polypeptides (right).

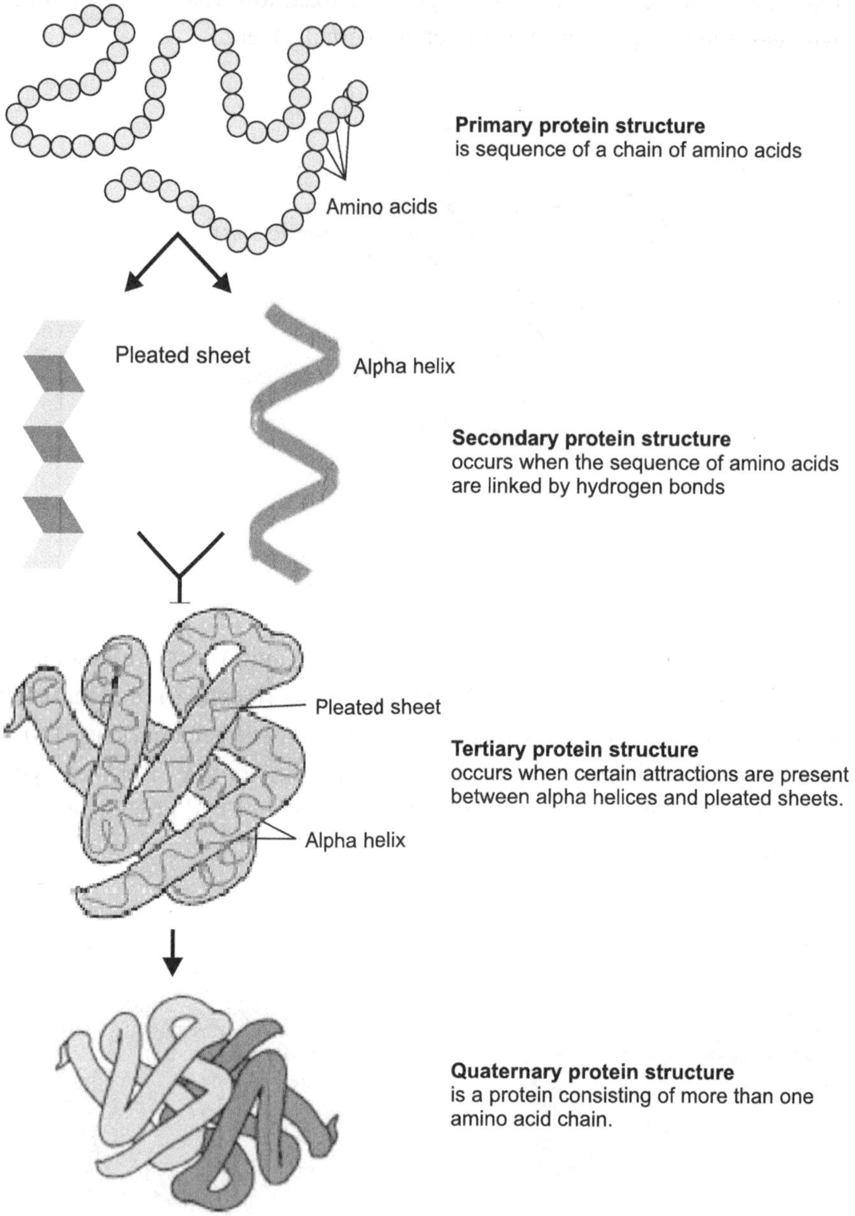

Fig. 3.9 Different levels of structure of proteins.

Role of Proteins

(*i*) **Structural** role, *e.g.* cell membrane, fibres, cartilage (made of collagen).

(*ii*) Many proteins are **enzymes,** *e.g.* ribonuclease, glutamine synthetase.

(*iii*) Many **hormones** are peptides or proteins, *e.g.* insulin, glucagon, oxytocin, vasopressin, growth hormone, etc.

(*iv*) **Blood proteins,** *e.g* haemoglobin (carries O_2), albumin (maintains osmotic pressure of blood), α-, β- and **γ-globulins** (immunoglobulins or antibodies, which are produced by B-lymphocytes), fibrinogen (helps in blood clotting).

(*v*) Many proteins help in **transport** *e.g.,*carriers in membranes, haemoglobin carries O_2, myoglobin carries and stores O_2 in muscles

(*vi*) **Movement** of muscles is performed by actin and myosin (contractile proteins).

(*vii*) Glutelin of wheat and rice performes **storage** function (cohesiveness and elasticity of dough is due to this protein, which is rich in -S-S- bonds). Egg albumin is also a storage protein.

QUESTIONS

1. Write characteristic features of amino acids. What are zwitter ions?
2. What are the different levels of protein organization?
3. Discuss the important roles of proteins.

Chapter 4

LIPIDS

The word lipid has been derived from the Greek word '*lipos*' and they are the **esters** of **fatty acids** with **alcohols.** Generally, they are **triglycerides,** containing three fatty acids linked to the alcohol, glycerol. Lipids are hydrophobic and may contain some hydrophilic groups. They are made up of C, H and O, but sometimes P, N, S are also present. They are insoluble in water (polar solvent) but, soluble in organic solvents (non-polar solvents), *e.g.* benzene, petroleum ether, acetone, chloroform, hot alcohol, etc.

Lipids are of three types:

1. **Simple Lipids:** They are **esters** of **fatty acids** with **alcohols,** *e.g.* fats (solid lipids), oils (liquid lipids), waxes, cutin, suberin, etc. **Fats** (having higher m.p. and generally, solid at room temperature) and **oils** (having lower m.p. and liquid at room temp.) are the esters of fatty acids with alcohols, whereas, **waxes** are esters of high molecular weight fatty acids with alcohols other than glycerol.

   (*i*) ***Oils and Fats:*** They are generally **triglycerides,** where three fatty acids (that may be similar or dissimilar) are esterified with the three hydroxyl groups of the trihydroxy alcohol, glycerol and are the main energy storing compounds in animals. In monoglycerides

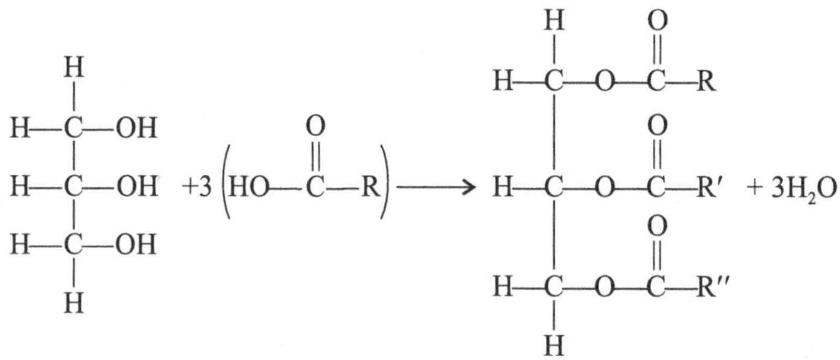

Glycerol — Fatty Acids (where R represents C—C chains) — Fat Molecule (R's may/or may not be the same)

and diglycerides, only one or two hydroxyls of the glycerol are esterified with one or two fatty acids, respectively. During each ester group formation between the —COOH of fatty acids and the —OH of alcohol a molecule of H_2O is given out.

The C chain length in fatty acids ranges from 4-24 and these fatty acids can be saturated (without any double or triple bonds) or unsaturated (with one or more double or triple bonds), *e.g.,*

Saturated fatty acids (have high m.p.):

(*i*) Butyric acid (4C chain)

(*ii*) Palmitic acid (16 C chain)

(*iii*) Stearic acid (18 C chain) (Fig. 4.1)

(*iv*) Lignoceric acid (24 C chain)

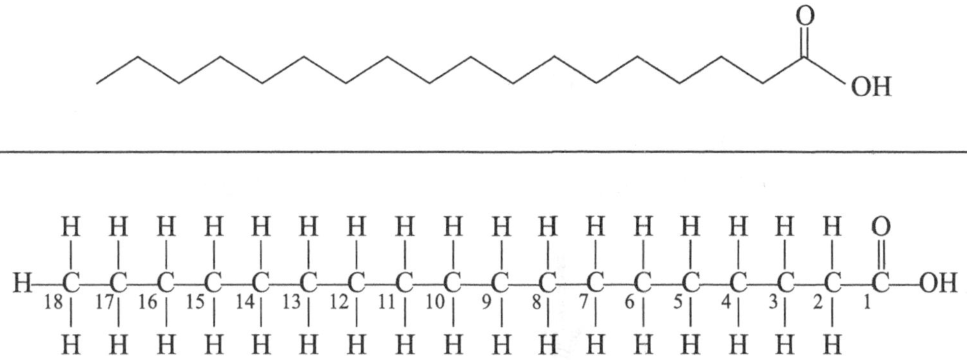

Fig. 4.1 Stearic acid, a saturated fatty acid.

Unsaturated fatty acids (have low m.p.):

(*i*) Oleic acid (18 C chain with 1 double bond, $C_9 = C_{10}$) (Fig. 4.2)

(*ii*) Linoleic acid (18 C chain with 2 double bonds)

(*iii*) Linolenic acid (18 C chain with 3 double bonds)

Trans oleic acid

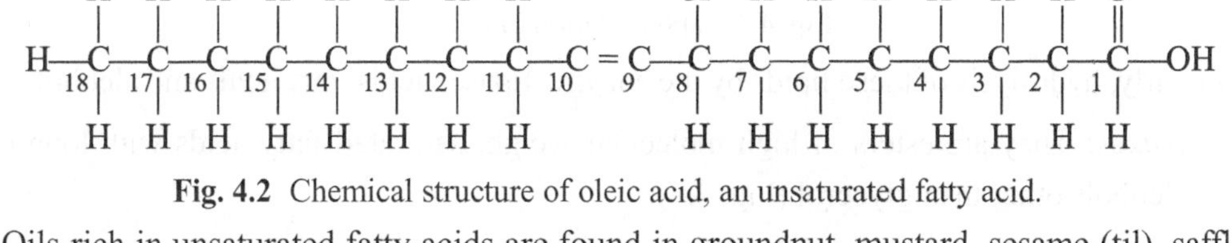

Fig. 4.2 Chemical structure of oleic acid, an unsaturated fatty acid.

Oils rich in unsaturated fatty acids are found in groundnut, mustard, sesame (til), safflower, etc. seeds, that provide nourishment to young plant during their germination.

As the number of double bonds in fatty acids increases, their m.p. decreases. Fatty acids with 16 C and 18 C chain length are most common, particularly the unsaturated ones, in living system. Hydrocarbon chains of the fatty acids are non-polar (hydrophobic).

Upon hydrolysis, these lipids **by alkali** of (saponification) form soaps (salts of fatty acids) and alcohol (glycerol) (Fig. 4.3).

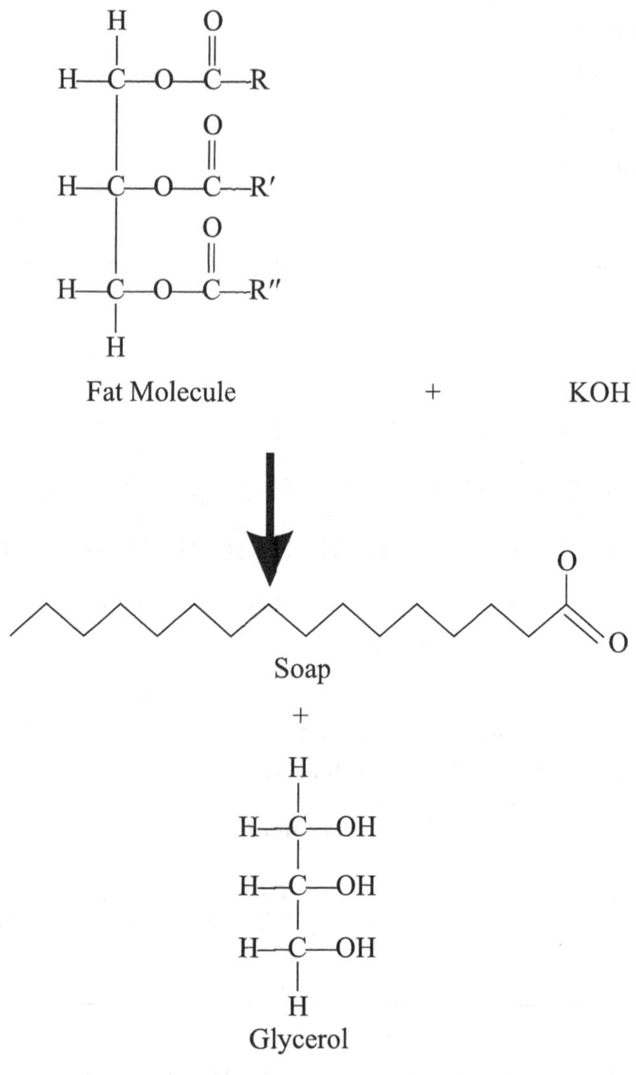

Fig. 4.3 Saponification of oil.

Similarly, hydrolysis of these lipids by the enzyme lipase yields fatty acids and alcohol.

(*ii*) ***Waxes:*** They are esters of high molecular weight saturated fatty acids with long chain alcohols other than glycerol (Fig. 4.4).

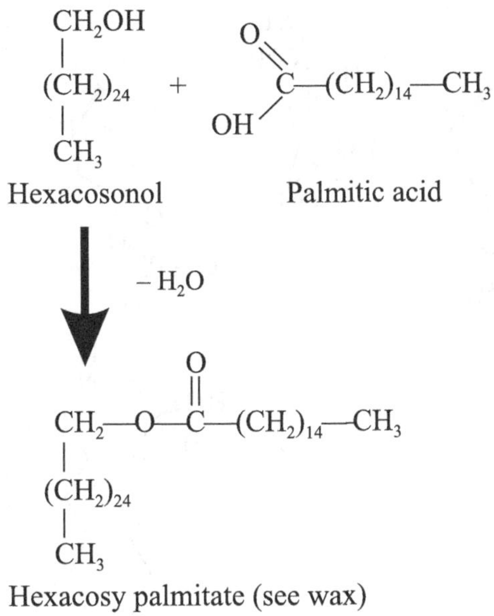

Fig. 4.4 Bee wax.

They are found as water barrier in exoskeleton of insects, in honey beehives, feathers of birds, furs of animals, as protective coating over human skin, fruits and leaves, etc. In many bacteria the wax present over the cell walls imparts pathogenicity as they are not engulfed by WBC of blood, *e.g. Mycobacterium tuberculosis* causing tuberculosis, leprosy bacterium etc.

2. Compound Lipids: They have **additional groups** in addition to fatty acids and alcohols, e.g. phospholipids, glycolipids, etc. (Fig. 4.5). In **phospholipids,** the phosphoric acid and two fatty acids are esterified to glycerol. They are amphipathic having both polar and non-polar groups, and in aqueous media they arrange in the form of a lipid bilayer with fatty acid tails towards the inside and phosphate groups towards the outside in contact with aqueous medium. They are found in cell membrane, egg yolk, etc.

In **glycolipids,** fatty acids and one or more sugars (usually galactose) are esterified to glycerol.

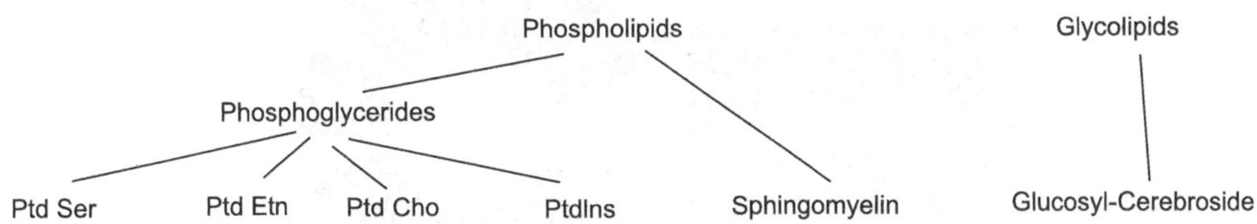

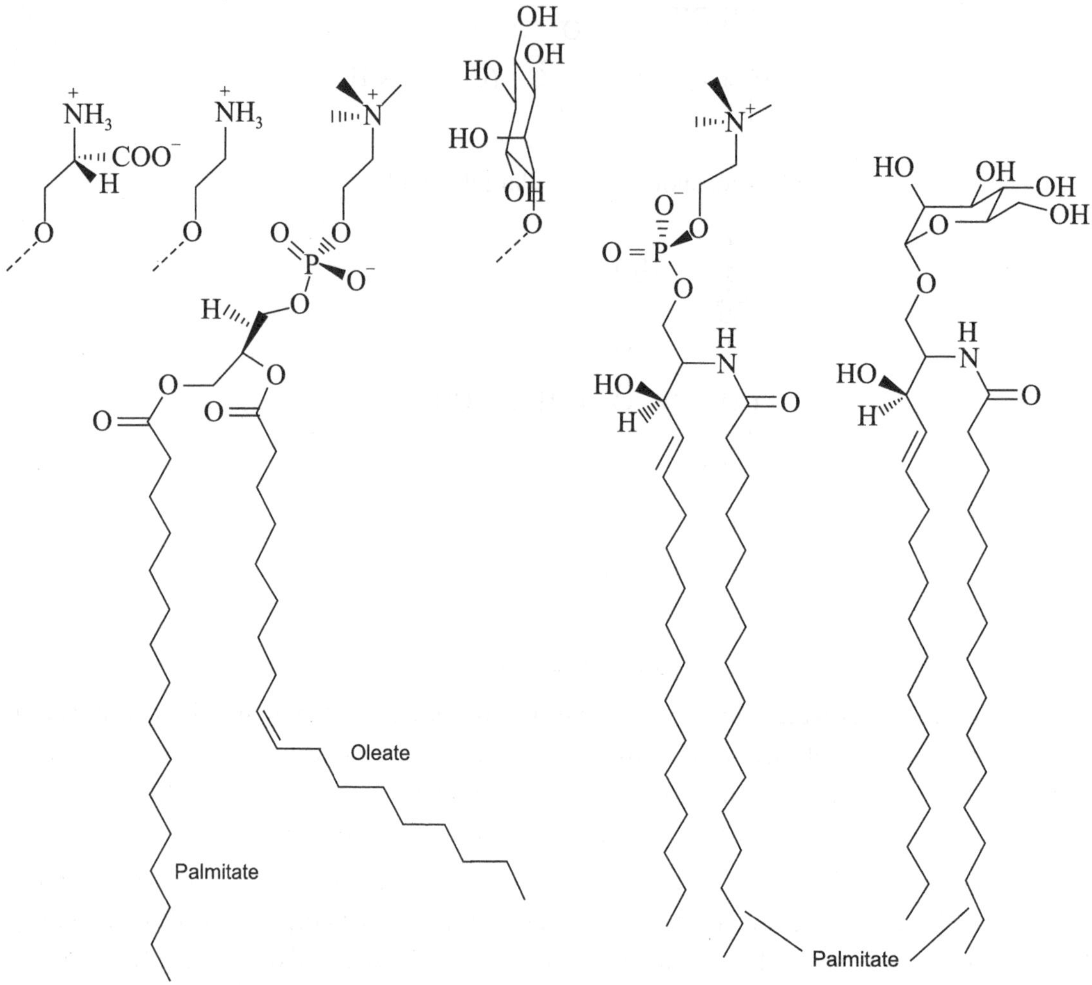

(1)

(2)

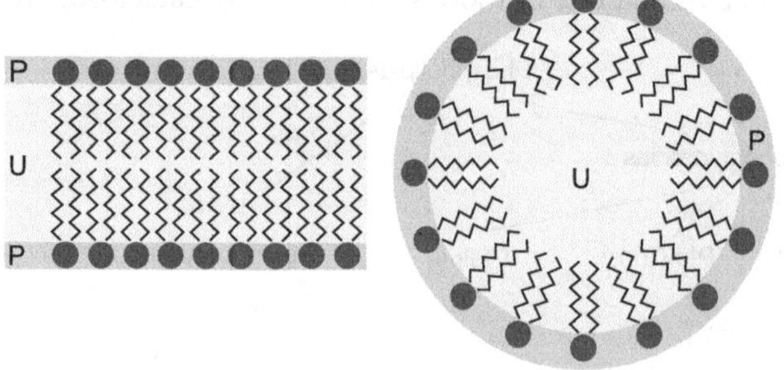

Phospholipid bilayer

A micelle

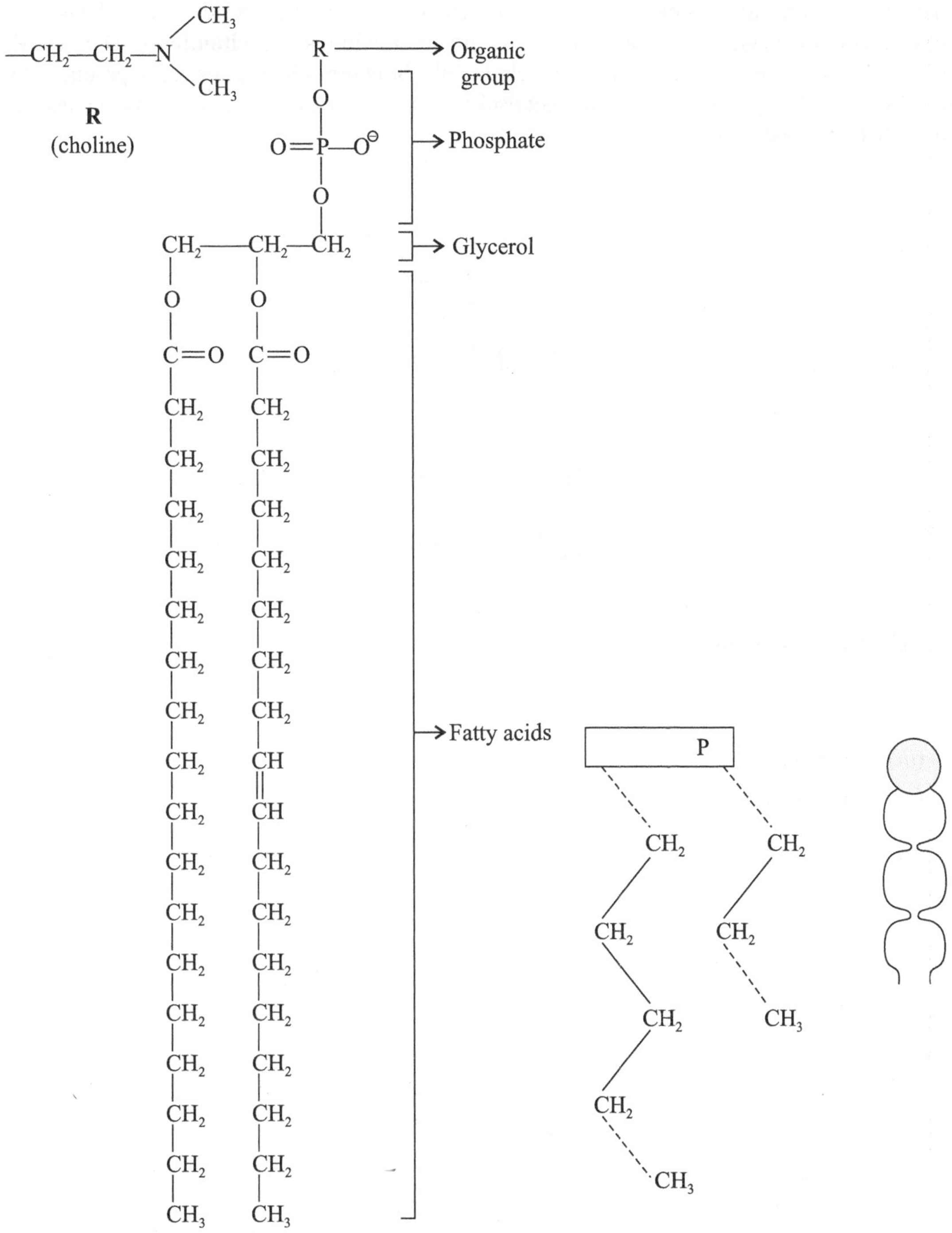

Phospholipids

Fig. 4.5 Compound lipids.

3. Derived Lipids: Though, they have similar physical properties like lipids, but, chemically they have fused hydrocarbon **four-ring** structure (sterols) with hydrocarbon side chains. Thus, they are not esters of fatty acids with alcohols and, thus, are not true lipids, *e.g.* **vitamins** A, D, E and K, cholesterol (Fig. 4.6) found in cell membrane, blood, bile, **hormones** like testosterone produced by testes, estrogen and progesterone by ovary, **cortisol** by adrenal cortex that regulates metabolism of fats, carbohydrates and proteins, etc.

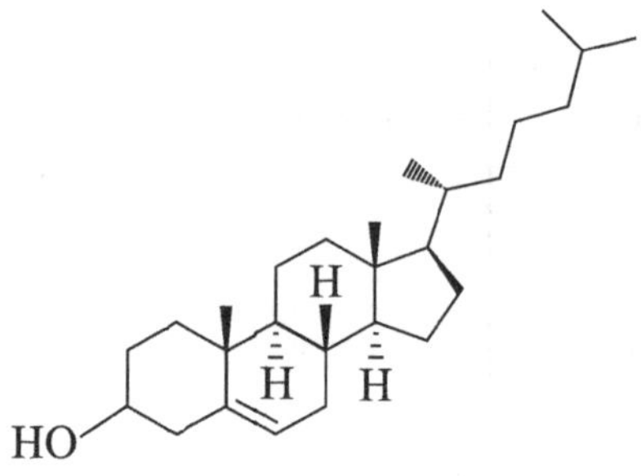

Fig. 4.6 Cholesterol.

QUESTIONS

1. What are lipids and discuss their role in living system.
2. Write short notes on:
   (*i*) Simple lipids
   (*ii*) Compound lipids
   (*iii*) Honey bee wax
   (*iv*) Oils and fats

Chapter 5

CELL DIVISION

For growth and reproduction, different types of cells undergo division process by various ways. Thus, the **prokaryotic cells** simply divide through **binary fission** method, in which the parent cell divides into two identical daughter cells for both growth as well as asexual reproduction. Here, after DNA replication, the two DNA molecules get distributed preciously into two daughter cells with the help of mesosome (an invagination of cell membrane).

Unlike prokaryotes, the **eukaryotic cells** divide by two types of more complex methods of cell division:

I. MITOSIS

This type of cell division is necessary for the maintenance and perpetuation of life. The growing tissues and asexually reproducing cells divide by mitosis, where the parent cell gives rise to two daughter cells that are identical not only to each other but also to their parent cell. Thus, the growth of organisms occurs due to mitosis and increase in size of cells. In advanced multicellular organisms, all the divisions from zygote (formed by the fusion of two gametes) to the adult are of mitosis type, giving rise to genetically identical cells of the individual. Generally, the organisms reproducing asexually give rise to **genetically similar organisms** (clones) within a species, except for variations arising due to mutations or other means, causing somaclonal variation.

Though, the cell division is a continuous process, but for convenience, it has been divided into various stages. The mitotic cell cycle has been divided into following stages:

1. Interphase: During this phase, since the level of coiling is very low, the individuality of the chromosomes is not prominent and they appear in the form of a mass, known as **chromatin** (Fig. 5.1).

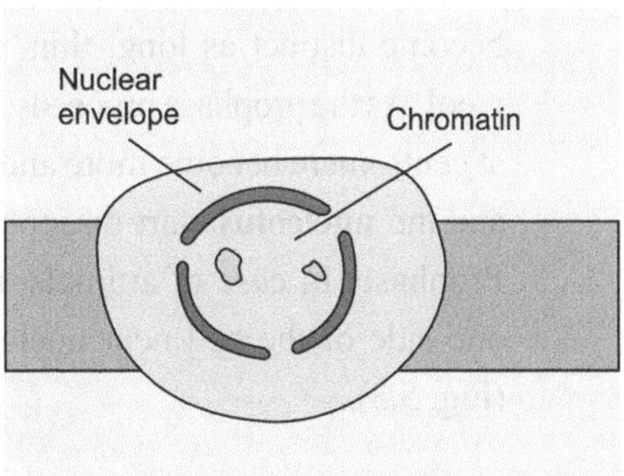

Fig. 5.1 Interphase.

During this phase, the cell prepares itself for division and it is divided into three phases (Fig. 5.2):

(*i*) ***G1 Phase:*** During this phase, the cell grows (G means growth or gap) in size and synthesizes the necessary RNAs and proteins for division.

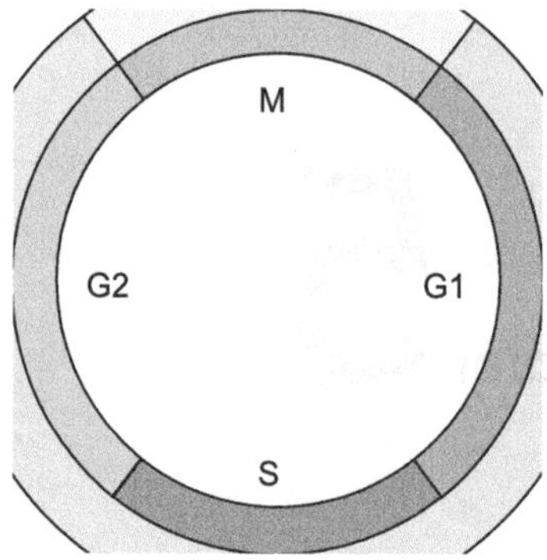

Fig. 5.2 Cell cycle

(*ii*) ***S Phase:*** The **DNA synthesis** (S means synthesis of DNA) takes place during this phase so that each chromosome, now, contains two DNA molecules still held together at centromere. Thus, though the chromosome number (the individuality of the chromosomes is not prominent and they appear as a mass, known as chromatin) remains the same, but due to DNA replication, the DNA content becomes double, *e.g.* if the diploid parent cell (having two sets of chromosomes) has $2n = 4$ chromosomes (each chromosome consists of 1 DNA molecule) with DNA content = 4C, after S phase the chromosome number $2n = 4$ remains the same (each chromosome, now, consists of 2 DNA molecules with 2 chromatids held at centromere) but, the DNA content becomes double (8C).

(*iii*) ***G2 Phase:*** Again, the cell synthesizes the proteins and RNAs required for mitosis. Thus, G1 and G2 are the resting (or gap) phases between DNA replication and the mitosis.

The Interphase varies from organism to organism, in different tissues, in different culture media, etc. from about 21–24 hours. It may be short in rapidly dividing cells.

2. Mitosis: In comparison to Interphase, the actual cell division process, Mitosis, is of short duration, about less than 1 hour. For convenience, Mitosis is again divided into four phases:

(*i*) ***Prophase:*** Due to continually increasing level of coiling, the individual chromosomes become distinct as long, thin, thread-like structures and appear as mass of interwoven wool. As the prophase proceeds, chromosomes (consisting of two sister chromatids attached at centromere become more and more thicker and shorter in size), the **nuclear membrane** and the **nucleolus** start disappearing and ultimately **disappear** completely at the end of Prophase. In case of animals, the two centrioles, lying at right angle to each other on one side of the cell near nucleus, separate and get arranged at the two opposite poles (Fig. 5.3).

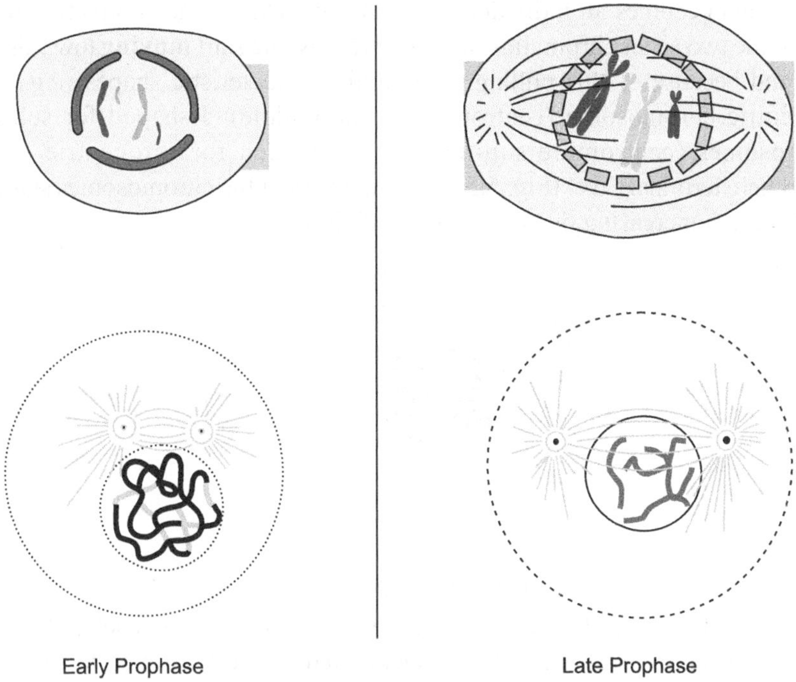

Fig. 5.3 Prophase.

(*ii*) ***Metaphase:*** The spindle fibres start appearing and get attached to the centromere of the chromosomes. In animals, the **spindle fibres** (tubules) are produced from a pair of centrioles, lying at the two opposite poles of the cell, but in plants where centrioles are absent, the mechanism of formation of spindle tubules is unknown. The chromosomes get arranged at the **equatorial plate** in the centre of the cell. At this stage, the degree of coiling is highest and the chromosomes become thickest and shortest, therefore, chromosome counting is generally performed at this stage (Fig. 5.4). [Colchicine obtained from *Colchicum autumnale*, inhibits the spindle fibre formation, causing polyploidy.]

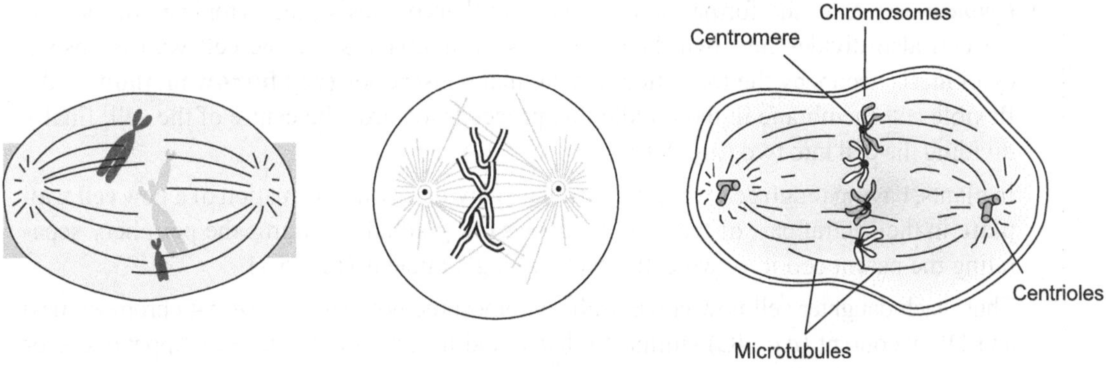

Fig. 5.4 Metaphase.

(*iii*) ***Anaphase:*** Each **chromosome splits** at the centromere and the two sister chromatids separate and becomes an individual chromosome. Due to the contraction of the spindle tubules, the two sister chromatids of each chromosome start **moving** towards the opposite poles and because of this pulling, they attain characteristic shapes, *e.g.,* V-shape for metacentric chromosome (centromere in the middle), J-shaped for sub-metacentric chromosomes (centromere sub-terminal), I-shaped for acrocentric chromosomes (centromere terminal), etc. (Fig. 5.5). The thickness of the chromosomes starts decreasing and the length increasing due to uncoiling of chromosomes.

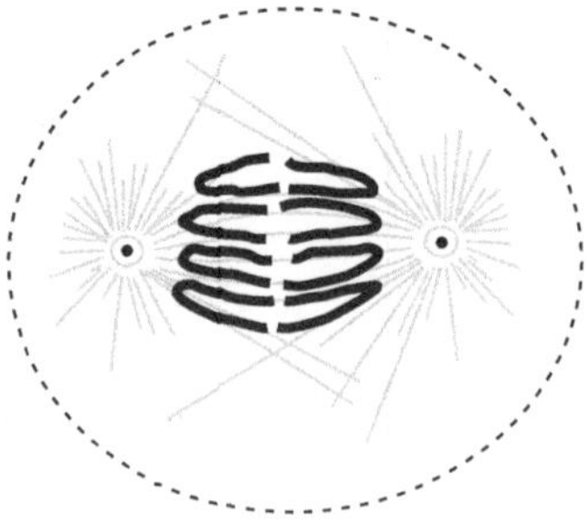

Fig. 5.5 Anaphase.

(*iv*) ***Telophase:*** Ultimately, the two complete sets of chromosomes reach at the two opposite poles of the cell, around which the **nuclear membrane** is formed to form the nucleus, that contains **nucleolus** also. The chromosomes become thin and thread-like, as in Prophase (Fig. 5.6).

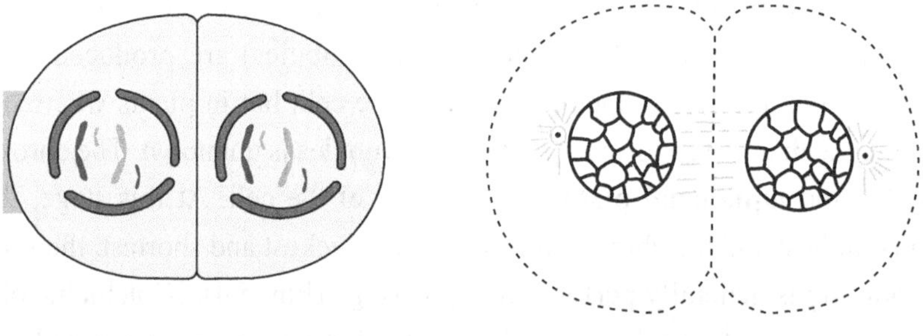

Fig. 5.6 Telophase.

(*v*) ***Cytokinesis:*** After the formation of two nuclei (karyokinesis), the cytoplasm of the parent cell also divides into two daughter cells. In animals, since the cell wall is absent, cytokinesis occurs by the formation of a circular constriction **(cell furrow method)** in the flexible cell membrane in the middle that proceeds towards the centre of the cell, finally dividing the cell into two (Fig. 5.8).

In plants, having tough cell wall, the cytoplasm divides by the formation of a new **cell wall plate** in the central part of the cell that grows out gradually towards the periphery separating the parent cell into two cells, each having a nucleus (Fig. 5.7).

Thus, each daughter cell now contains the chromosome number (*e.g.* $2n = 4$ chromosomes) and DNA content (*e.g.* 4C) similar to that found in the parent cell (See Appendix—for Fig. 5.7) and (Fig. 5.8).

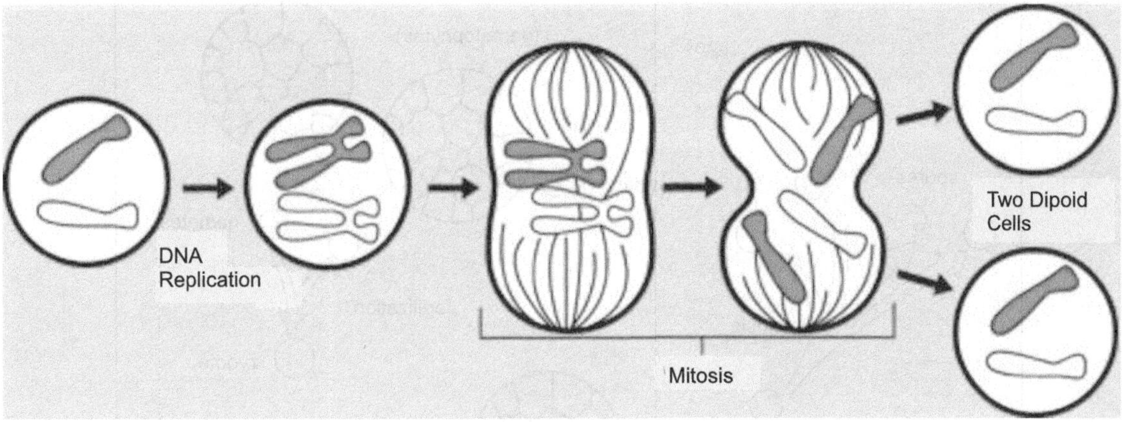

Fig. 5.8 Mitosis in brief (in animal cell).

The various stages of mitosis can be observed in rapidly dividing tissues, *e.g*, onion root tips. For the preparation of slides, about 5 mm long root tips are fixed in a solution of **acetic acid** and **methanol** present in 1 : 3 ratios. These tips are then treated with **1N HCl** and heated at 60°C for 15 minutes in a watch glass. After washing with tap water, the tips are put on a slide and stained with **acetocarmine** dye that colours the chromosomes. After covering the tips with coverslip, they are squashed with the thumb and warmed over the flame for a few seconds and finally observed under the microscope.

II. MEIOSIS

This type of cell division is found only in **sexually reproducing organisms,** where the male and female gametes (or positive and negative gametes in primitive organisms) fuse to form the zygote. This union of gametes in each generation would lead to successive doubling of chromosome number, causing unlimited number of chromosomes in the nucleus, which is undesirable. In such sexually reproducing organisms from generation to generation, the constant normal chromosome number is maintained by the specific type of division, meiosis, in which the **chromosome number** in daughter cells (gametic cells) is reduced to **half.** Thus, the meiosis involves two divisions of the parent cell, whereas, the DNA of the chromosomes replicates only once, during S-phase of Interphase. This produces **4 daughter cells** having **haploid** number of chromosomes (n) which differ not only among themselves with respect to the distribution of paternal and maternal chromosomes and exchange of genetic material (crossing over or genetic recombination), but also from the parent cell that has diploid number of chromosomes ($2n$). The crossing over causes the variation among various individuals within a species, that is necessary the evolution process.

In advanced organisms (*e.g.* higher plants, mammals, etc.), meiosis occurs during the formation of haploid **gametes** (*e.g.* eggs in females and sperms in males) that unite to form the diploid zygote having similar chromosome number as in somatic cells of normal male and female individuals. The zygote, thereafter, divides by mitosis to form the adult individual (Fig. 5.9).

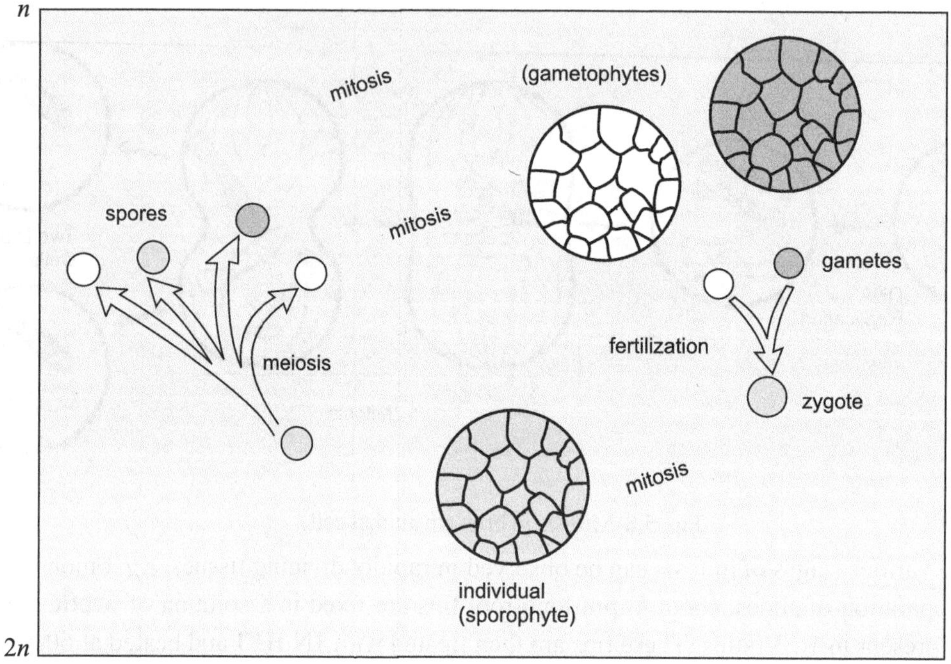

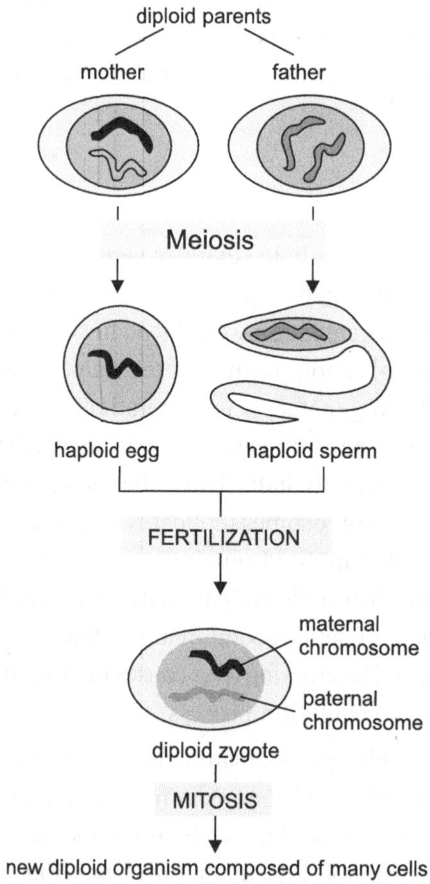

Fig. 5.9 Meiosis occurring at the time of gamete formation.

Whereas, in primitive organisms (*e.g.* green alga—*Chlamydomonas*, fungus—*Neurospora*, Bryophytes, Pteridophytes, etc.), where the normal body is haploid (n), at the time of sexual reproduction they produce male and female gametes (n) that unite to form the diploid zygote ($2n$). To attain the normal haploid chromosome number, the **zygote** undergoes meiosis (Fig. 5.10).

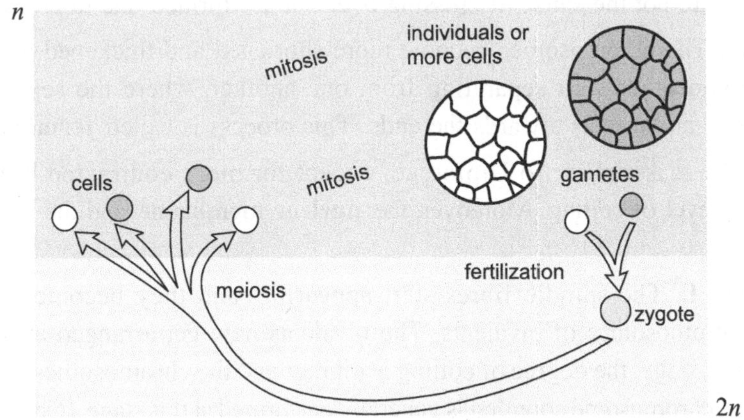

Fig. 5.10 Meiosis occurring after zygote formation.

Thus, meiosis involves **two divisions,** where the **first division** causes reduction in chromosome number in the two daughter nuclei (without dividing the sister chromatids of each chromosome) and the **second division** is similar to the mitosis that leads to the splitting and separating of the sister chromatids of individual chromosomes to the separate cells.

Like mitosis, meiosis is also a continuous process and has been divided into various stages. The meiotic cell cycle has been divided into following stages:

1. **Interphase:** It is similar to that of mitosis, divided into G1, S and G2 phases, where the DNA replication takes place during S-phase.

2. **Meiosis:** The cell division process, Meiosis involves two divisions:

I Division

This complex division is the reduction division that is again divided into four phases:

(*i*) **Prophase I:** In animals, the two centrioles separate and get arranged at the two opposite poles. The complex Prophase I is again divided into five stages:

(*a*) ***Leptotene:*** The individual chromosomes as long, thin, thread-like structures due to continually increasing level of coiling appear as loosely entangled **mass of wool.**

(*b*) ***Zygotene:*** The homologous chromosomes (similar chromosomes one maternal and the other paternal) undergo pairing (**synapsis**) in a zipper-like fashion to form bivalents having four chromatids (tetrads). During S-Phase of Interphase, total DNA replication does not occur and 0.3% DNA replicates during Zygotene that controls the synapsis.

(*c*) ***Pachytene:*** The chromosomes become thicker and shorter and exchange of segments (**crossing over** or recombination) of non-sister chromatids in some of the bivalents may take place by breakage and reunion, resulting in the formation of cross-like structures, called **chiasmata** (singular is chiasma) (Fig. 5.11 see Appendix). Number of chiasmata in a bivalent is directly proportional to the length of paired homologous chromosomes. Chiasmata are not the cause of crossing over but are formed due to crossing over.

(*d*) ***Diplotene*:** The chromosomes become more shortened and thickened and the homologous chromosomes start separating from one another, where the separation starts at centromeres and travels towards the ends. This process is called **terminalization.**

(*e*) ***Diakinesis:*** It is similar to Diplotene, except for more contracted bivalents due to increasing level of coiling. Moreover, the **nuclear membrane** and the **nucleolus** disappear.

(*ii*) **Metaphase I:** The spindle fibres start appearing, and they become attached to the centromeres of the chromosomes of bivalents. The bivalents now get arranged at the **equatorial plate** in the cell. At this stage, the degree of coiling is highest and the chromosomes become thickest and shortest, therefore, chromosome counting is generally performed at this stage. [Colchicine obtained from *Colchicum autumnale,* inhibits the spindle fibre formation, causing polyploidy.]

(*iii*) **Anaphase I:** The maternal and paternal chromosomes of each bivalent start moving towards the opposite poles, where each chromosome is made up of two chromatids. (Unlike mitotic Metaphase, the two sister chromatids of each chromosome remain attached to the centromere)

(*iv*) **Telophase I:** Two complete sets of chromosomes at the two opposite poles of the cell get surrounded by the **nuclear membrane** to form nuclei, each containing a nucleolus. Each nucleus, though, contains half the **chromosome number,** but each chromosome consists of **two** DNA molecules **(chromatids).** The chromosomes become thin and thread-like.

(*v*) **Cytokinesis I:** It may or may not be present.

After the first division, the resting phase Interphase may or may not be present, but if present, does not involve any DNA replication.

II Division

This division is simply the mitotic division where the two cells of Meiosis I divide to produce four cells:

(*i*) **Prophase II:** The chromosomes become distinct as long, thin, thread-like structures and the **nuclear membrane** and the **nucleolus disappear** at the end.

(*ii*) **Metaphase II:** The spindle fibres start appearing and the chromosomes, each made up of two chromatids, get arranged at the **equatorial plate.**

(*iii*) **Anaphase II:** Each **chromosome** is **splits** at the centromere and the two sister chromatids start **moving** towards the opposite poles.

(*iv*) **Telophase II:** Ultimately, the four complete sets of chromosomes separately get surrounded by a **nuclear membrane** to form four nuclei, each containing a **nucleolus.** The chromosomes become thin and thread-like.

(*v*) **Cytokinesis II:** The cytoplasm of the parent cell divides into four daughter cells, each containing half the number of chromosomes (*e.g.* $\boldsymbol{n = 2}$ chromosomes) and DNA content (*e.g.* **2C**) in comparison to the parental cell (*e.g.* with $2n = 4$ chromosomes and DNA content = 4C, which at S-Phase of Interphase I becomes 8C due to DNA replication) (Fig. 5.12 see Appendix).

At the time of Metaphase I, the arrangement of maternal and paternal chromosomes in a pair of homologous chromosomes is independent of the other pairs. Thus, $\boldsymbol{2^n}$ combinations become possible to give 2^n types of daughter cells, where n is the number of pairs of homologous chromosomes, *e.g.*, a cell with one pair of homologous chromosomes would give $2^n = 2 \times 1 = 2$ types of products, with 2 pairs giving $2^n = 2^2 = 2 \times 2 = 4$ combinations, with 3 pairs giving $2^n = 2^3 = 2 \times 2 \times 2 = 8$ combinations, etc., (Fig. 5.13).

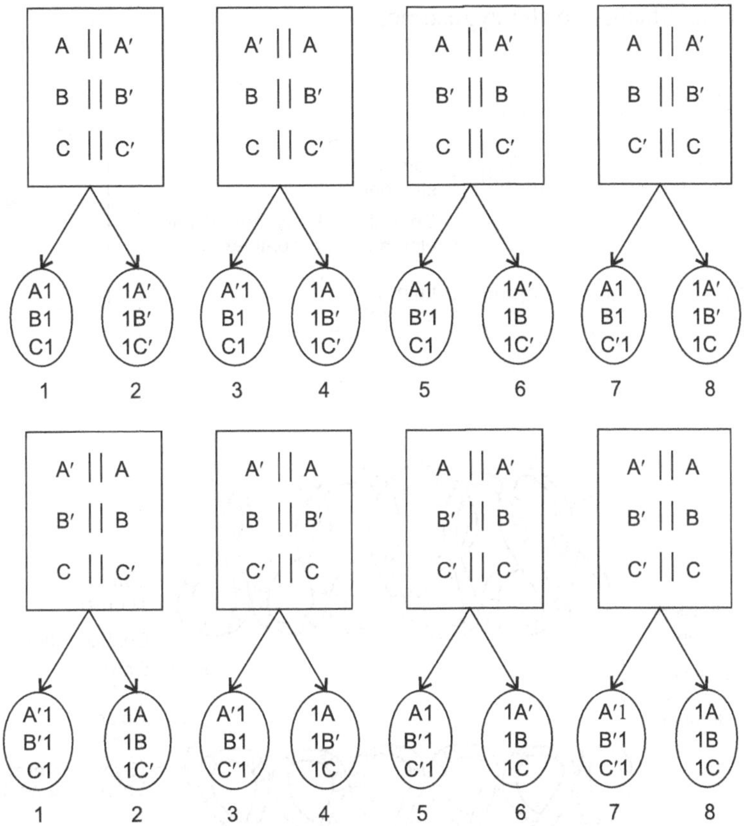

Fig. 5.13 Different arrangements of chromosomes at Metaphase I.

Synaptonemal Complex

It is a three-layered structure formed between the paired homologous chromosomes of a bivalent at the time of Prophase I (Zygotene and Pachytene) of meiosis. It facilitates the proper alignment of the paired homologous chromosomes that is necessary for the exchange of DNA segments (recombination). At the time of Zygotene, 0.3% replication of DNA is somehow involved in the formation of synaptonemal complex and, thus, if the cell is treated with the inhibitor of DNA synthesis at this time, it arrests the proper formation and functioning of the synaptonemal complex.

This complex was discovered by **Moses** (1956) and consists of three electron dense layers (two lateral elements and one central element) alternating with two electron lighter layers. The two **lateral elements** are made up of mainly DNA with some RNA and proteins, whereas, the **central element** consists of mainly RNA and proteins with a little DNA.

King (1970) gave a hypothesis, called **synaptomere-zygosome hypothesis,** where the two lateral elements are made up of about 10 nm (100 Å) fibres, called synaptomeres, whereas, the central element is ladder-like consisting of zygosomes.

Each **synaptomere** in the two **lateral elements** has three segments A, B and C, where A and C segments are involved in pairing with the similar types of segments of the adjacent synaptomeres, *e.g.* A segment of one synaptomere pairs with the A segment of other synaptomere, similarly C of one synaptomere pairs with C segment of other synaptomere. The B segments face towards the central element and are attached to the zygosomes.

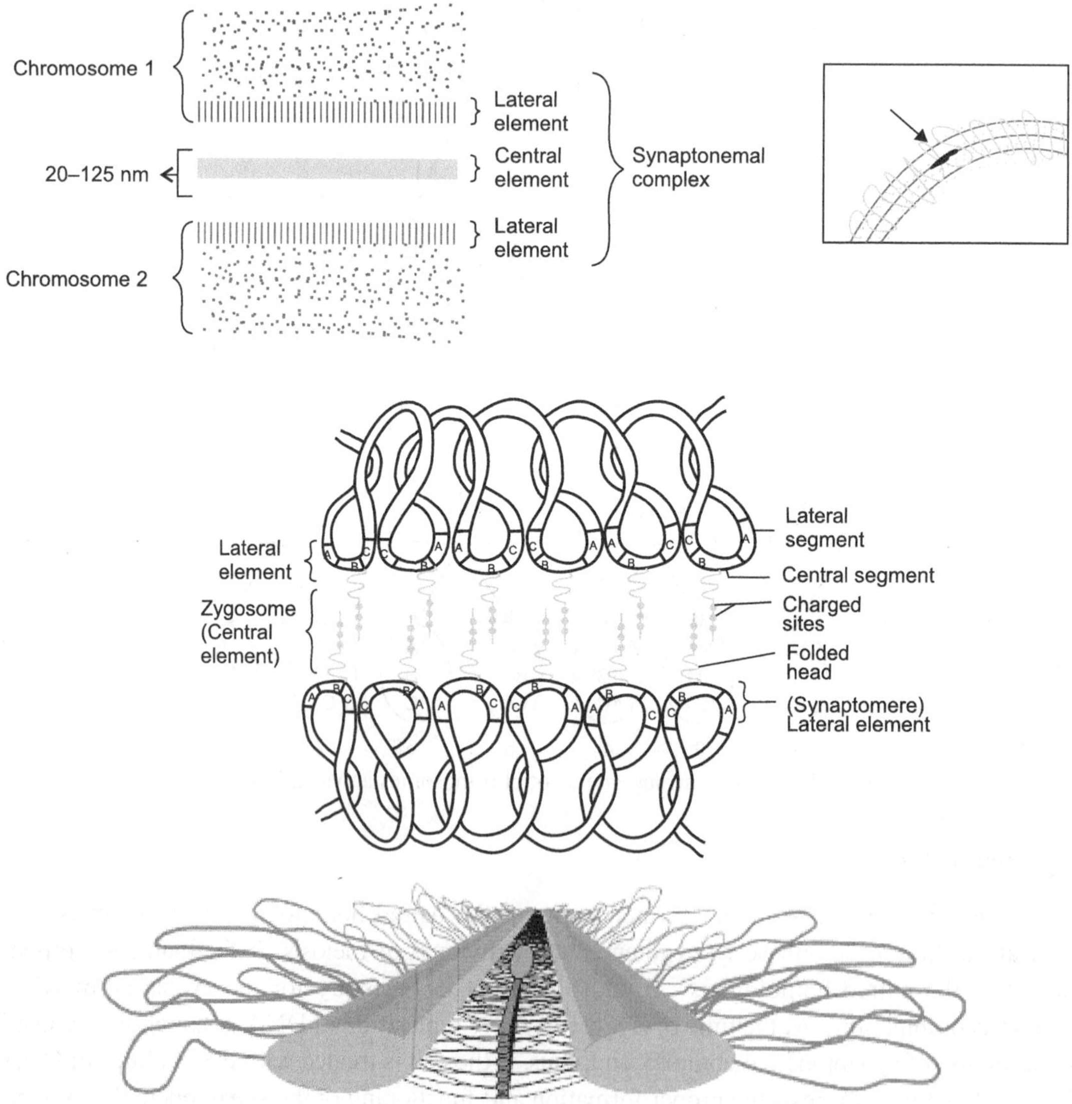

Fig. 5.14 Synaptonemal complex between homologous cromosomes during Prophase I.

The **zygosomes** in the **central element** are rod-shaped, each having a head that is attached to the B-segment of synaptomere and a tail with charged sites that bind laterally in a ladder-like fashion with thc tail of opposite zygosome (Fig. 5.14).

Comparison of Mitosis and Meiosis

Differences between mitosis and meiosis (Fig. 5.15):

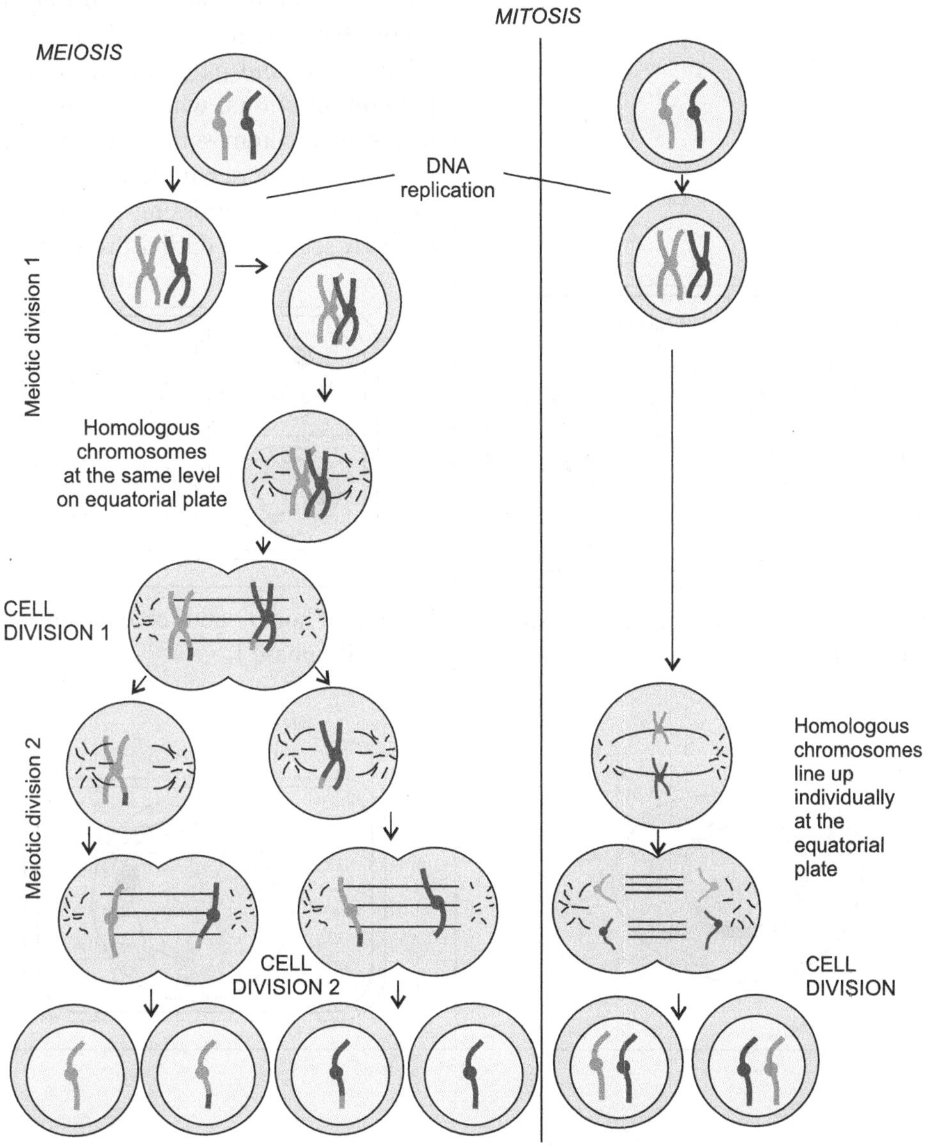

Fig. 5.15 Comparison of mitosis and meiosis.

S. No.	*Mitosis*	*Meiosis*
1.	It occurs in **somatic cells** for **growth** or **asexual reproduction.**	It occurs in **gametic cells** or zygote in **sexually reproducing** organisms.
2.	The product of mitosis are **two cells** that are not only **identical** with each other but also to its parent cell. 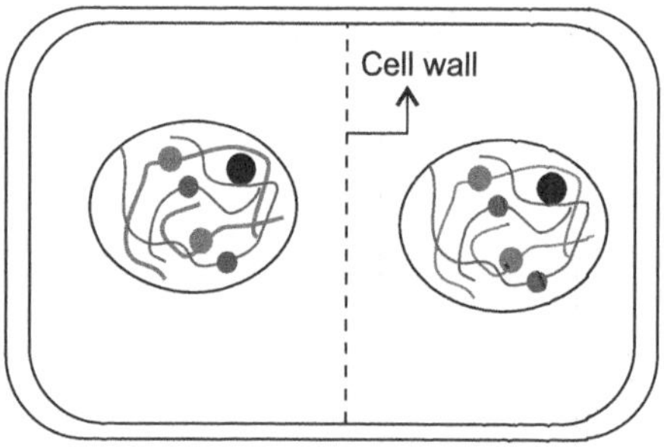	This type of cell division gives rise to **four daughter cells** that have **half the chromosome number** (n) as the parent cell ($2n$). These four daughter cells, though have same number of chromosomes, but **differ among themselves** due to independent orientation of maternal and paternal chromosomes (2^n combinations, where n = number of homologous chromosome pairs) at Metaphase I and then their separation at Anaphase I as well as due to crossing-over (exchange of DNA segments) during Prophase I. 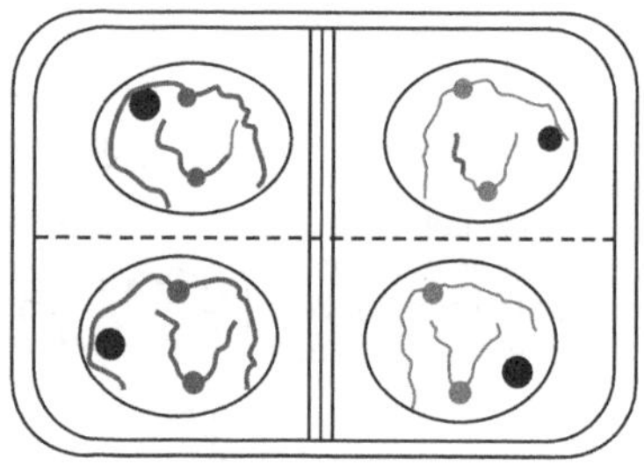
3.	No pairing of homologous chromosomes occurs here. 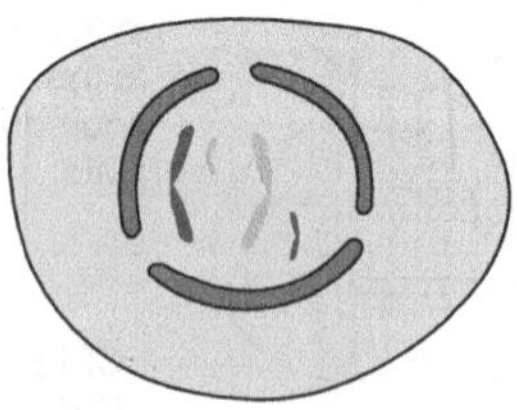	**Pairing** of homologous chromosomes occurs during Prophase I. 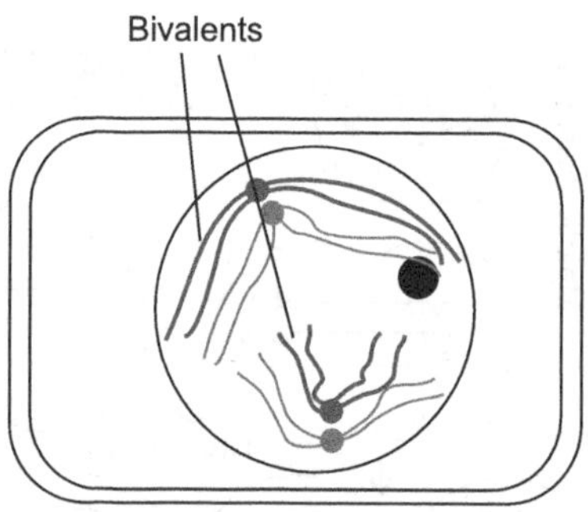

4.	Crossing-over does not takes place.	**Crossing-over** and, therefore, chiasma formation may take place during Prophase I. 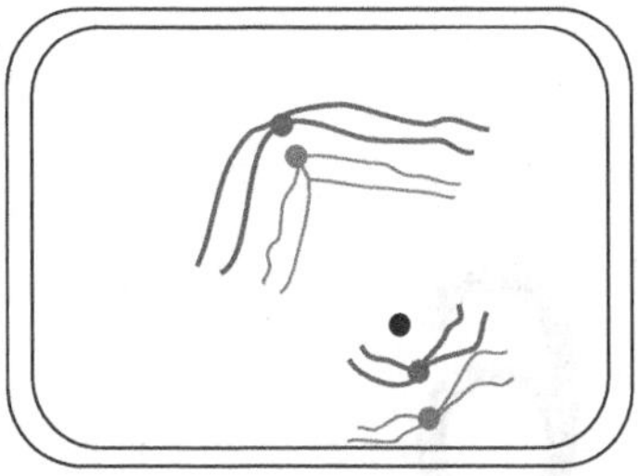
5.	During Anaphase, the centromeres divide and the chromatids after separation move towards the opposite poles. Movement of chromosomes, here, is called **equational separation.** 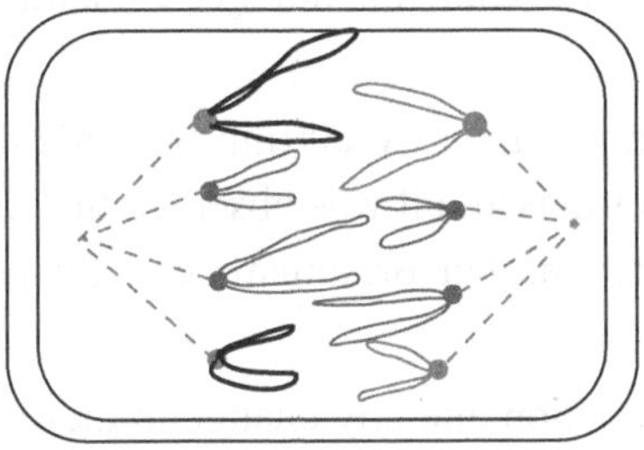	During Anaphase I, centromeres do not divide and the two chromosomes of each homologous pair start moving towards opposite poles. Movement of chromosomes, here, is called **reductional separation** (partly equational in the regions of crossing-over). 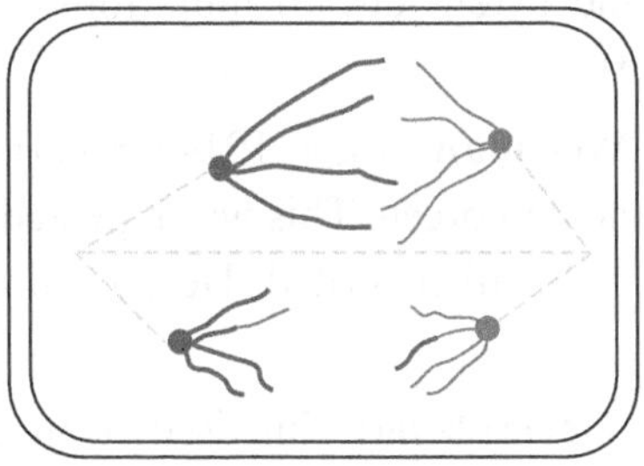

QUESTIONS

1. Describe the details of meiosis with particular emphasis on Prophase I. What is the significance of meiosis ?
2. What do you understand by reductional (disjunctional) and equational divisions. Demonstrate that first and second meiotic divisions are partly reductional and partly equational.
3. Explain why do you expect differences among cells arising from meiosis and not from mitosis.
4. Write a short note on synaptonemal complex.
5. Describe mitotic cell division and its significance.
6. Compare and contrast mitosis with meiosis.

Chapter 6 MENDEL'S LAWS OF INHERITANCE

The term 'genetics' was coined by **W. Bateson** (1905) and is the science that deals with heredity and variation. **Heredity** is the transmission of traits from parents to offspring, whereas variation can be hereditary (that are transmitted from generation to generation and arising mainly due to independent assortment of chromosomes and recombination in sexual reproduction and to some extent due to mutations) or environmental (that do not get transmitted to next generation).

A. Weisman (1834–1914) experimented on inheritance by cutting the tails of mice and allowing them to breed. This was repeated for 22 generations and it was observed that complete tail character was still inherited. He then proposed that the higher organisms contain two types of tissues:

(*i*) **Germplasm:** This tissue is meant for reproduction and any change in these cells affects the progeny, *e.g.,* any change in egg or sperm will be transmitted to the offspring.

(*ii*) **Somatoplasm:** Its cells do not enter the sex cells and the variations arising here are not transmitted to next generation, *e.g.* the tails of mice, therefore, the cut tail character was not transmitted to the offsprings of mice with cut tails and they were born with intact tails.

Johannsen (1909) gave the 'genotype-phenotype' concept, according to which:

(*i*) **Genotype** is the sum total of heredity and two or more organisms with the same genotype (*e.g.*, identical twins) can exhibit different phenotypes under different environmental conditions (*e.g.*, of the two identical twins, the one is nourished properly would have good physique in comparison to the other one who is malnourished).

(*ii*) **Phenotype** (*e.g.,* appearance) arises due to interaction between genotype and the environment (*e.g.*, of the two identical twins that have same genotype, if one for most of the time plays in the ground in sunlight would have dark complexion, whereas, the other mostly staying at home would have comparatively fair complexion).

Many scientists tried to understand the principle of inheritance but they could not formulate the laws of inheritance as they did not give the numerical treatment to their results, *e.g.*, **Kölreuter** (1733–1806) performed hybridization experiments on tall and dwarf varieties of *Tobacco* and got plants of intermediate height in F1 (first filial or progeny) generation. In F2 generation, obtained by self pollination of F1 generation, tall plants of varying heights and dwarf plants were obtained. But, he did not count thc number of dwarf and tall plants thus obtained.

Gregor John Mendel (1822–1884) was a teacher of Science and Math in Vienna and took interest in plant hybridization (Fig. 6.1). He is known as 'Father of Genetics' and formulated the laws of inheritance of traits in 1866 while experimenting on *Pisum sativum* (garden pea). But, at that time his work was not recognized until 1900 when **Hugo de Vries, C. Correns** and **E. von Tschermak** independently experimentally confirmed his laws. The secret of Mendel's success was that he considered one trait at a time and numerically treated his results by counting the number of different types of offsprings obtained in various crosses.

Fig. 6.1 Gregor John Mendel, the 'Father of Genetics'.

Mendel selected garden pea as it has **bisexual** flowers (with both male and female reproductive parts in the same flower) and it reproduces mainly by **self fertilization** method (the male gamete of the same flower fertilizes the female gamete), so that homozygous **pure lines** (*e.g.* DD tall and dd dwarf plants) were easily obtained due to natural self fertilization. Moreover, Mendel could select 7 pairs of **contrasting traits** for **hybridization** which was quite easier. Fortunately, when he considered two (dihybrid cross) or more than two traits (trihybrid cross, and so on) at a time, many these traits were located on **different chromosomes** (garden pea has 7 pairs of chromosomes, *i.e.* $2n = 14$) to successfully show the independent assortment of these traits (Fig. 6.2 see Appendix). These are also tabulated in Table 6.1.

Table 6.1: Traits selected by Mendel in garden pea for hybridization experiments

S. No.	*Traits*			
			Dominant	**Recessive**
1.	Seed	(*i*) Shape	Round	Wrinkled
		(*ii*) Cotyledon colour	Yellow	Green
2.	Flower	(*iii*) Colour	Purple	White
3.	Pod	(*iv*) Shape	Inflated	Constricted
		(*v*) Colour	Green	Yellow
4.	Stem	(*vi*) Position of flower	Axial	Terminal
		(*vii*) Plant height	Tall	Dwarf

The two different forms of a trait (*e.g.,* D and d in case of plant height, and the alphabet is selected on the basis of initial letter of the mutant trait) are called **alleles** or **allelomorphs** (later, it was found that, actually, these alleles are the different forms of the same gene, *e.g.*, d allele arose due to mutation in the normal allele D for plant height). There can also be possibility of occurence of more than two alleles of a gene when a gene mutates in more than one form. Most of the mutants are recessive, except for, a few *e.g.*, bar eye character in *Drosophila.* Since at that time no one was aware of the term gene, Mendel called these trait determining genes as **factors).**

The **dominant** traits that can express in both **homozygous** (*e.g.*, DD tall plant, the two doses of D indicate the diploid nature of plant, where one D comes from father and the other from mother) as well as **heterozygous** (*e.g.*, Dd tall plant) conditions. Here, DD are the normal wild type, whereas, the **recessive** (dd) ones are the small population of mutants that can express only in homozygous condition (*e.g.,* dd dwarf plant) in the absence of its dominant allele.

The crosses where only one trait is considered at a time (*e.g.*, plant height) are known as **monohybrid crosses,** whereas, those crosses are called **dihybrid, trihybrid,** etc. where two, three, or more traits are considered together.

While crossing the two contrasting characters of a trait, the flower that was used as female was **emasculated** (Fig. 6.3 see Appendix) (anthers removed before reaching maturity) to avoid self fertilization and covered with a polythene or muslin cloth bag to inhibit undesired cross-pollination. The cross was performed by collecting the pollen grains of the flower to be treated as male and dusting them over the stigma of emasculated flower.

When Mendel crossed the seven pairs of contrasting characters, considering one pair at a time (monohybrid cross), it was found that **reciprocal crosses** (*e.g.,* in one cross if DD taken as male and *dd* as female, in another cross DD was selected as female and *dd* as male) gave the same results, showing that both male and female parents make equal genetic contribution to the progeny.

MENDEL'S LAWS OF INHERITANCE

By performing various crosses of the seven pairs of traits, Mendel formulated two laws of inheritance:

1. Law of Purity of Gametes or Principle of Segregation

When the homozygous dominant (*e.g.,* DD tall) and recessive (*dd* dwarf) plants for a trait were crossed, all the F1 progeny (Dd heterozygous tall), thus, obtained exhibited only the dominant phenotype.

When such F1 plants were self-fertilized (pollens of the same flower pollinated its stigma), about 75% of F2 progeny were tall, whereas about 25% were dwarf. This shows that the trait that appeared in F1 was **dominant** over the trait that did not appear in F_1 but reappeared in 25% of F_2 progeny **(recessive).**

This result would only be possible if F_1 produces two types of gametes for a trait (both male as well as female), D and d or it can be said that the gametes are pure for a trait and would contain either D or d allele **(Law of purity of gametes).** [Actually, at the time of gamete formation, the two doses of D and d in F1 plants go to different gametes due to meiosis]. Thus, the two alleles D and d get segregated or separated at the time of gamete formation **(Principle of segregation).**

Again, the self fertilization of F2 dwarf (dd) gave rise to only dwarf plants in F3 generation, whereas, among the F2 tall plants 33% gave only tall plants (TT) and rest of the 66% gave both tall (75%) as well as dwarf (25%) plants. This confirms that all the F1 plants had the genotype Dd and among the F2 individuals, 25% had DD (homozygous tall), 50 had Dd (heterozygous all) and 25% had dd (homozygous dwarf) genotypes. Thus, in F2 generation though the **phenotypic ratio** (on the basis of external appearance) of tall and dwarf plants was **3 : 1,** the **genotypic ratio** was **1 : 2 : 1** (1 DD:2 Dd:1 dd) (Fig. 6.4).

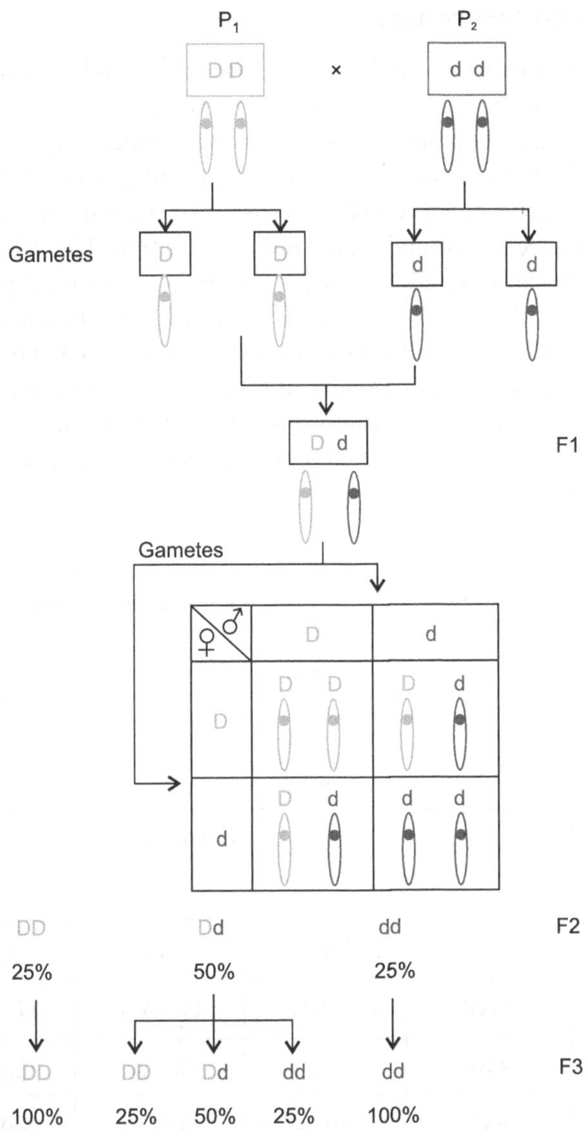

Genotype	Genotypic ratio	Phenotypic ratio	Phenotype
DD	1	3	Tall
Dd	2		
dd	1	1	Dwarf
	1 : 2 : 1	3 : 1	

Fig. 6.4 Monohybrid cross, selecting plant height as a trait, in garden pea.

2. Principle of Independent Assortment

When two (dihybrid crosses) or more than two (trihybrid and so on) traits are considered together (*e.g.* yellow and green and round and wrinkled seed traits in a dihybrid cross), one pair of trait (*e.g.* yellow and green seed colour) behaves completely independent of the other trait (*e.g.* round and wrinkled seeds). Such crosses of homozygous dominant (*e.g.* GGWW yellow-round seeds) and recessive parents (*e.g.* ggww green-wrinkled seeds), like the monohybrid crosses, gave only dominant phenotype (GgWw yellow-round seeds) in F1 generation. This F1 upon self fertilization gave 9 : 3 : 3 : 1 phenotypic ratio (9 yellow-round : 3 yellow-wrinkled : 3 green-round : 1 green-wrinckled) with gnotypic ratio of 4 : 2 : 2 : 2 : 2 : 1 : 1 : 1 : 1 in F2 generation, which is possible only when F1 would give rise to all the possible four types of gametes (GW, Gw, gW and gw) due to independent assortment of various alleles (actually, at the time of gamete formation during meiosis, the various types of maternal and paternal chromosomes can get arranged independent of each other in all the possible combinations to give different types of gametic cells). This is known as **principle of independent assortment** (Fig. 6.5).

P1 GGWW × ggww P2
Yellow-round seeds Green-wrinkled seeds

GgWw F1
Yellow-round seeds

F1: ↓

Male → / Female ↓		Gametes			
		GW	**Gw**	**gW**	**gw**
Gametes	**GW**	GGWW	GGWw	GgWW	GgWw
	Gw	GGWw	GGww	GgWw	Ggww
	gW	GgWW	GgWw	ggWW	ggWw
	gw	GgWw	Ggww	ggWw	ggww

Genotype	Genotypic ratio	Phenotypic ratio	Phenotype
GgWw	4	9	Yellow-round
GGWw	2		
GgWW	2		
GGWW	1		
Ggww	2	3	Yellow- wrinkled
GGww	1		
ggWw	2	3	Green-round
ggWW	1		
ggww	1	1	Green-wrinkled
	4 : 2 : 2 : 2 : 2 : 1 : 1 : 1 : 1	9 : 3 : 3 : 1	

Fig. 6.5 Dihybrid cross in garden pea considering two traits, seed colour and shape at a time.

Similarly, in a trihybrid cross, the F1 gives rise to eight types of possible gametes due to independent assortment of three pairs of traits. These gametes after fertilization gave a 27 : 9 : 9 : 9 : 3 : 3 : 3 : 1 phenotypic ratio in F2 generation.

Thus, when one pair of trait is considered, F1 produces two types of gametes ($n = 1, 2^1 = 2$), when two pairs of traits are considered F1 gives four types of gametes ($n = 2, 2^2 = 4$), when three pairs of traits are considered, F1 gives 8 types of gametes ($n = 3, 2^3 = 8$), and so on or it can be said that F1 can give 2^n possible types of gametes, where n is the number of pairs of traits considered at a time. (Independent assortment is not shown when the genes for two or more traits are located on the same chromosome (linked) as they would always move together at the time of meiosis)

Back Cross and Test Cross

When F1 (*e.g.*, Dd) is crossed either with homozygous dominant (*e.g.* DD) or with homozygous recessive (*e.g.*, dd) parent, it is called **back cross.**

When F_1 is crossed with homozygous dominant parent, all the resulting progeny exhibit only dominant phenotype (*e.g.*, tall plants with 50% DD and 50% Dd genotype).

But, when F1 is crossed with homozygous recessive parent, 50% of progeny show dominant phenotype (*e.g.*, Dd) and rest of the 50% recessive (*e.g.*, dd genotype) phenotype. Since this cross is performed to know whether an individual is homozygous dominant or heterozygous dominant, it is known as **test cross.** Thus, when a homozygous dominant (*e.g.*, DD) individual is back crossed with homozygous recessive individual, 100% of the resulting progeny would be of dominant phenotype (*e.g.*, 50% DD and 50% Dd). But when a heterozygous dominant (*e.g.*, Dd) individual is back crossed with a homozygous recessive individual, 50% of progeny shows dominant phenotype (*e.g.*, Dd) whereas 50% recessive (*e.g.*, dd) phenotype (Fig. 6.6).

(*a*) Dd × DD

F1 ↓ P1

Gametes

	Male →	**D**	**D**
	Female ↓		
Gametes	D	DD	DD
	d	Dd	Dd

50% DD + 50% Dd → 100% tall

Back cross with homozygous dominant parent

(*b*) Dd × dd

F1 ↓ P2

Gametes

	Male →	**d**	**d**
	Female ↓		
Gametes	D	Dd	Dd
	d	dd	dd

50% Dd + 50% dd

Tall Dwarf

Back cross (test cross) with homozygous recessive parent.

Fig. 6.6 Monohybrid back crosses.

Similarly in dihybrid crosses, where two traits are considered together, the back cross of F1 (*e.g.*, GgWw yellow-round seeds) with homozygous dominant parent (*e.g.*, GGWW yellow-round seeds) would give 100% dominant phenotype.

Whereas, test crossing of F1 with homozygous recessive parent (*e.g.*, ggww green-wrinkled seeds) gives 1 : 1 : 1 : 1 ratio in the progeny (Fig. 6.7).

(*a*) GgWw × GGWW

F1 P1

Gametes

	Male →	**GW**	**GW**
	Female ↓		
Gametes	**GW**	GGWW	GGWW
	Gw	GGWw	GGWw
	gW	GgWW	GgWW
	gw	GgWw	GgWw

100% Yellow-round seeds

Back cross with homozygous dominant parent.

(*b*) GgWw × ggww

F1 ↓ P1

Gametes

	Male →	**gw**	**gw**
	Female ↓		
Gametes	**GW**	GgWw	GgWw
	Gw	Ggww	GgWw
	gW	ggWw	ggWw
	Gw	ggww	ggww

Phenotypic ratio = 1 : 1 : 1 : 1

Back cross (test cross) with homozygous recessive parent.

Fig. 6.7 Dihybrid back crosses.

Deviations from Mendel's Laws

The results of Mendel, as found in garden pea, are not universal, except for the law of purity of gametes, and various traits in different organisms exhibit different behaviour, *e.g.*,

1. **Incomplete Dominance:** In many cases the dominant allele for a trait is not completely dominant over its recessive allele, *e.g.*, in *Mirabilis jalapa* (4 O' clock plant) the red flower colour-imparting allele R does not exhibit its complete dominance over its recessive

allele r that does not characterize any colour to the flower. Thus, the plants with Rr genotype bear pink flowers that is an intermediate phenotype between red and pink (Fig. 6.8).

RR × rr

Red flower   White flower

Rr

Pink flower   F1

F2

RR 25%   Rr 50%   rr 25%

Fig. 6.8 Incomplete dominance in *Mirabilis jalapa.*

2. **Interaction of Genes:** Sometimes, many genes interact in different ways to control a trait, *e.g.* epistasis, where one gene marks the effect of other gene.
3. **Linkage:** Independent assortment is exhibited only when the two or more pairs of traits are located on different chromosomes and is not shown when they are located on the same chromosome (linked) as they would always move together at the time of meiosis. For example, the gene for yellow/green seed trait was located on one pair of chromosomes whereas, the gene for round/wrinckled trait on different chromosome pair, therefore, they exhibited independent assortment. But, it they were located on the same chromosome pair they would not show any independent assortment of traits (Fig. 6.9).

G| |g   G W| |g w

W| |w

F1   F1

G/g and W/w located on different homologous chromosome pairs   G/g and W/w located on same homologoues

↓   ↓

G,g,W and w would exhibit independent assortment   G,g,W and w would not exhibit independent assortment as GW would move together and gw again move together

Fig. 6.9 Inheritance of linked (right) and non-linked (left) genes.

CHROMOSOMAL THEORY OF INHERITANCE

Sutton (1903), while working on meiosis in grasshopper, and **Boveri** (1903), on sea urchin, independently proposed the chromosomal theory of inheritance, according to which the inheritance is due to the transfer of chromosomes from one generation to other. Thus, there is parallelism between behaviour of chromosomes during meiosis and Mendelian traits.

1. Similar to the separation of homologous chromosomes during Anaphase I of meiosis, the two alleles (*e.g.*, D and d) for a trait in a hybrid (*e.g.*, Dd F1) get separated into two gametes at the time of gamete formation **(law of purity of gametes or law of segregation).** This shows that Mendelian factors (genes) are located on chromosomes.

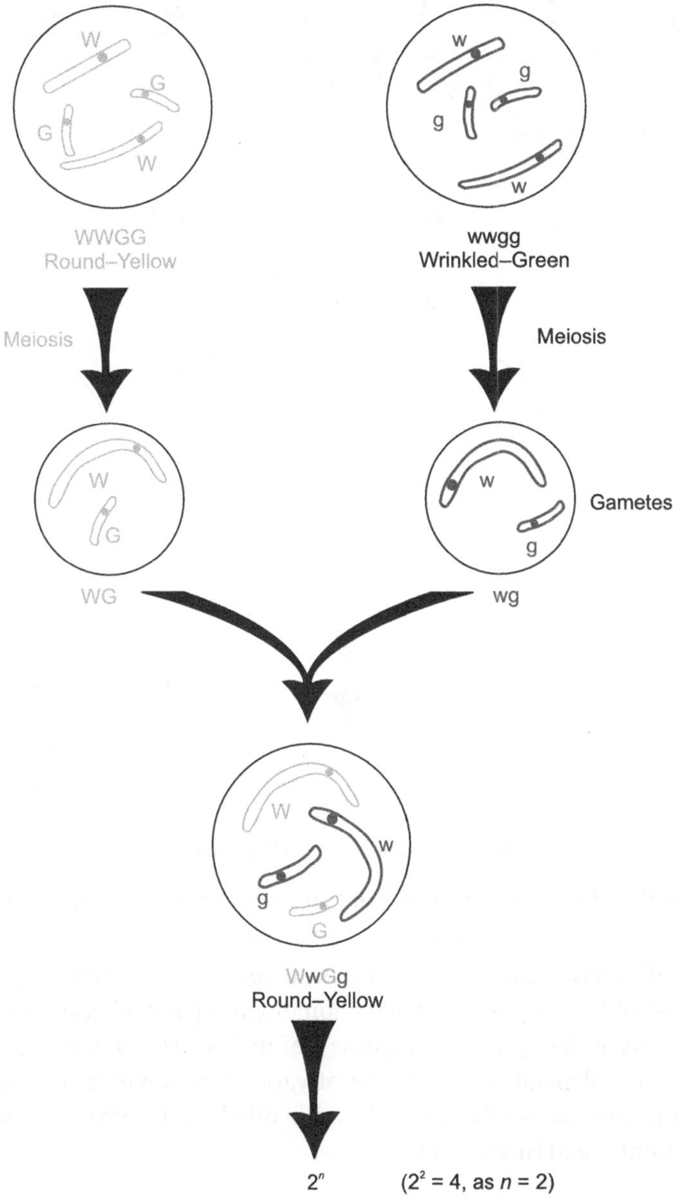

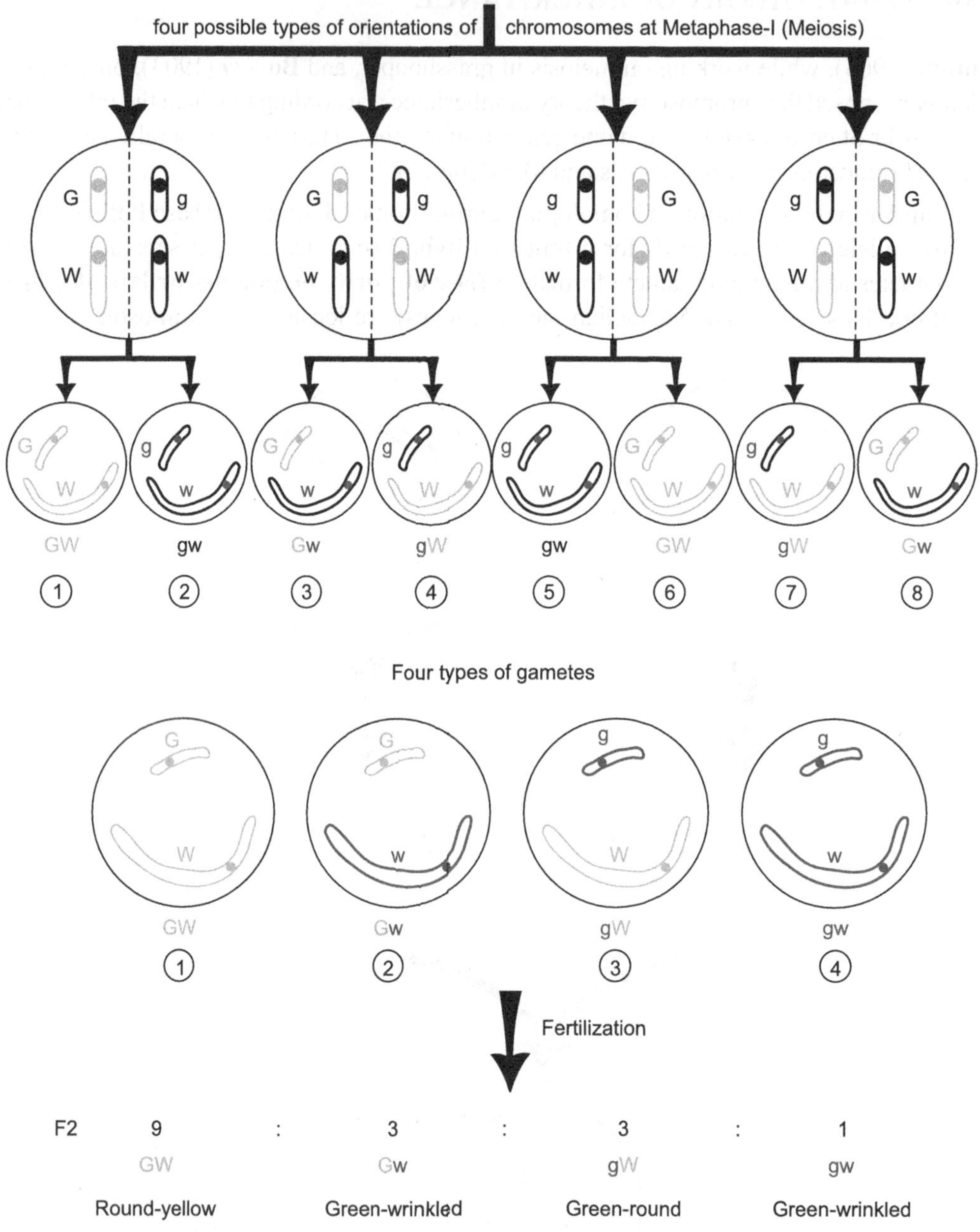

Fig. 6.10 Independent assortment of two pairs of alleles located on two pairs of homologous chromosomes.

2. Orientation of maternal and paternal chromosomes in a pair of homologous chromosomes is independent of the orientation of other homologous pairs of chromosomes, thus, to give 2^n arrangements at the time of Metaphase I of meiosis (n = number of pairs of chromosomes), like the independent assortment of various traits where one pair of trait is independent of the other pairs of traits as shown in dihybrid, trihybrid and so on crosses **(law of independent assortment).** Thus,

(*i*) when one pair of trait/homologous chromosomes is considered, F1 produces two types of gametes ($n = 1$, $2^1 = 2$) due to two types of orientations of chromosomes at Metaphase I,

(*ii*) when two pairs of traits are considered, F1 gives four types of gametes ($n = 2$, $2^2 = 4$), due to four types of possible arrangements of chromosomes (Fig. 6.10)

(*iii*) when 3 pairs of traits are considered together, F_1 gives eight types of gametes ($n = 3$, $2^3 = 8$), and so on (Fig. 6.11).

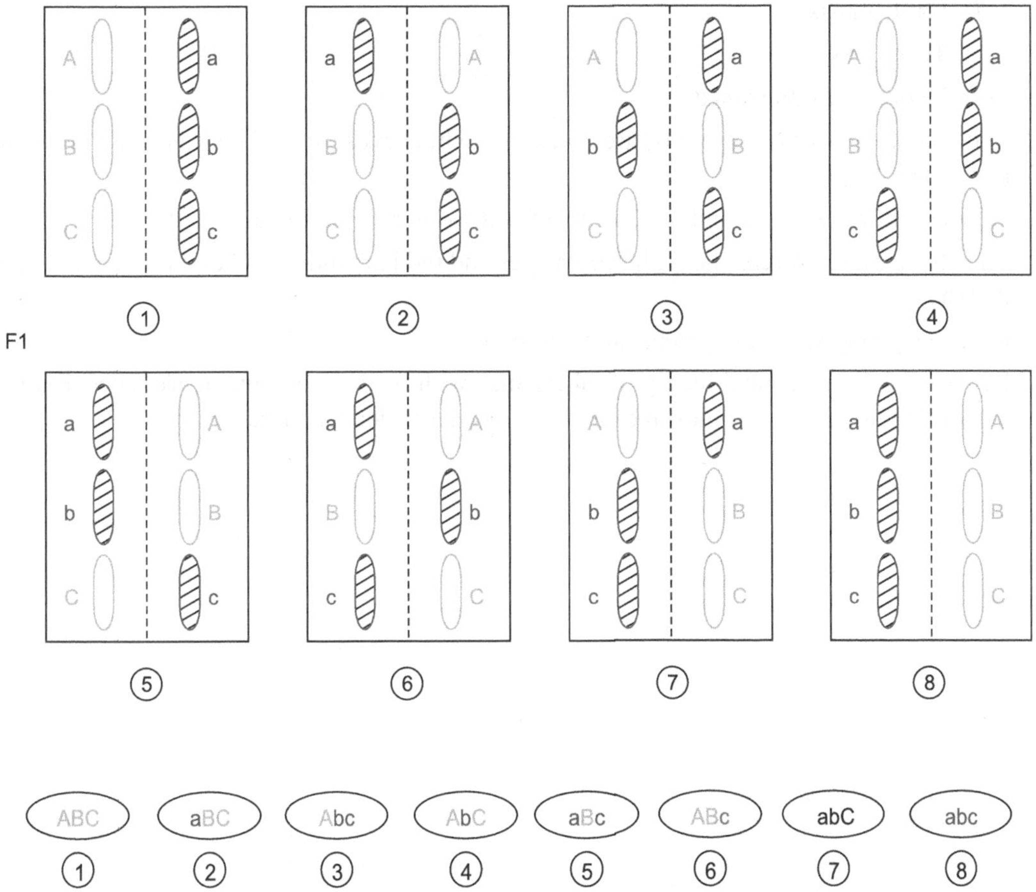

Eight types of possible gametes produced by F1.

↓

After self fertilization, give 27 : 9 : 9 : 9 : 3 : 3 : 3 : 1 ratio in F2 generation.

Fig. 6.11 Independent orientation of three pairs of homologous chromosomes at Metaphase I of meiosis.

QUESTIONS

1. Why did Mendel use pea as the experimental material in his experiments. Give a brief outline of the experiments conducted by him.

   Or

   Discuss Mendel's laws of inheritance.
2. Write short notes on:
   (*i*) Reciprocal crosses
   (*ii*) **Back cross**
   (*iii*) **Test cross**
   (*iv*) Incomplete dominance
3. In human beings brown eyes (B) are dominant to blue eyes (b); and dark hair (R) dominant to red hair (r).

   A man with brown eyes and red hair marries a woman with blue eyes and dark hair.

   They have two children, one with brown eyes and red hair and the other with blue eyes and dark hair.

   Give the genotypes of the parents and children.
4. Discuss the chromosomal theory of inheritance with the help of suitable sketches. Explain Mendel's laws using chromosomes as the carriers of Mendelian factors.

Chapter 7 NUCLEIC ACIDS

Nucleic acids are long chain polymers (actually macromolecules as they consist of four types of nucleotides and they are not repeated in an orderly manner) of **nucleotides** (monomers). Each nucleotide is made up of a **sugar** (either ribose or deoxy-ribose), a **base** (thymine, cytosine, uracil, adenine or guanine) and a **phosphate group** (where the sugar and the base together constitute a nucleoside) (Fig. 7.1).

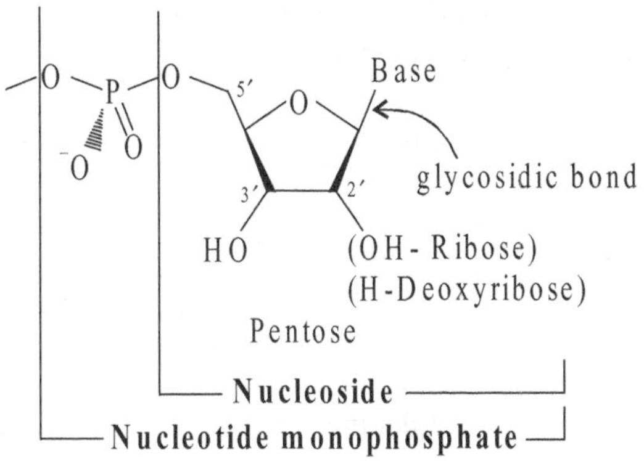

Fig. 7.1 Nucleotides.

There are two types of nucleic acids, DNA (Deoxy-ribo Nucleic Acid) and RNA (Ribo Nucleic Acid) that differ mainly with respect to the sugar, which is **2-deoxy-ribose** ($C_5H_{10}O_4$, where an O atom is missing at position 2 in the ring) in **DNA** and **ribose** ($C_5H_{10}O_5$) in **RNA** (Fig. 7.1 and 7.2). The sugar is found in the ring form where the **phosphate** is attached to the sugar either at **3′ —OH** or **5′ —OH** by elimination of a molecule of water, whereas the **base** (pyrimidine with N1 and purine with N9) at **C1** (See Fig. 7.5).

Ribose 2-Deoxy-ribose

Fig. 7.2 Structure of ribose and deoxy-ribose sugars.

DNA

In addition, the **DNA** contains four types of bases, **thymine** (nucleotide deoxy-thymidylic acid), cytosine (nucleotide deoxy-cytidylic acid), adenine (nucleotide deoxy-adenylic acid), and guanine (nucleotide deoxy-guanylic acid), whereas the **RNA** contains cytosine (nucleotide cytidylic acid), **uracil** (nucleotide uridylic acid), adenine (nucleotide adenylic acid), and guanine (nucleotide guanylic acid).

Actually the bases are of two types:

(*i*) **Pyrimidines:** Thymine, cytosine and uracil (Fig.7.3).

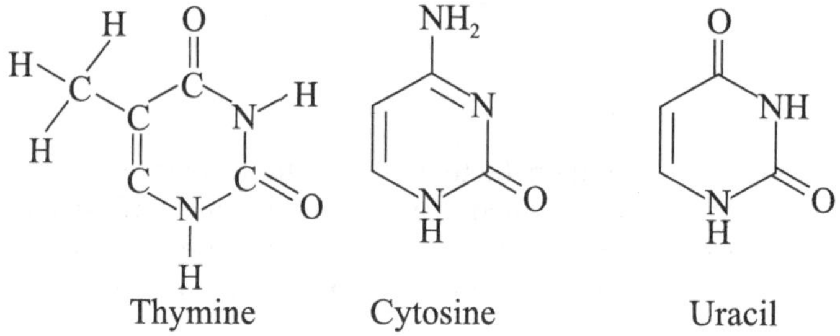

Fig. 7.3 Pyrimidines.

(*ii*) **Purines:** Adenine and guanine (Fig. 7.4).

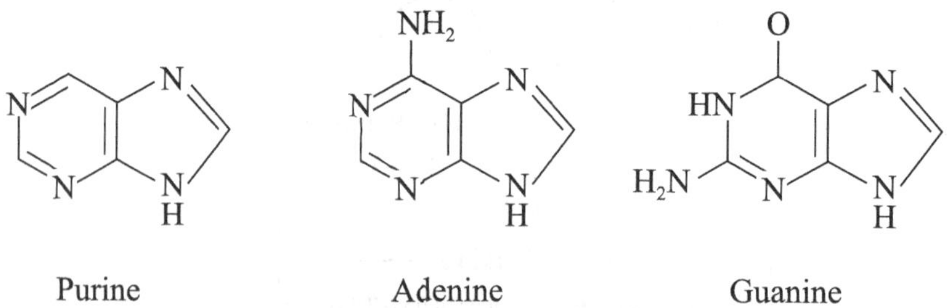

Fig. 7.4 Purines.

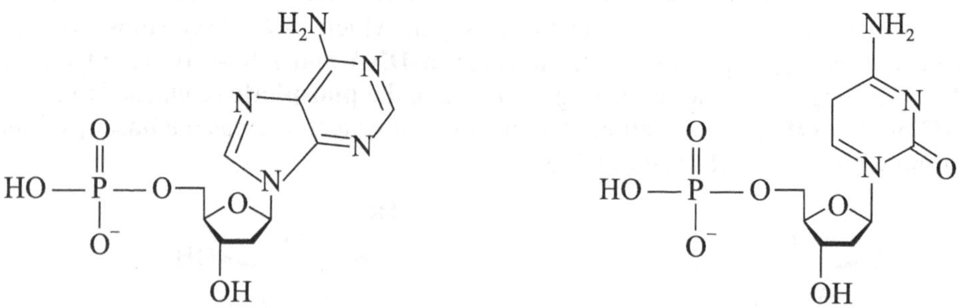

Deoxy-adenosine monophosphate (dAMP)

Deoxy-cytosine monophosphate (dCMP)

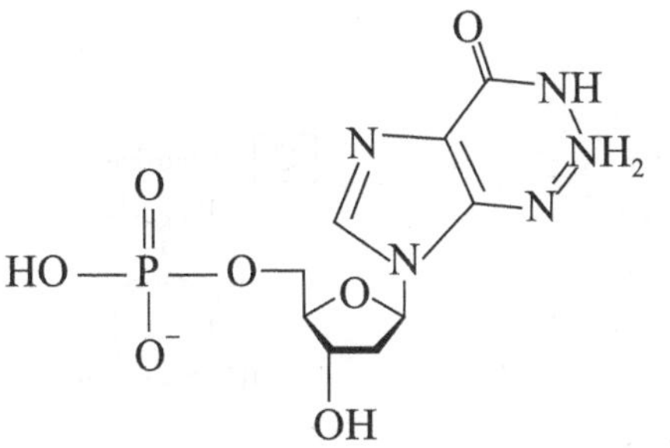

Deoxy-guonosine monophosphate (dGMP)

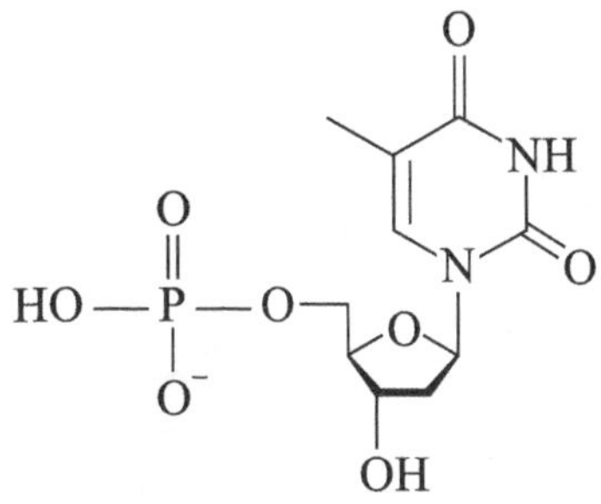

Deoxy-thymosine monophosphate (dTMP)

Fig. 7.5 Deoxy-ribo-nucleotides (dNTP).

RNA

The **RNA** is generally **single-stranded** (exceptions, *e.g.* certain viruses, like reovirus, wound tumour virus, etc., where it is double-stranded), whereas the **DNA** is **double-stranded** (exceptions, *e.g.* certain viruses, like ΦX174). In the nucleic acid strands, the backbone is made up of alternating units of sugar and phosphate, whereas the bases lie at right angle to the backbone and are attached to the sugar. The adjacent nucleotides are linked by **phosphodiester bonds** that are formed by the elimination of water molecules (Fig. 7.6 and 7.7).

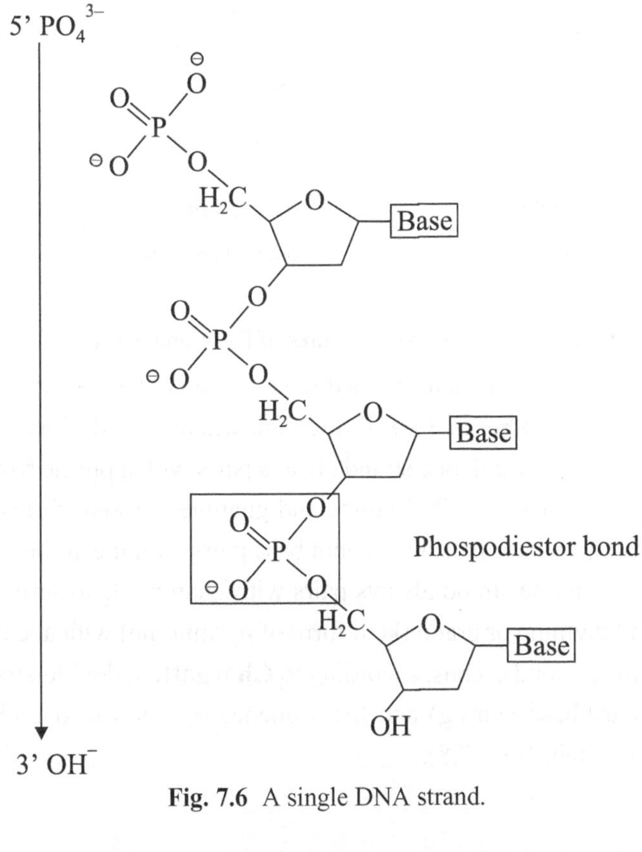

Fig. 7.6 A single DNA strand.

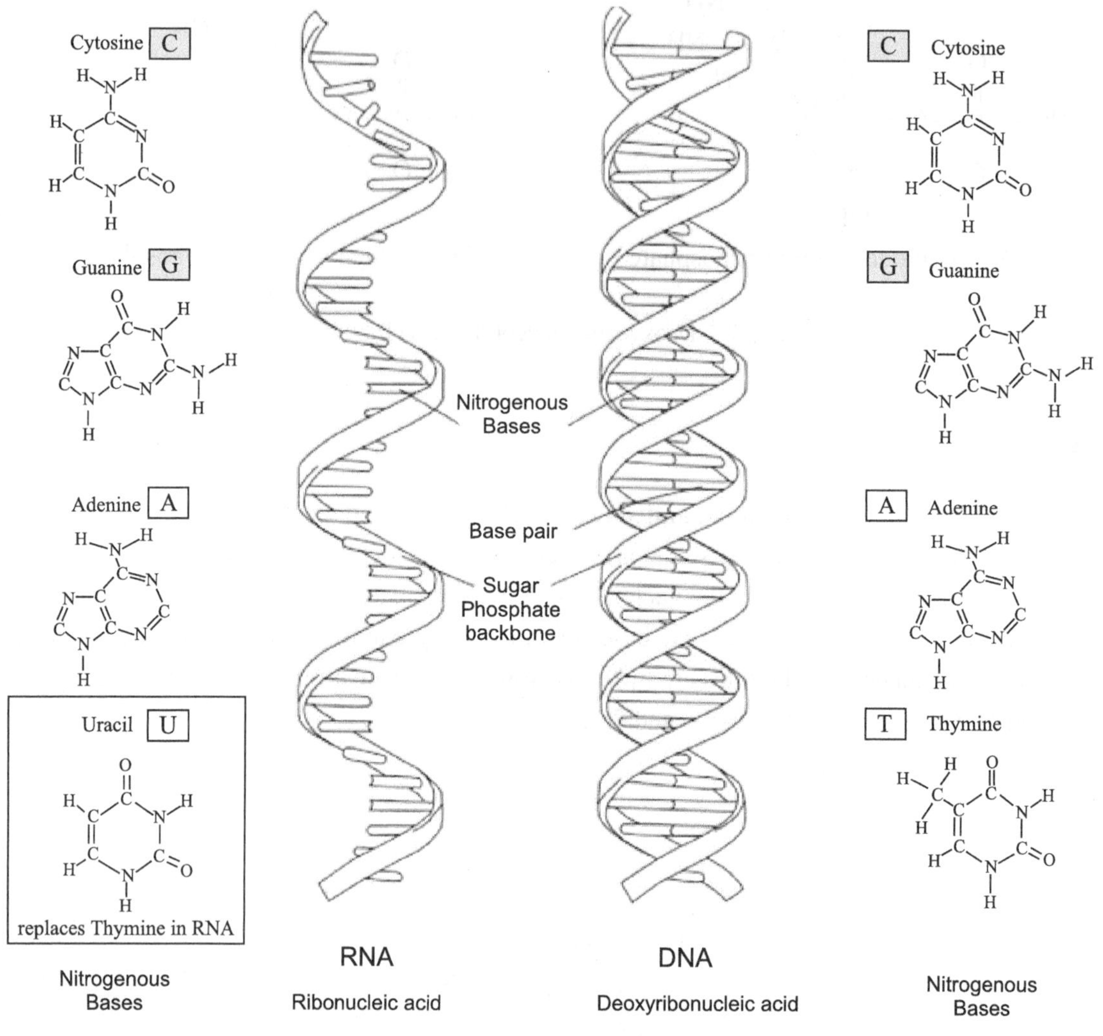

Fig. 7.7 Comparative structures of DNA and RNA.

Thus, whenever the nucleic acid is found as double-stranded, the two strands run anti-parallel (one in 5'→ 3' direction and the other in 3'→ 5' direction where 5' and 3' are either free PO_4^{3-} or OH^- groups) and the pyrimidine base in one strand always pairs with a purine base in another strand through hydrogen bonds. The bases uracil, thymine and guanine are **keto** forms, whereas cytosine and adenine are **amino** bases. In base pairing, an enol base pairs with the amino base, thus, cytosine (amino form of pyrimidine) in one strand always pairs with guanine (keto form of purine) through three hydrogen bonds, and thymine or uracil (keto form of pyrimidine) with adenine (amino form of purine) through two hydrogen bonds. Thus, according to **Chargaff,** in double-stranded DNA, **A = T** and **C = G** (complementary base pairing) and the sequence of bases in one chain determine the sequence of bases in other chain (Fig. 7.8).

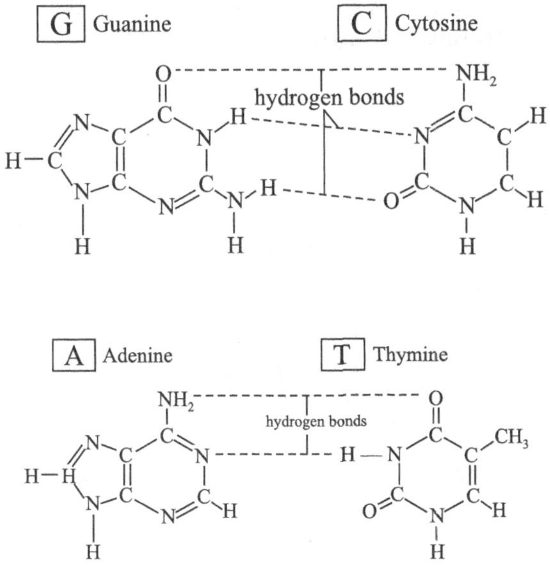

Fig. 7.8 Pairing between complementary bases.

J.D. Watson and **F.H.C. Crick** (1953) (Fig. 7.9) gave the model for double helical structure of DNA that was published in the journal, Nature. According to their model, the two polynucleotide chains connected by hydrogen bonds run in **opposite directions** and they are coiled around a central axis in a **right-handed** manner, so that they can be separated only by uncoiling and not by lateral separation. Thus, the backbones of the two strands, consisting of alternating sugar and phosphate units, are like the railing of a staircase whereas, the hydrogen-bonded base pairs, found at right

Fig. 7.9 James D. Watson (left) and Francis Crick (right).

angle to the long axis, lie at the steps. The diameter of the DNA double helix is **2 nm** (20 Å) with a pitch (1 round) of **3.4 nm** (34 Å), so that the distance between two adjacent bases becomes **0.34 nm** (3.4 Å), as **10 bases** constitute one turn. This is known as **B-form** of DNA (Fig. 7.10 and 7.11).

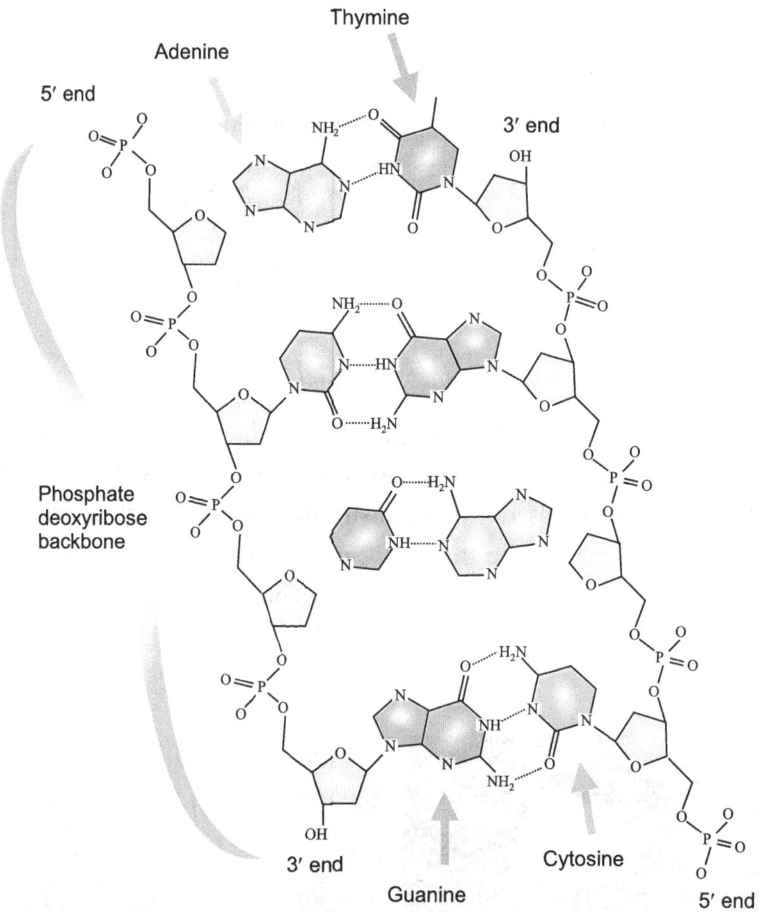

Fig. 7.10

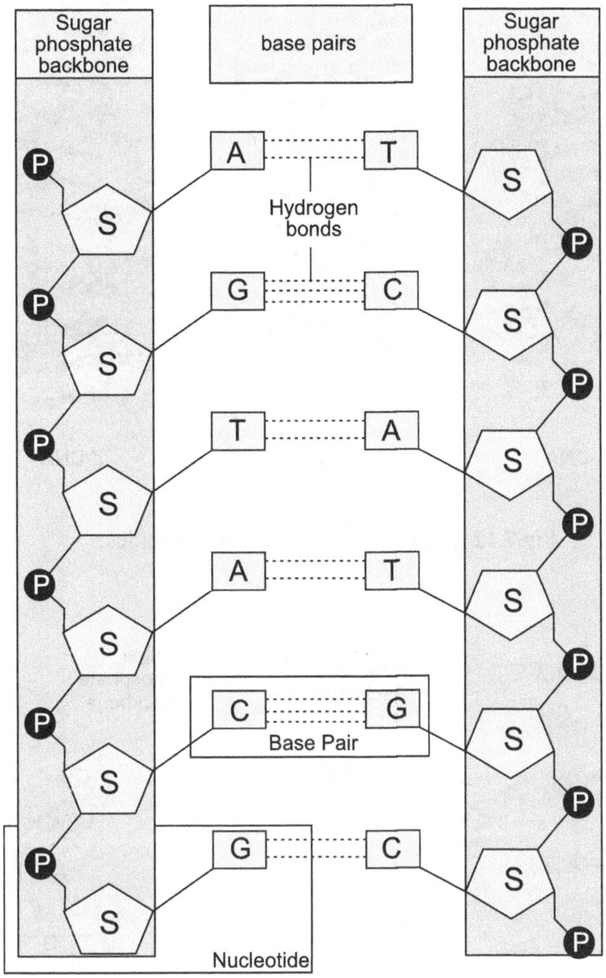

Fig. 7.11 Base pairing in double-stranded DNA molecules.

A- and Z-forms (1979) of DNA have been obtained in crystalline form. The occurrence of A-form of DNA (right-handed coiled) in the cells is uncertain, but evidences indicate the presence of short fragments of Z-form in both prokaryotes and eukaryotes that may play some role in regulation of gene expression or in genetic recombination. The pitch in Z-form DNA is 4.5 nm with 12 base pairs/turn, diameter 1.8 nm, distance between two adjacent bass 0.38 nm, sugar-phosphate backbone zig-zag, repeating units being dinucleotides, and coiling of the two strands left-handed (Fig. 7.12).

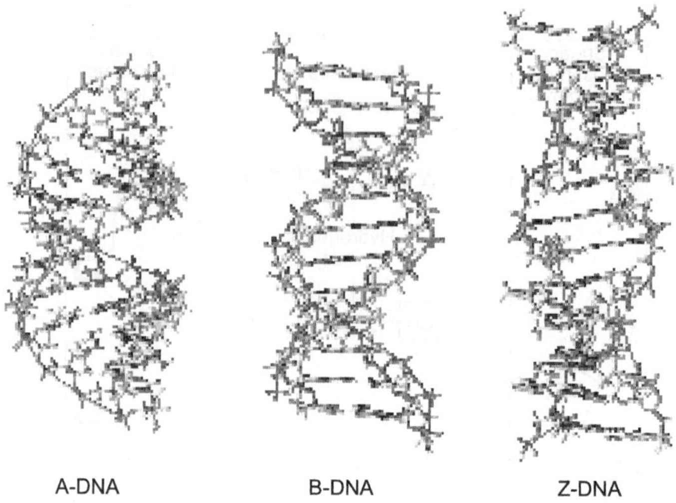

Fig.7.12 Different types of DNA molecules.

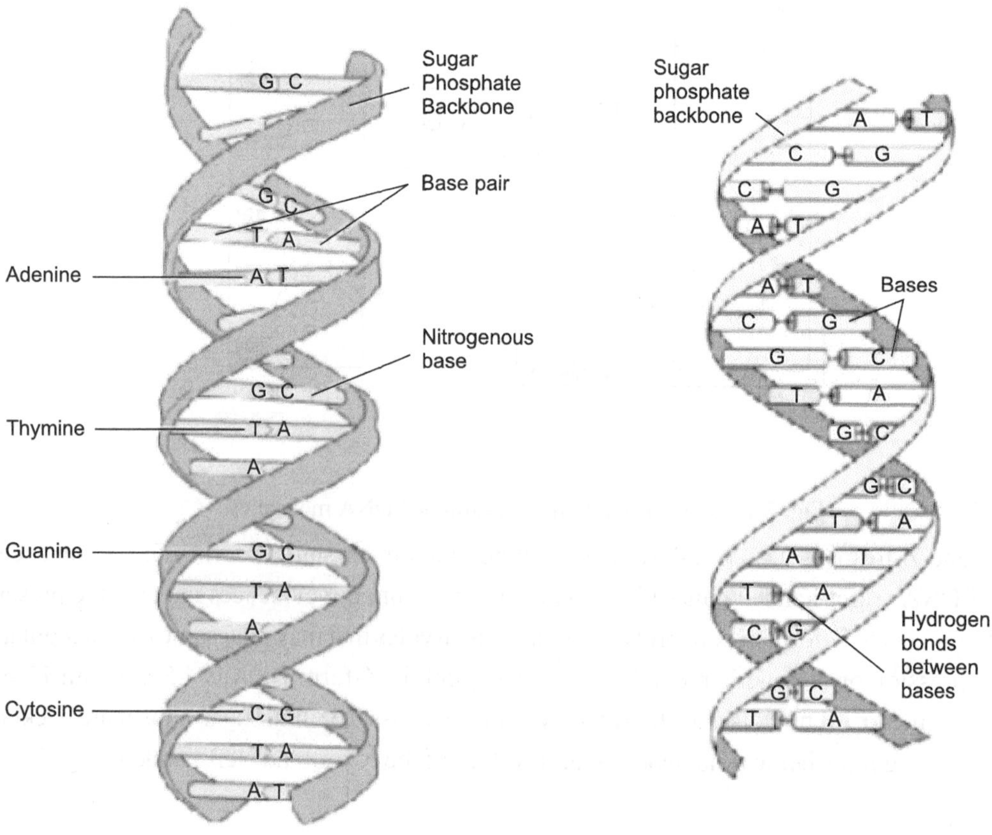

Fig. 7.13 Structure of DNA.

Cells, having both DNA as well as RNA, **DNA** constitutes the **genetic material** containing the information for development and existence of an organism, whereas RNA constitutes the genetic material only in certain **RNA** viruses, *e.g.* retroviruses, reoviruses, etc. In **eukaryotes,** the DNA is mainly found in the nucleus, but is also present in the organelles, chloroplasts and mitochondria. In certain eukaryotes it is also found in the cytoplasm as plasmids that are double-stranded circular molecules of DNA. In **prokaryotes,** the nucleoid contains a circular double-stranded DNA attached to mesosome. Many prokaryotes also contain one to many plasmids in the cytoplasm.

RNA in the cells is of three types:

(*i*) mRNA

(*ii*) tRNA

(*iii*) rRNA

(Details given in the chapter "Protein Synthesis")

Hammerling's grafting experiments (1953) on the green alga *Acetabularia* demonstrated that nucleus contains the hereditary information. The body of alga is made up of a rhizoid that contains the nucleus, a stalk and a cap that is generally umbrella-shaped reproductive body. He used two species, *Acetabularia mediterranea* and *A. crenulata* that differ with respect to the shape of

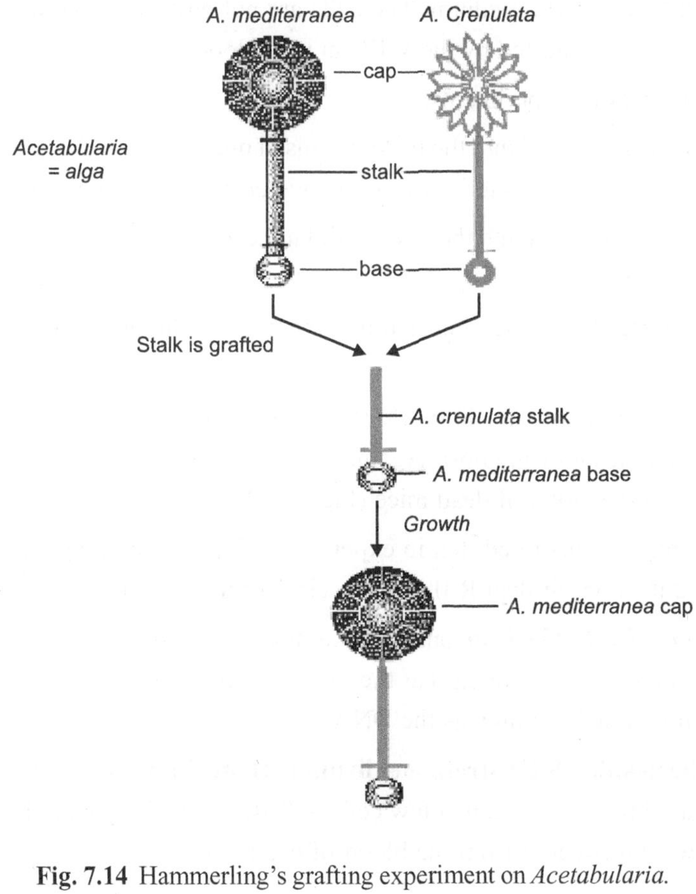

Fig. 7.14 Hammerling's grafting experiment on *Acetabularia.*

caps. He cut the body of the organisms of the two species in 3 pieces - cap, stalk and the rhizoid. Then, he grafted the stalk of one species over the rhizoid of the other, thus, the cap regenerated resembled the species that donated the rhizoidal portion containing the nucleus (Fig. 7.13). (For details see the chapter 'Introduction to Cell Structure and Viruses').

Many experiments have demonstrated that **nucleic acids** contain the **genetic information:**

Transformation Experiment of Griffith

Griffith (1928) worked on two strains of the bacterium, *Streptococcus pneumoniae* (*Pneumococcus pneumoniae* or *Diplococcus pneumoniae*) where the cells are found in aggregates of two (diplococci):

(*a*) **S III strain** that is virulent (pneumonia disease causing) and forms smooth colonies on solidified nutrient medium. The cells are enveloped within a polysaccharide capsule that prevents the bacterial cells from being engulfed by the WBC in host's (mouse) blood, and thus is responsible for the virulence.

(*b*) **R II strain** that is non-virulent (did not cause pneumonia) and forms rough colonies on solidified nutrient medium. The cells are not enveloped within any capsule and, thus, are easily engulfed by the WBC in host blood.

He performed a series of experiments:

1. **S III strain** when injected into the mice, it caused pneumonia and death of the host. Cells of the S III strain of the bacterium were recovered from the blood of dead mice.
2. **R II strain** when injected into the mice, it did not cause pneumonia and, thus, no death of the host occurred.
3. **Heat-killed S III** strain when injected into the mice, it did not cause pneumonia and no death of the host.
4. **Heat-killed S III strain** and **living R II strain** when injected into the mice, it caused pneumonia and death of the host. Again, the cells of S III strain of the bacterium were recovered from the blood of dead mice (Fig. 7.15).

These experiments demonstrated that in experiment No. 4, something from the heat-killed S III strain transformed the non-virulent R II into the virulent S III strain.

Later, **O.T. Avery, C.M. Macleod** and **Mc Carthy** (1944) through their *in vitro* (outside the intact cells) experiments demonstrated that the virulence-transforming agent from heat-killed S III strain into non-virulent R II strain was the DNAs.

1. Extract of **heat-killed S III strain** and **living R II strain** when injected into the mice, it caused death of the host. Again, a few cells of S III strain along with the RII strain of the bacterium were recovered from the blood of dead mice.

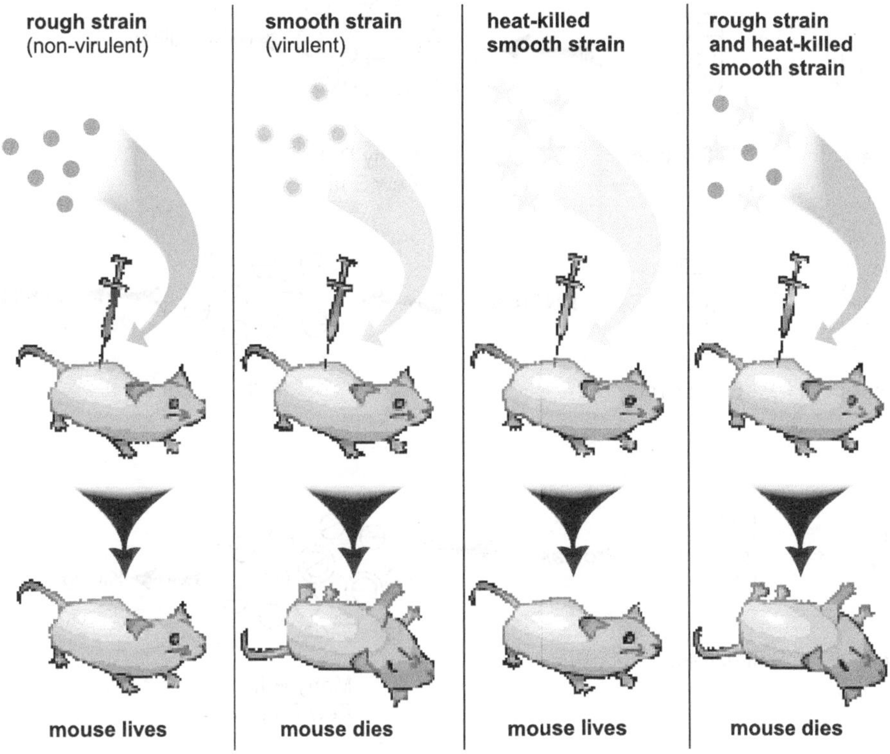

Fig. 7.15 Griffith's experiment of transformation in *Pneumococcus pneumoniae*.

2. To know the transforming agent in the heat killed S III strain, extract of **heat-killed S III strain, living R II strain** along with the **DNase** enzyme, was injected into the mice, and it did not cause death of the host. Thus, the DNase enzyme, here, destroyed the transforming agent DNA from the heat-killed S III strain that transformed a few cells of non-virulent R II strain into the virulent S III strain in experiment No. 1.
3. But, when the injection of extract of **heat-killed S III strain, living R II strain** along with the **protease** enzyme, was given to the mice, it caused death. Thus, the protease enzyme, here, destroyed the proteins in the heat-killed S III strain and the actual transforming agent, DNA, remained intact. A portion of this DNA entered into a few non-virulent R II strain cells and got incorporated into their genetic material, thus, transforming them into the virulent S III strain that were recovered from the blood of the dead mice (Fig. 7.16).

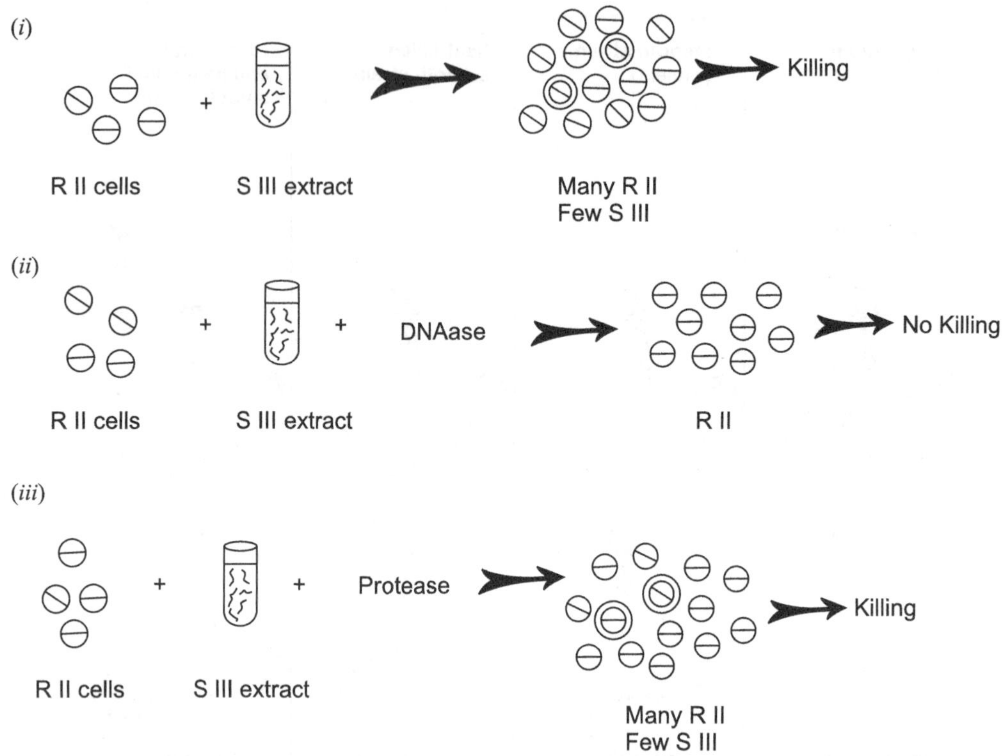

Fig. 7.16 Experiment of Avery, Macleod and Mc Carthy on *Streptococcus pneumoniae.*

Experiment of Hershy & Chase

A.D. Hershy and **M.J. Chase** (1952) performed experiments with **T2 bacteriophage** that infects *E. coli* bacterium. The T2 phage particle consists of a proteinaceous head and a tail, where the head contains a double-stranded DNA molecule. Since proteins contain S but no P and the DNA contains P but no S, in order to know whether the protein or the DNA contains the information for the formation of intact progeny phages during the infection of a host cell, Hershey & Chase separately radio labeled the protein and DNA of T2 phage.

The **protein** coat of T2 bacteriophage was labeled with radioactive ^{35}**S** by growing T2-infected *E. coli* cells in $^{35}SO_4^{2-}$ containing nutrient medium, whereas the DNA of a separate lot of T2 bacteriophage was labeled with radioactive ^{32}**P** by allowing growth of T2-infected *E. coli* cells in $^{32}PO_4^{3-}$ containing nutrient medium.

When the fresh cells of *E. coli* were infected with ^{35}**S-labeled T2 phage,** the protein coat remained outside the host cell and no radioactivity was observed in any of the progeny phages inside the host cells (after agitaton the protein coat detached from the infected host cells and the pellet obtained after centrifugation contained the infected host cells). Here, radioactivity was detected in the extracellular medium obtained after the centrifugation that contained the lighter T2 protein coat.

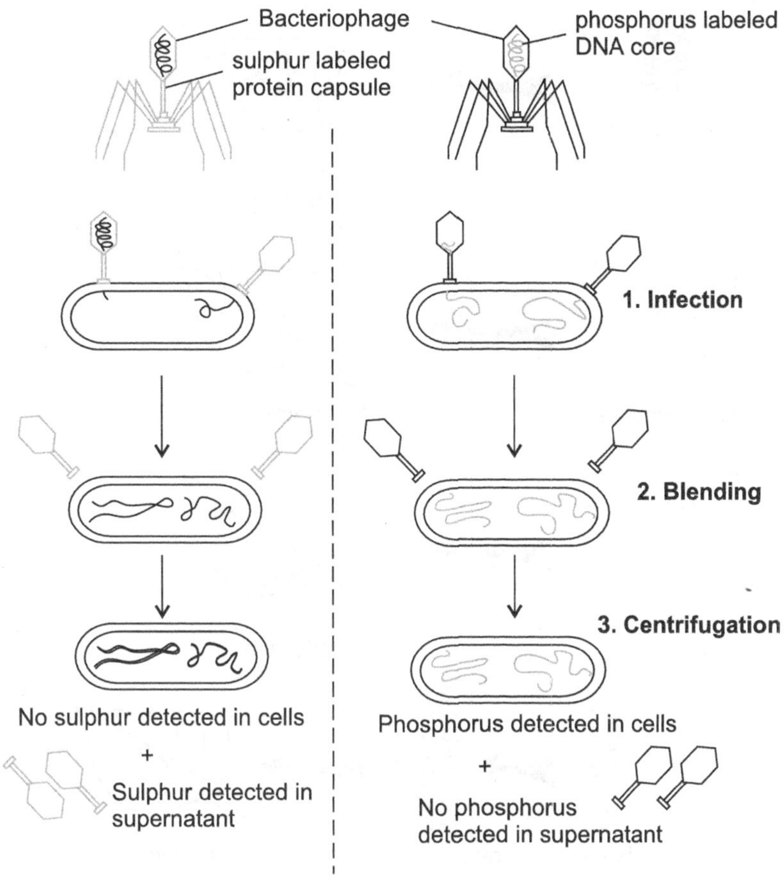

Fig. 7.17 Experiment of Hershy and Chase.

On the other hand, when fresh cells of *E. coli* were infected with **^{32}P-labeled T2 phage,** radioactivity was detected in the pellet containing infected host cells with progeny phages inside the cells obtained after the centrifugation, whereas the supernatant containing lighter T2 protein coat did not show any radioactivity.

This experiment confirms that at the time of infection, ^{32}P-labeled DNA enters the host cell and directs the formation of progeny phage particles. Thus, the DNA contains the information for the assembly of phage particles and not the proteins (Fig 7.18).

Experiment of Fraenkel-Conrat

H. Fraenkel-Conrat worked on a plant virus, Tobacco Mosaic Virus (TMV), where the genetic material is RNA and it infects the tobacco plant to cause mosaic on leaves. The virus is rod-shaped; consisting of a protein coat (made up of spirally-arranged capsomeres), enclosing a spirally-coiled single-stranded RNA molecule. They performed various experiments that showed that the genetic material RNA, here, contains the information for the formation of progeny phages:

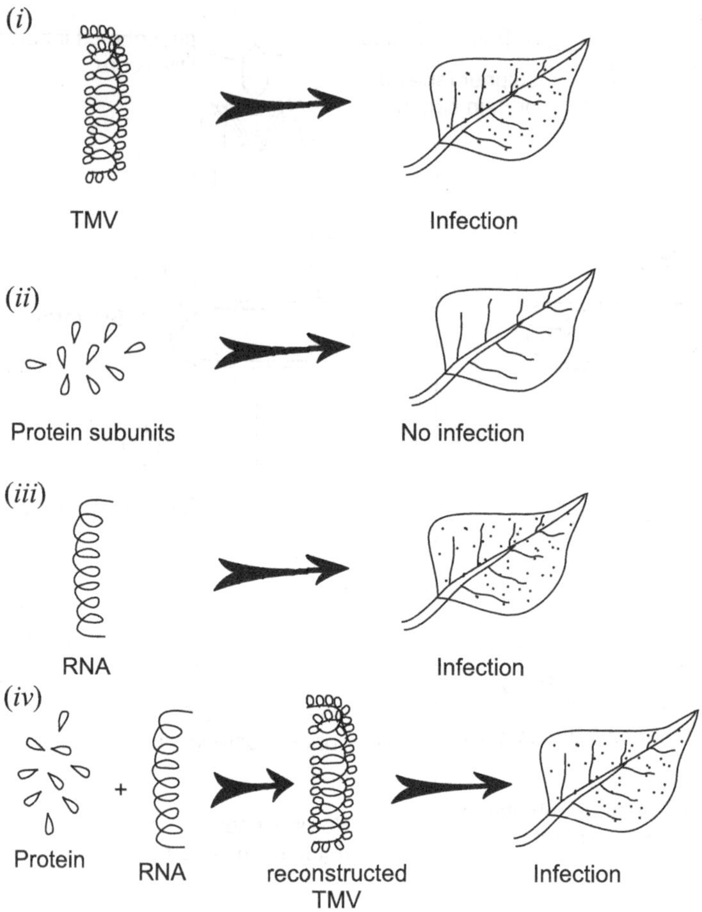

Fig. 7.18 Fraenkel-Conrat's experiment on TMV.

1. When tobacco leaves were infected with **TMV,** it caused mosaic on leaves and progeny viruses were recovered from the infected parts of the leaves.
2. The isolated **RNA** of TMV also successfully infected the tobacco leaves and progeny virus particles could be obtained from the infected leaves.
3. The isolated **protein** of TMV could not infect the tobacco leaves.
4. Reconstructed virus from the isolated RNA and protein again caused infection (Fig. 7.18).

DNA Replication

DNA (and genetic RNA in RNA viruses) at the time of cell division (multiplication in case of viruses) undergo self replication. In case of double-stranded DNA, the replication is **semi-conservative** in nature, where the two strands separate and each parental strand synthesizes its complementary strand. Thus, in each of the two newly synthesised double-stranded DNA molecules, one strand is the older parental DNA molecule, whereas the other strand is newly assembled.

M. Meselson and **F.W. Stahl** (1958) were the first who gave evidence for semi-conservative replication of DNA in *E. coli.* They labeled the DNA of *E. coli* by growing the cells in nutrient medium containing ^{15}N (heavy isotope of nitrogen in the form of $^{15}NH_4Cl$) for 14 generations. After harvesting the cells from ^{15}N-containing medium, they were grown in ^{14}N-containing medium, so that the newly synthesized DNA, now, contained the normal ^{14}N isotope of nitrogen. The isolated DNA from the cells in successive generations accordingly showed variable heaviness in CsCl density gradient ultracentrifugation:

1. 0 generation → 100 % heavy DNA
2. 1st generation → 100 % intermediate DNA
3. 2nd generation → 50 % light + 50 % intermediate DNA
4. 3rd generation → 75 % light + 25 % intermediate DNA

This shows that DNA replicates in a semi-conservative manner (Fig. 7.19)

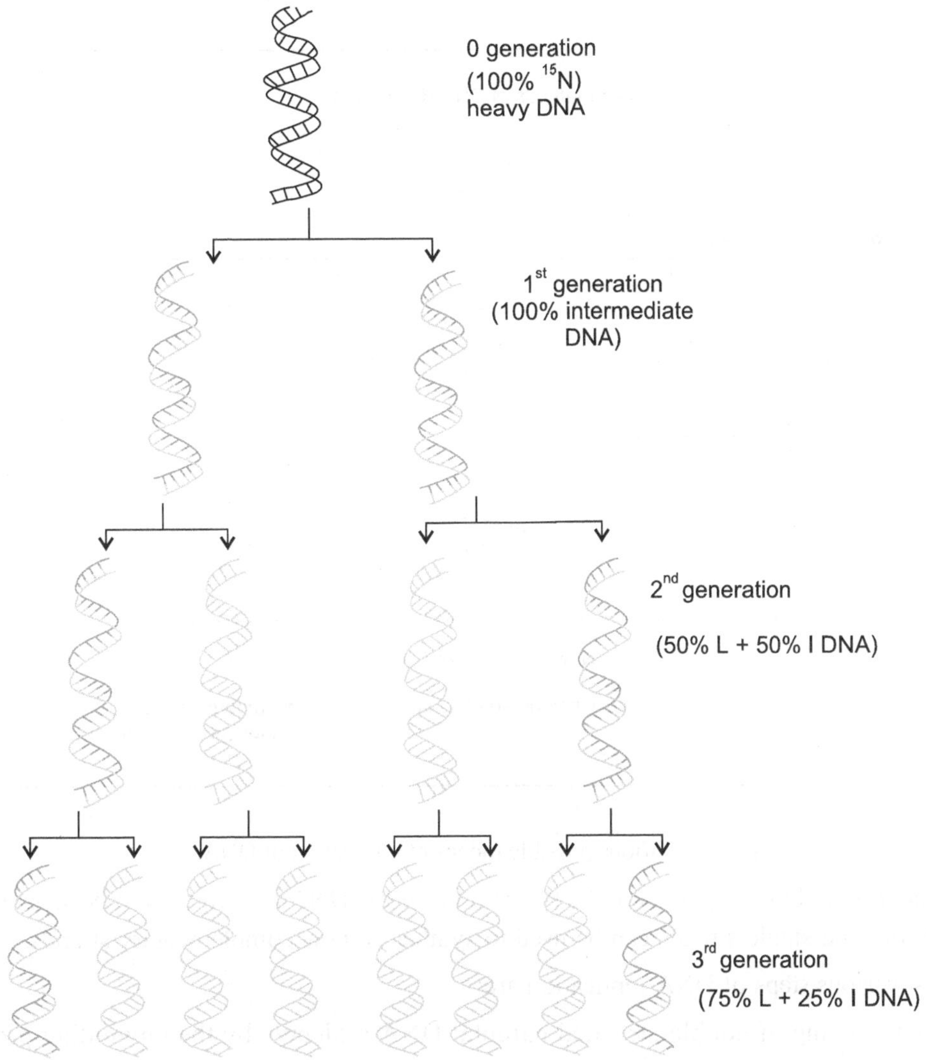

Fig. 7.19 Semi-conservative replication of DNA (L = light DNA, I = intermediate DNA).

If DNA replicated in the conservative manner to produce two double-stranded DNA molecules, where both the strands of one daughter DNA molecule remaining conserved and consisted of the older strands, whereas the other DNA molecule consisted of entirely newly synthesized DNA strands, they would show the result:

1. 0 generation → 100 % heavy DNA
2. 1st generation → 50 % heavy DNA + 50 % light DNA
3. 2nd generation → 25 % heavy + 75 % light DNA
4. 3rd generation → 12.5 % heavy + 87.5 % light DNA (Fig. 7.20).

As these results were not obtained, this confirms that DNA replicates in a semi-conservative manner.

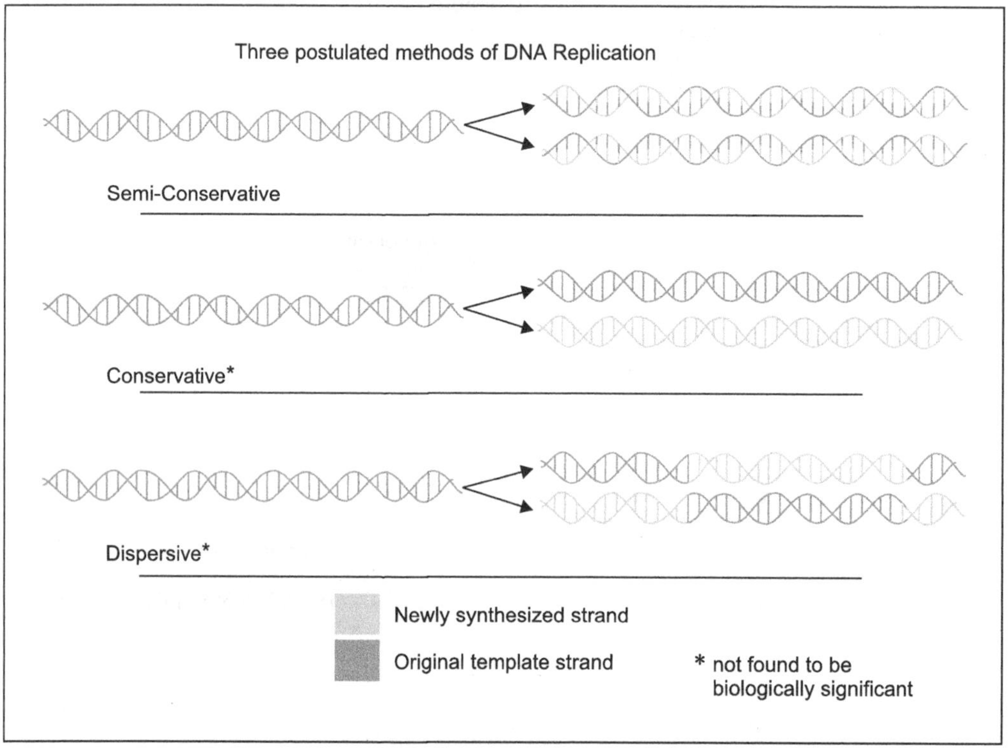

Fig. 7.20 Various possible modes of replication of DNA.

At the time of DNA replication, the two strands of the DNA are required to be separated, so that each of the two single strands can be used to synthesize new complementary strands.

Thus, various steps of DNA replication are:

1. Unwinding of double-stranded parental DNA molecule by the unwinding proteins, *e.g.*, **helicases.**

2. Since, the two strands of DNA are right-handed coiled around each other that cannot be separated laterally, the unwinding creates a coiling tension ahead of moving replication fork. This coiling tension is reduced by **topoisomerases** that help in the separation of two strands by relaxing super-coils by making cuts in one (*e.g.* ω protein in *E. coli*) or both the DNA strands (*e.g.* DNA gyrase in *E. coli*) and then after the release of tension, causing rejoining of the broken strands.
3. Since the two strands of a double-stranded DNA molecule are complementary, the **single-strand binding proteins** (SSB) stabilize the single strands thus produced and prevent their rejoining by complementary base pairing.
4. The DNA replication requires a short segment of RNA primer that is synthesized by a RNA polymerase, **primase.**
5. Now, the polymerization process occurs over the RNA primer by the enzyme **DNA polymerase III.** In *E. coli* three types of DNA polymerases are found:

   (*i*) **DNA polymerase I:** It was reported for the first time by **A. Kornberg** (1960) and has both exonuclease (5' → 3' and 3' → 5') as well as polymerase (5' → 3') activities. The rate of polymerization of this enzyme is about 1,000/min and each cell contains about 400 molecules. Using its exonuclease and polymerase activities, it is mainly involved in:

   (*a*) DNA repairing by removing mismatch bases (by 3' → 5' exonuclease activity) and then filling the gaps thus produced (by 5' → 3' polymerase activity) (Fig. 7.21).

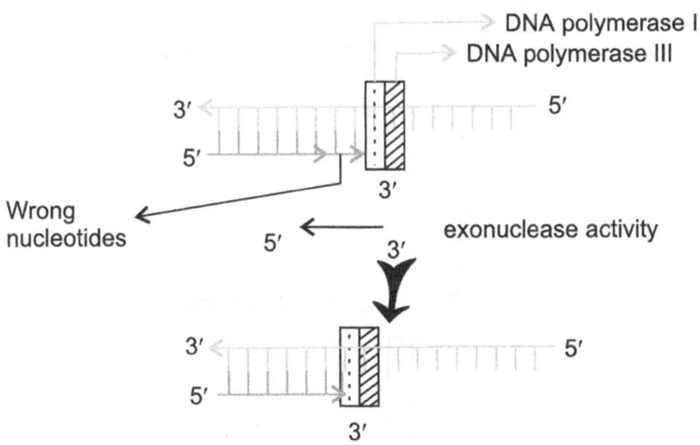

Fig. 7.21 Removal of wrong nucleotides during DNA replication.

   (*b*) removal of RNA primer (by 3' → 5' exonuclease activity) and filling up of the gap by adding dNTPs (by 5' → 3' polymerase activity) (Fig. 7.22)

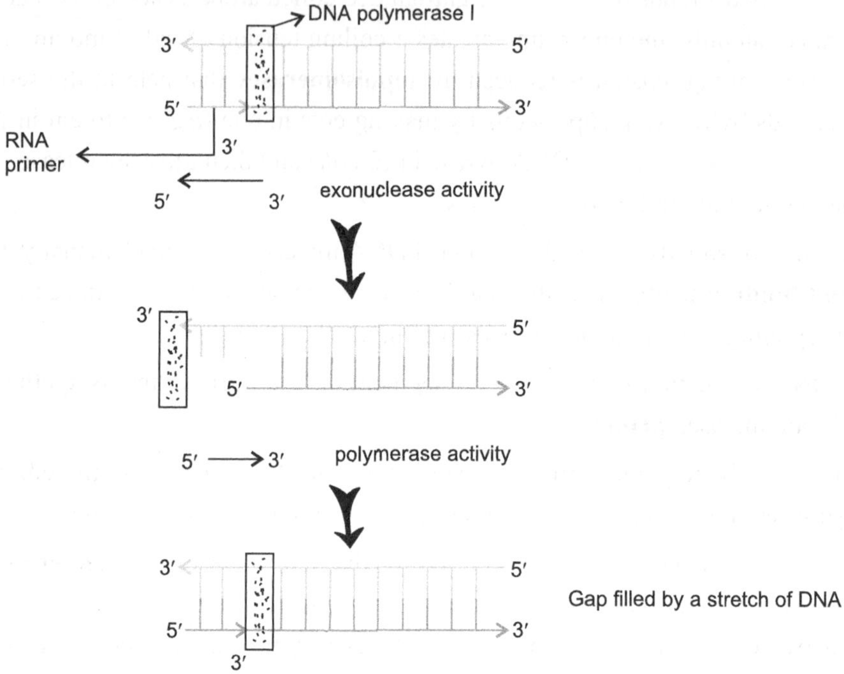

Fig. 7.22 Removal of RNA primer during DNA replication.

(*c*) removing obstructions (*e.g.* wrong DNA segments) encountered during DNA synthesis (by 5' → 3' exonuclease activity) (Fig. 7.23), etc.

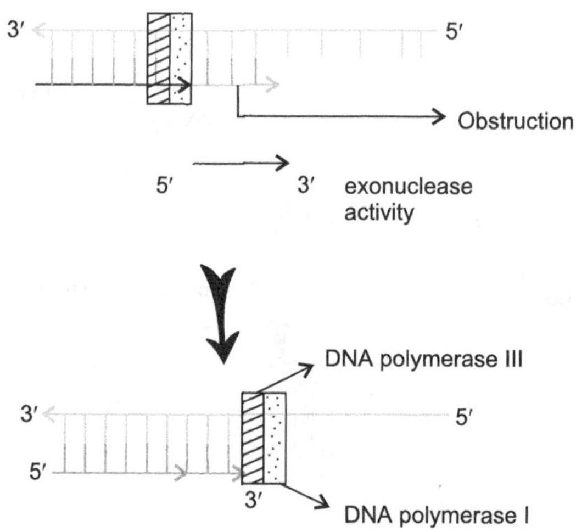

Fig. 7.23 Removal of the obstruction during DNA synthesis by DNA polymerase I.

(*ii*) **DNA polymerase II:** It has 3' → 5' exonuclease (no 5' → 3' exonuclease activity) and 5' → 3' polymerase activities. The rate of polymerization of this enzyme is about 50/min and each cell contains about 100 molecules. Its exact role is not known.

(*iii*) **DNA polymerase III:** It has 5' → 3' and 3' → 5' exonuclease and 5' → 3' polymerase activities. It is made up of 7 subunits (α, β, δ, ε, γ, θ, τ) and is the main enzyme involved in DNA replication. The rate of polymerization of this enzyme is about 15,000/min and each cell contains about 10 molecules. (τ = tau)

In eukaryotes, the three types of DNA polymerases are:

1. **DNA polymerase α:** It is found in nucleus but leaks into the cytoplasm and is mainly responsible for DNA replication. It does not contain any exonuclease activity.
2. **DNA polymerase β:** It is found in the nucleus and is mainly involved in DNA repair and recombination. It does not have any exonuclease activity.
3. **DNA polymerase γ:** It is found in the nucleus and mitochondria and resembles bacterial polymerases. It is mainly involved in DNA repair and recombination. It has 5' → 3' and 3' → 5' exonuclease activities.

After the separation of two strands of a double-stranded DNA molecule, each strand acts as a template and DNA polymerase III reads its nucleotides and adds complementary nucleotides in the newly synthesizing DNA strand. Since, the two strands of a double-stranded DNA molecule are anti-parallel (⇄) and the same DNA polymerase III synthesizes the two new strands simultaneously using the two template DNA strands and since, DNA polymerization always occurs in **5' → 3'** direction, DNA replication is **semi-discontinuous,** where the synthesis of one of the two newly synthesizing DNA strands occurs continuously and that of the other discontinuously. The DNA fragments synthesized discontinuously are called **Okazaki fragments** (a Japanese scientist **Okazaki** discovered them). The template where the DNA synthesis occurs continuously is known as the **leading strand,** whereas the other template where the DNA synthesis occurs discontinuously is known as the **lagging strand** (Fig. 7.24).

[Exception: In adenovirus, both the new DNA strands are synthesized continuously.]

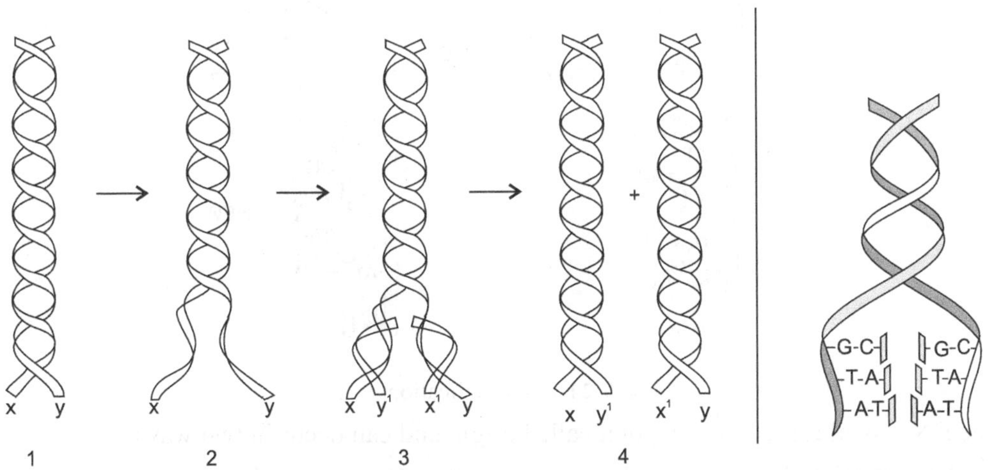

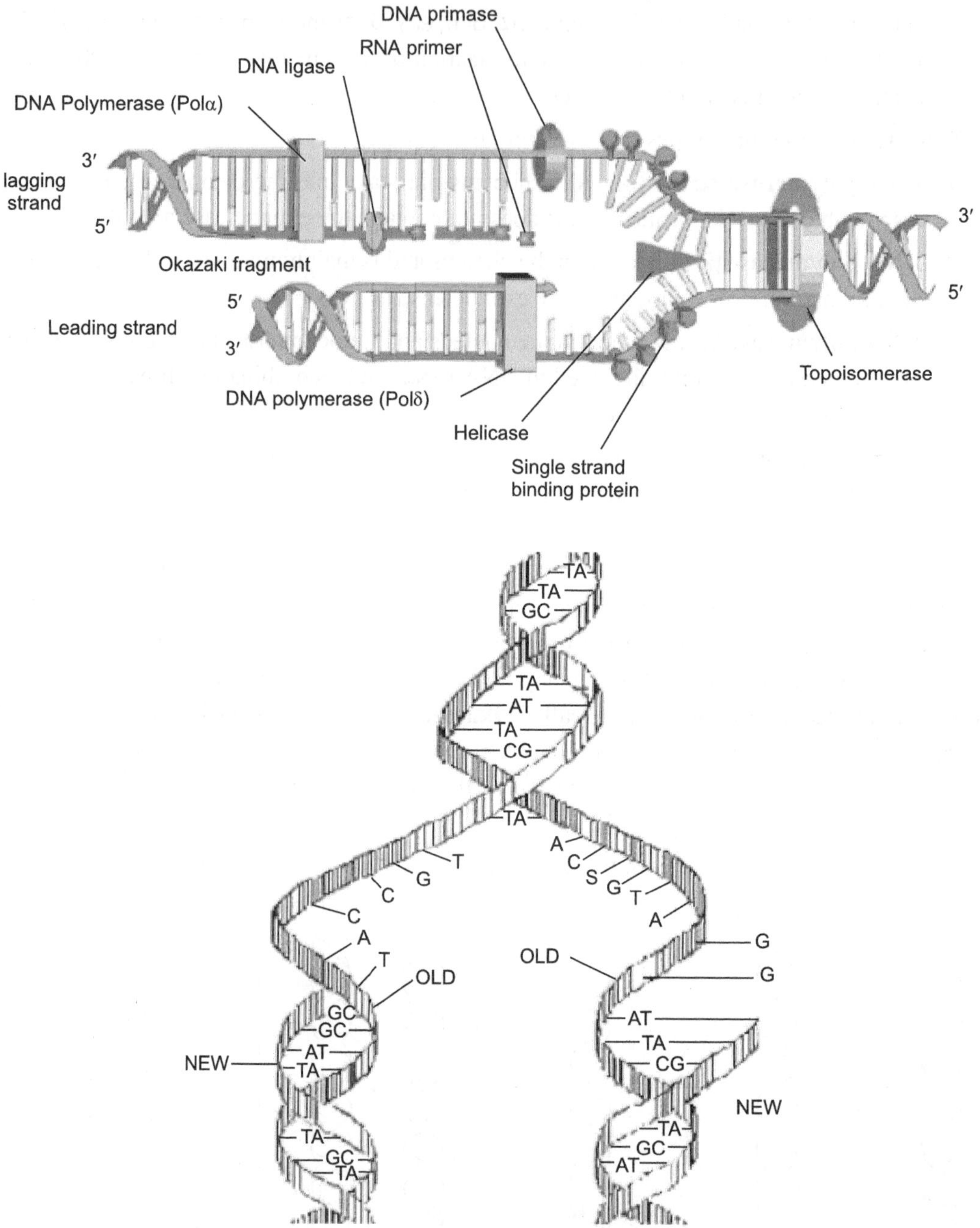

Fig. 7.24 DNA replication.

The DNA synthesis starts at a point called origin and can occur in two ways:

(*a*) **Unidirectional,** where a single enzyme complex (consisting of various enzymes required for the synthesis of DNA) binds at the origin and moves along the replication fork to synthesize the two new strands (Fig. 7.25).

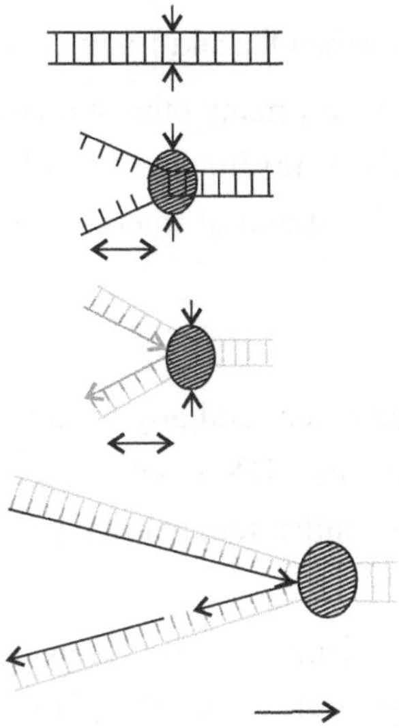

Fig. 7.25 Unidirectional DNA replication.

(*b*) **Bidirectional,** where two enzyme complexes bind at the origin and move in opposite directions synthesizing the new DNA strands (Fig. 7.26).

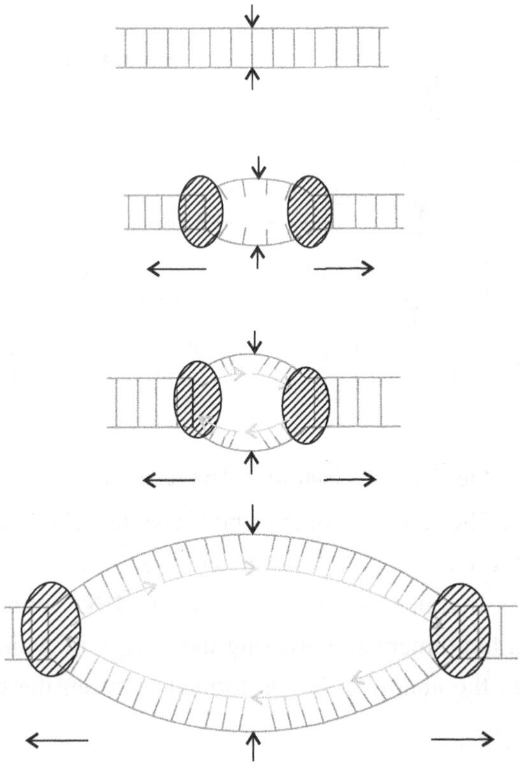

Fig. 7.26 Bidirectional DNA replication.

In eukaryotes, the DNA replication starts at many sites and can be unidirectional or bidirectional.

6. The **DNA ligase** enzyme joins the Okazaki fragments by forming phosphodiester bonds.
7. In addition to these main enzymes, many other enzymes are also involved in DNA replication, *e.g.* **exonucleases** (removing nucleotides at the ends of DNA strand), **endonucleases** (cleaving DNA strand at internal positions), etc.

DNA Repairing

DNA replication occurs quite accurately and any mistake done during this process is proof read by a very strong proof reading enzyme, DNA polymerase I. Despite this accuracy, if some mistake is done in DNA or during DNA replication, many repair systems are involved and some of the important ones are:

(*i*) **Dark Repair:** This repair mechanism is also known as **excision repair** mechanism. It does not need any light energy and can occur in dark as well as in light. Here, the wrong nucleotide segment is removed with the help of **endonucleases** and the correct DNA segment is synthesized by **DNA polymerase I** enzyme (Fig. 7.27).

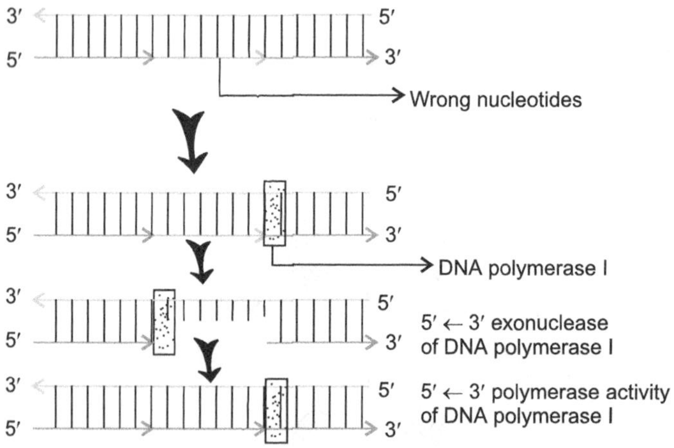

Fig. 7.27 Mechanism of dark repairing.

(*ii*) **Photoreactivation:** The sunlight contains both visible light as well as UV radiation and whenever the DNA is exposed to UV, it induces the formation of intrastrand **pyrimidine dimers** (particularly thymine dimers). The enzyme **DNA photolyase** repairs the dimmers by binding at the dimers and utilizing the energy of white light (blue being most effective) it cleaves the abnormal bonds formed between the two adjacent pyrimidine dimers (Fig. 7.28).

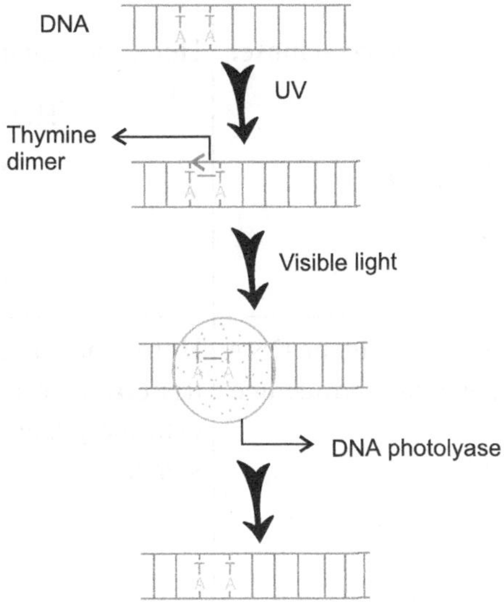

Fig. 7.28 Mechanism of photoreactivation.

Important differences between DNA and RNA

S. No.	*DNA*	*RNA*
1.	Generally, DNA is the **genetic** material and it transcribes the genetic information to RNA.	RNA is not the genetic material, except in certain viruses, *e.g.*, TMV containing single-stranded RNA. Of the three types of RNA molecules, mRNA is generally being translated to polypeptides, whereas, tRNA and rRNA that are not translated.
2.	Usually it is found in the **nucleus** and some cytoplasmic organelles (mitochondria and chloroplast), except in prokaryotes, where it occurs as nucleoid and in some cases as plasmids (also found in a few eukaryotes, like yeast) in the cytoplasm.	Most of the RNA is found in **cytoplasm** and very small amount (newly synthesized) being present in the nucleus. There are mainly three types of RNA molecules, mRNA, tRNA and rRNA.
3.	Generally, it is **double-stranded,** except in some viruses, *e.g.*, bacteriophage ΦX174 that contains a circular single-stranded DNA molecule as the genetic material. 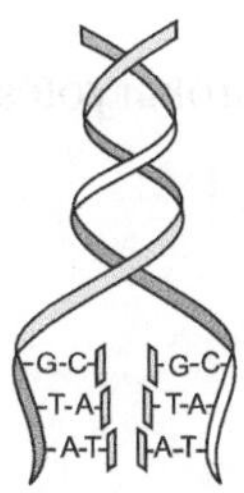	It is mostly **single-stranded,** except in some viruses like reo virus, wound tumour virus where it is double-stranded. 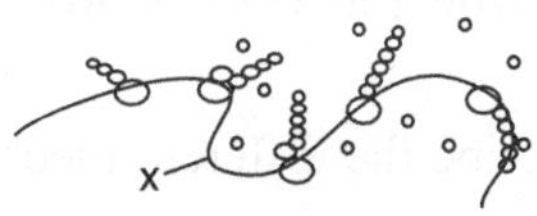 mRNA with newly synthesized polypeptides.

4.	The sugar found here is **2-deoxy-ribose.** 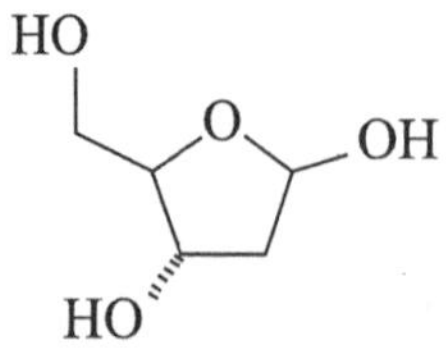	Here, the sugar is **ribose.** 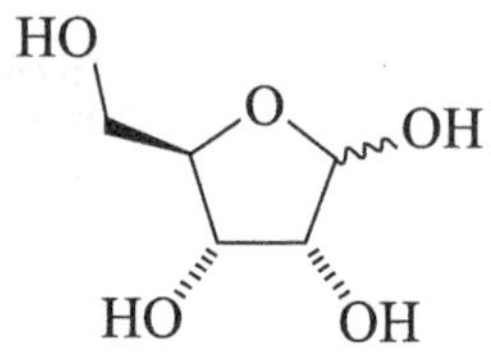
5.	The four types of bases present in DNA are **thymine,** cytosine, guanine and adenine. 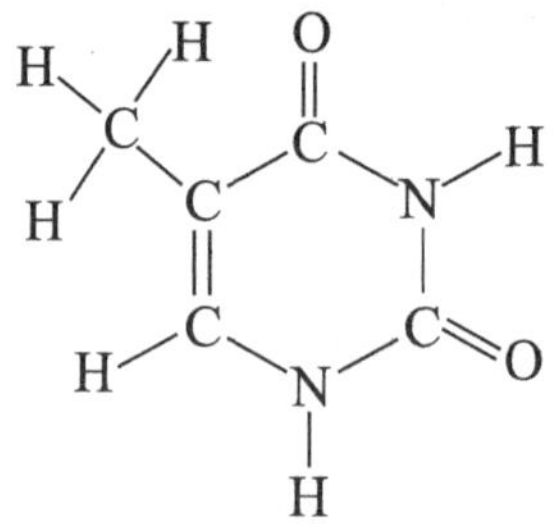	Here, the base thymine is replaced by **uracil,** whereas the other three bases, cytosine, guanine and adenine, remaining the same. 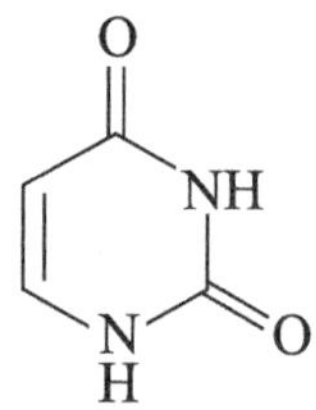
6.	Unusual bases are very few or absent.	Generally contains many **unusual bases.**
7.	It is self-replicating and after replication, it gives rise to two daughter DNA molecules.	It is not self-replicating and is produced by the transcription of DNA that produces large number of RNA molecules.

QUESTIONS

1. Describe the experiment which initially demonstrated that DNA is the genetic material.
2. Discuss an experiment which showed that RNA carries the genetic information.
3. **Discuss the structure of DNA.**
4. **What are the main differences between DNA and RNA.**
5. Give a brief account of different kinds of RNAs known in the living system.
6. Discuss an experiment which demonstrated the semi-conservative mode of DNA replication.
7. **Describe the DNA synthesis in living organisms, giving emphasis on *Escherichia coli.***
8. Describe the different mechanisms of DNA repair in prokaryotes.

Chapter 8 PROTEIN SYNTHESIS

Genes (DNA) are generally expressed through **transcription** (transfer of information from DNA to RNA, *e.g.*, synthesis of mRNA, tRNA, rRNA) and **translation** (transfer of information from mRNA to proteins) of language of nucleic acids (mRNA) into language of proteins. Thus, the sequence of bases in DNA and RNA determine the sequence of amino acids in proteins. Some genes do not contain any information of proteins and are expressed only through transcription, *e.g.*, tRNA, rRNA genes. This one way flow of information from DNA to RNA and finally to proteins is known as **central dogma** (Fig. 8.1).

DNA ⟶ RNA ⟶ Protein

G T G C A T C T G A C T C C T G A G G A G A A G   DNA
C A C G T A G A C T G A G G A C T C C T C T T C   (transcription)

↓

GUG CAU CUG ACU CCU GAG GAG AAG   RNA

↓ ↓ ↓ ↓ ↓ ↓ ↓ ↓   (translation)

V H L T P E E K   protein

But, in 1970, **H. Temin** in RSV (Rous Sarcome Virus) and **D. Baltimore** in MLV (Mouse Leukaemia Virus) independently reported the presence of **reverse transcriptase** (RNA directed DNA polymerase or RNA dependent DNA polymerase) enzyme which synthesizes DNA from RNA (acts as template). These RNA viruses when infect the host cells, the genetic RNA after entering the cell first gets copied into **DNA** by utilizing the viral reverse transcriptase enzyme. This DNA then again directs the formation of **genetic RNA** (which gets encapsulated into the protein cover to form the virus particle) and the **mRNA** (required for protein synthesis for the formation of progeny phage particles).

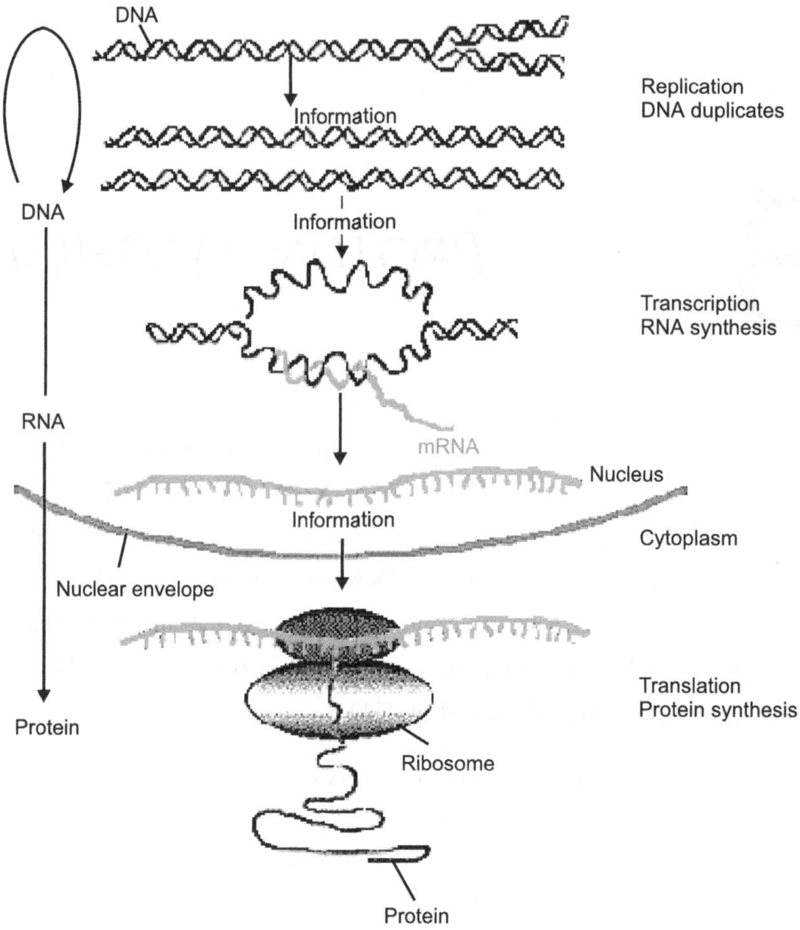

Fig. 8.1 Central dogma.

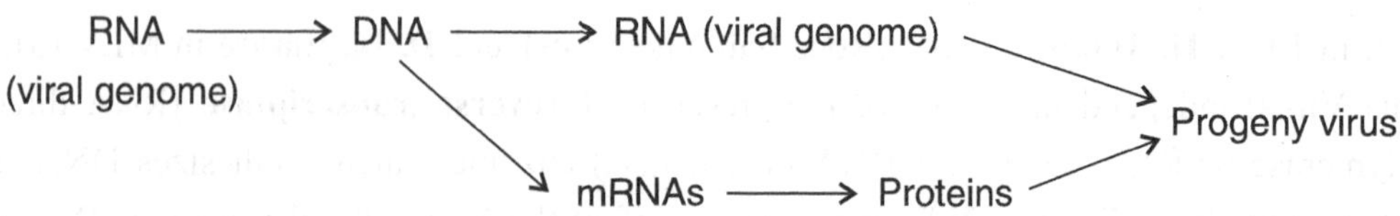

This backward flow of information has modified the central dogma. This reverse transcriptase enzyme has also been reported in many other retroviruses (tumour inducing RNA viruses) and HIV and is extensively utilized in molecular biology studies (*e.g.*, for the formation of cDNA from mRNA in the construction of cDNA library).

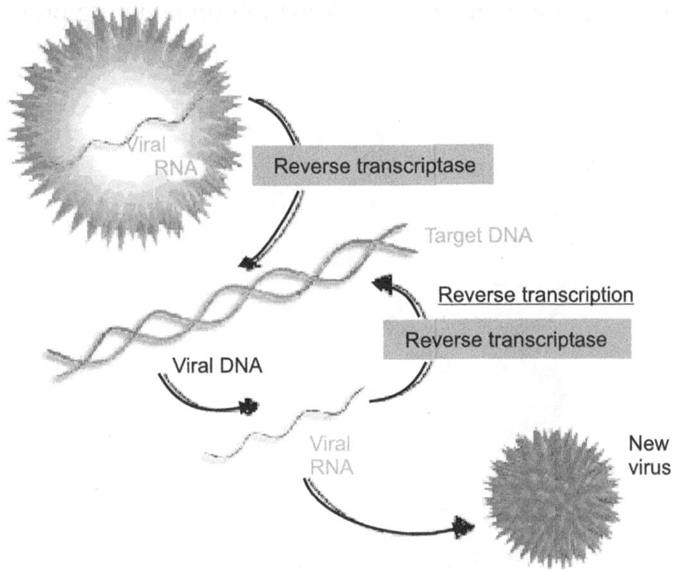

Fig. 8.2 Reverse transcription in retrovirus.

PROTEIN SYNTHESIS APPARATUS

For protein synthesis tRNA, mRNA, ribosomes, certain enzymes and proteins are required.

1. **tRNA:** Amino acids themselves cannot directly recognize the base sequences (codons) present in mRNA, but the anticodon of tRNA (that is complementary to codon in mRNA) helps in the recognition of the codons in mRNA. Thus, a tRNA (transfer RNA) carries a specific amino acid to its corresponding codon. The **nucleotide sequence** of tRNA for the first time was given by **Robert Holley** (1964) for alanyl-tRNA ($tRNA^{ala}$) of yeast and the **gene** for this $tRNA^{ala}$ was artificially synthesized by **Hargobind Khorana.** Several models of tRNA have been proposed and according to **clover leaf model,** the single-stranded tRNA molecule is folded like a clover leaf having three double helical stems (secondary structure due to intrastrand base pairing as A = U and G ≡ C) each terminating into a single stranded loop. All tRNA molecules have **G** (guanosine) at 5′ end and for the attachment of specific amino acids **CCA** sequence at the 3′ end. The **anticodon loop** contains anticodon (sequence of 3 nucleotides) that recognizes its complementary codon (sequence of 3 nucleotides) in mRNA. Not only the anticodon, but the 4 nucleotides present on the two sides of the anticodon and the base pairs of the attached stem are also important for the correct recognition of the respective codon, and this is known as extended anticodon hypothesis. The **amino acyl tRNA synthetase binding loop** recognizes the

specific amino acyl tRNA synthetase that in turn recognizes the specific amino acid. The **ribosomal binding loop** binds with the ribosome during protein synthesis (Fig. 8.3 and 8.4).

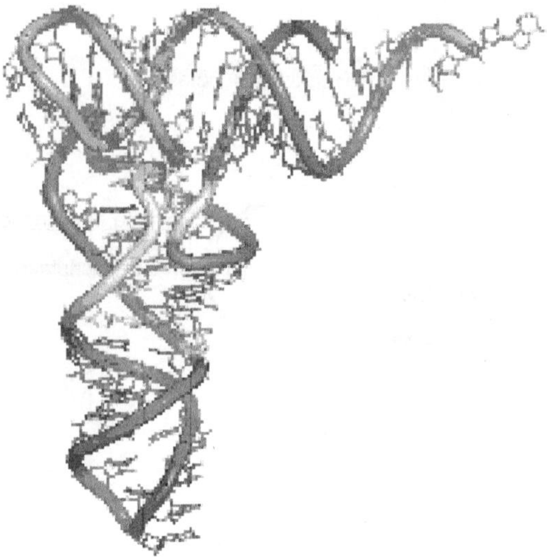

Fig. 8.3 Three dimensional structure of tRNA.

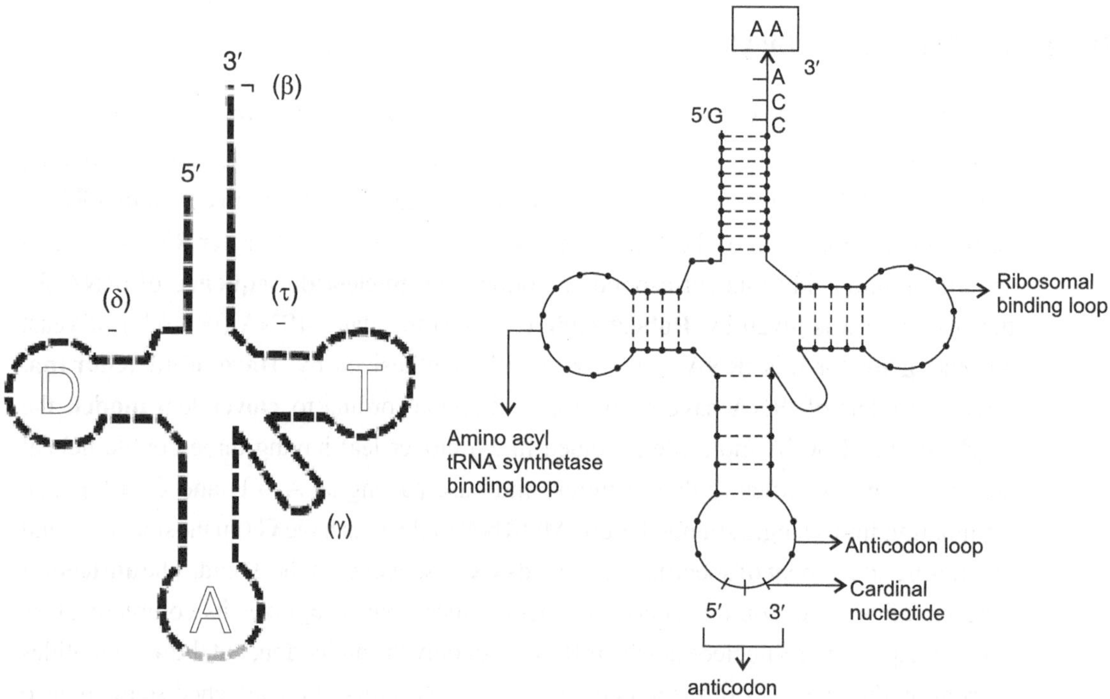

Fig. 8.4 Clover-leaf model of tRNA.

2. **mRNA:** After transcription, the information of protein present in DNA is transferred to mRNA (messenger RNA), which contains various codons (set of three nucleotides) for different amino acids constituting a particular protein. The initiation codon lies at (in prokaryotes and many mRNAs of eukaryotes) or near (many mRNAs of eukaryotes). 5′ end of mRNA and termination codon at or near the 3' end. These codons (except for the termination codons) are successively read by the anticodons present in their respective tRNAs (Fig. 8.5).

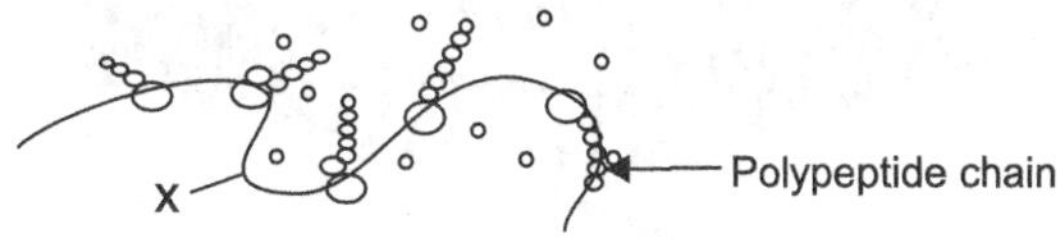

Fig. 8.5 mRNA with ribosomes and growing polypeptide chains.

3. **Ribosomes:** They are made up of rRNAs (ribosomal RNAs) and proteins and have about 23 nm average diameters. They have two subunits, one larger and the other smaller and at the time of translation process of protein synthesis, the two subunits bind with the mRNA. The mRNA attached with a number of ribosomes that are involved in the synthesis of the same protein constitutes a **polysome** (polyribosome).

In **prokaryotes,** the ribosomes have 66% rRNAs and 34% proteins and are of **70S** (S is the Swedberg unit indicating the sedimentation coefficient) type with 30S smaller (consisting of 16S rRNA and about 21 proteins) and 50S larger (consisting of 23S rRNA, 5S rRNA and about 31 proteins) subunits.

In **eukaryotes,** the ribosomes in the cytoplasm contain about 60% rRNA and 40% proteins and are of **80S** type with smaller 40S (made up of 18S rRNA and about 33 proteins) and larger 60S (consisting of 28S rRNA, 5.8S rRNA, 5S rRNA and about 45 proteins) subunits.

These ribosomes are found either freely in the cytoplasm or attached with the endoplasmic reticulum (rough endoplasmic reticulum). In the organelles, mitochondria and chloroplasts (in photosynthetic organisms), the ribosomes are generally of 70S type, as in prokaryotes.

MECHANISM OF PROTEIN SYNTHESIS

The protein synthesis occurs in two steps: Transcription and Translation.

Transcription

The process of synthesis of RNAs (mRNA, tRNA and rRNA) from DNA by the enzyme **RNA polymerase** is known as transcription. At the time of transcription, the RNA polymerase binds with double stranded DNA (gene) at a particular site (in prokaryotes known as promoter site) and after unwinding of the two strands of DNA by the rotation of the DNA, it starts copying one of the two strands, known as **coding strand** (sense strand or template strand). The other strand of the DNA, which is not copied for the RNA synthesis, is known as **non-coding strand** (antisense strand) (Fig. 8.6).

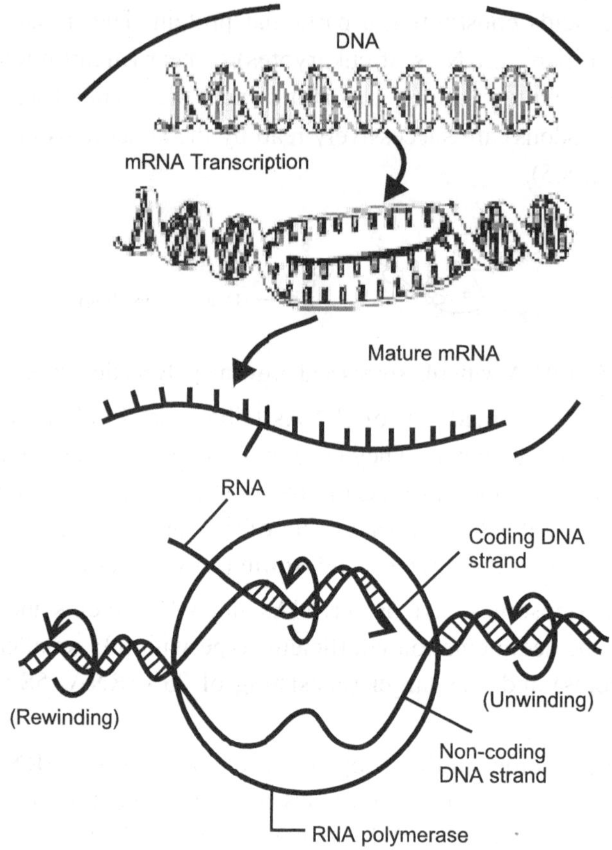

Fig. 8.6 Transcription in prokaryotes.

In **eukaryotes** there are four types of RNA polymerases for the synthesis of different types of RNA molecules:

(*i*) *RNA polymerase I* (or A) found in nucleolus for the synthesis of rRNAs (28S and 18S).

(*ii*) *RNA polymerase II* (or B) found in nucleoplasm for the synthesis of HnRNA (heterogeneous nuclear RNA) which, after processing gives rise to mRNA.

(*iii*) *RNA polymerase III* (C) found in nucleoplasm for the synthesis of tRNA and 5S rRNA.

(*iv*) *Organeller RNA polymerase* found in organelles, mitochondria and chloroplasts, (in photosynthetic organisms) and resembles the prokaryotic RNA polymerase.

In **prokaryotes,** only one type of RNA polymerase synthesizes all the types of RNAs. It is made up of 5 subunits (proteins), where $\alpha_2\beta\beta'$ makes the core enzyme that is found loosely attached with a σ factor. The σ factor helps in the recognition of promoter site and after initiating the transcription, it gets dissociated from the holoenzyme ($\sigma\alpha_2\beta\beta'$). Now, the RNA is synthesized by the movement of core enzyme ($\alpha_2\beta\beta'$) along the DNA.

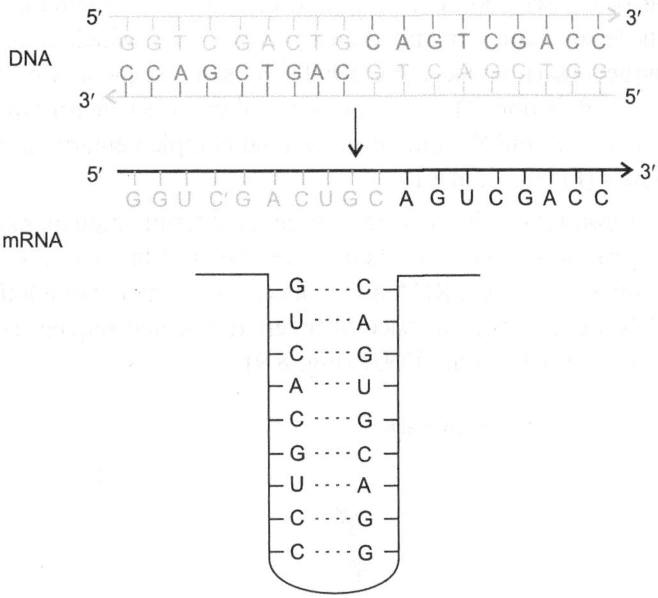

Fig. 8.7 Hairpin loop formation in mRNA.

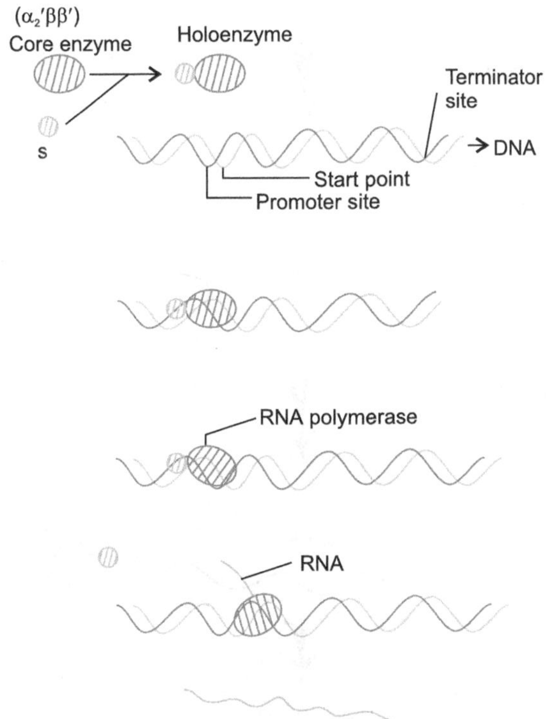

Fig. 8.8 Mechanism of transcription in prokaryotes

The **termination** of transcription occurs due to the presence of certain terminator sites on DNA. In prokaryotes, the termination is mainly done by in two ways, which involves the presence of certain **palindromic sequences** (Inverted repeats with the same sequences on the two strands of DNA when read in 5' → 3' direction. Thus, the presence of complementarity of bases on the same strand causing loop formation in mRNA due to intrastrand complementary base pairing) on DNA near the termination signals (Fig. 8.7 and 8.8).

(*i*) ***ρ-dependent Termination*:** In some prokaryotes, the termination of transcription is helped by a **ρ (rho) protein** that gets attached at the 5′ end of the newly synthesizing mRNA. The ρ then moves along the mRNA and induces the formation of a **hairpin loop** near the 3′ end of mRNA due to the presence of inverted repeated sequences. This helps in the detachment of mRNA from the DNA (Fig. 8.9).

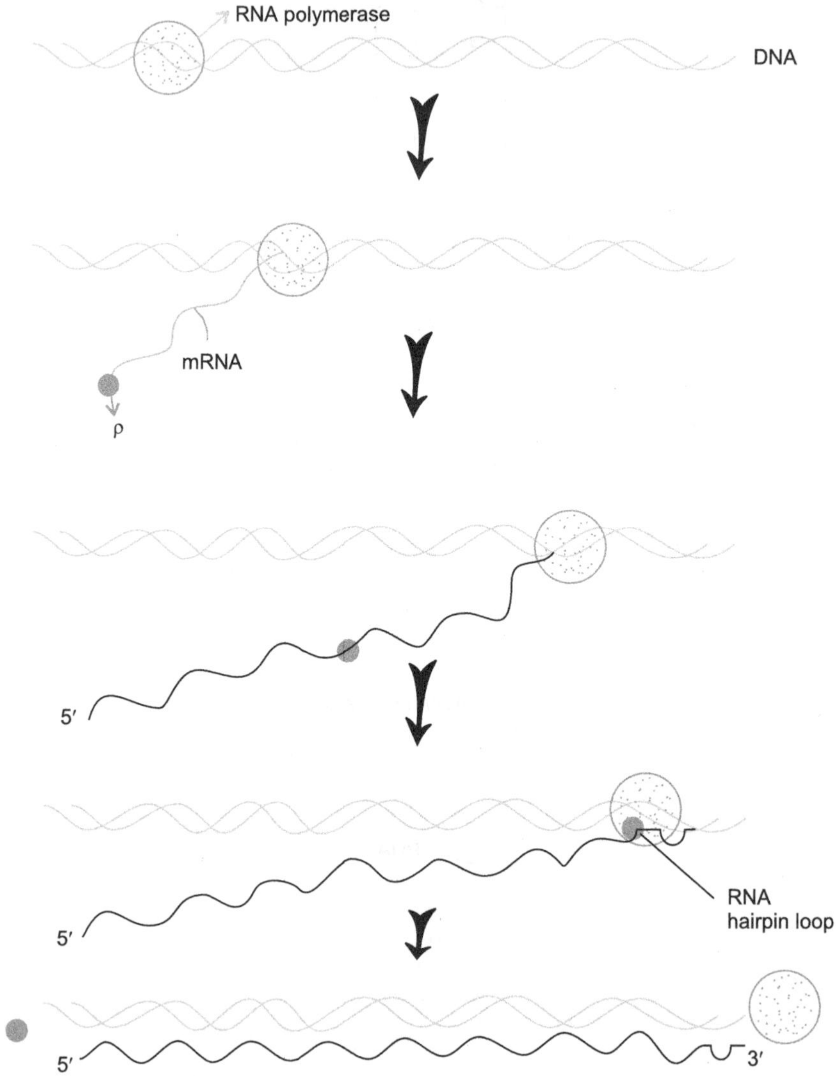

Fig. 8.9 ρ-dependent termination of transcription in prokaryotes.

(*ii*) ***ρ-independent Termination***: In many prokaryotes after the hairpin loop formation at 3′ end of mRNA, a **long run of U** (A in coding strand of DNA) is found and due to weaker U-A pairing in mRNA-DNA complex, the mRNA gets detached from the DNA (Fig. 8.10).

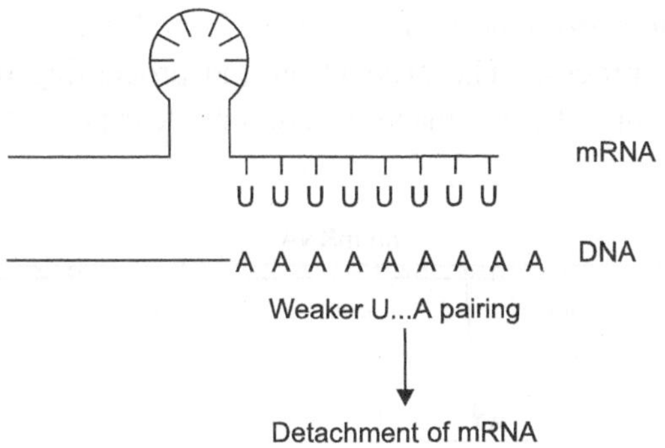

Fig. 8.10 ρ-independent termination of transcription in prokaryotes.

In prokaryotes, the transcription and translation processes occur **simultaneously** as there is no nuclear membrane to separate the genetic DNA and the cytoplasm.

In **eukaryotes,** the termination sites in DNA are present far away from the corresponding actual 3′ end of mRNA, thus, to produce **HnRNA** (heterogeneous nuclear RNA). The 3' end of the mRNA is generated after the processing of HnRNA by snurp (small nuclear RNA-protein complex). In addition to these extra nucleotides at 3′ end, the HnRNA may also contain extra nucleotide sequences at the 5′ end and at the internal positions. These extra nucleotide sequences at internal

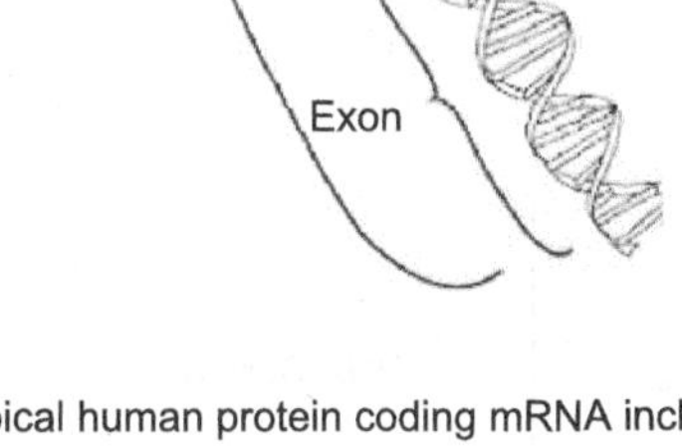

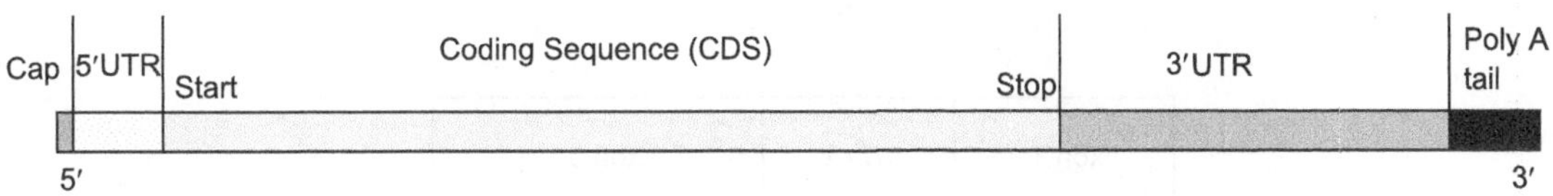

Fig. 8.11 The eukaryotic gene (above) and mRNA (below).

positions are called **introns,** whereas, the nucleotide sequences in between the introns that are present in mRNA and contain the information of proteins are known as **exons.** Thus, HnRNA produced after transcription is quite longer than the mRNA. Most of the extra nucleotide sequences, including introns, are cleaved by snurp. Moreover, after removal of the extra nucleotides from the 3′ end of the HnRNA, **poly A tail** is added that is required for the stability of the mRNA. Similarly, after removal of extra nucleotides from the 5′ end, a cap of 7-methyl guanosine **(7mG)** is added that is required for the translation process. The production and processing of HnRNA occurs in the nucleus from where it escapes into the cytoplasm through nuclear pores for the translation process (Fig 8.11 and 8.12).

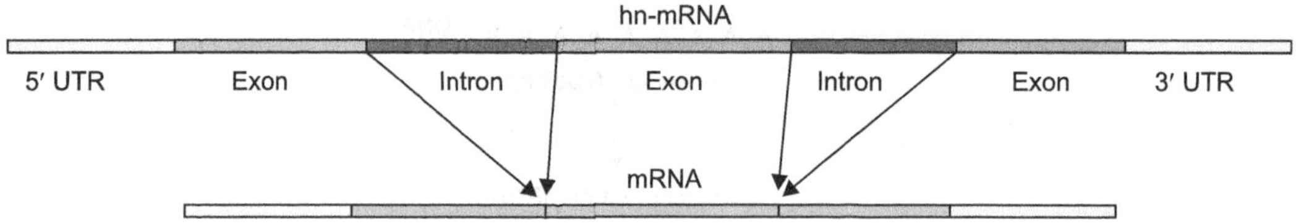

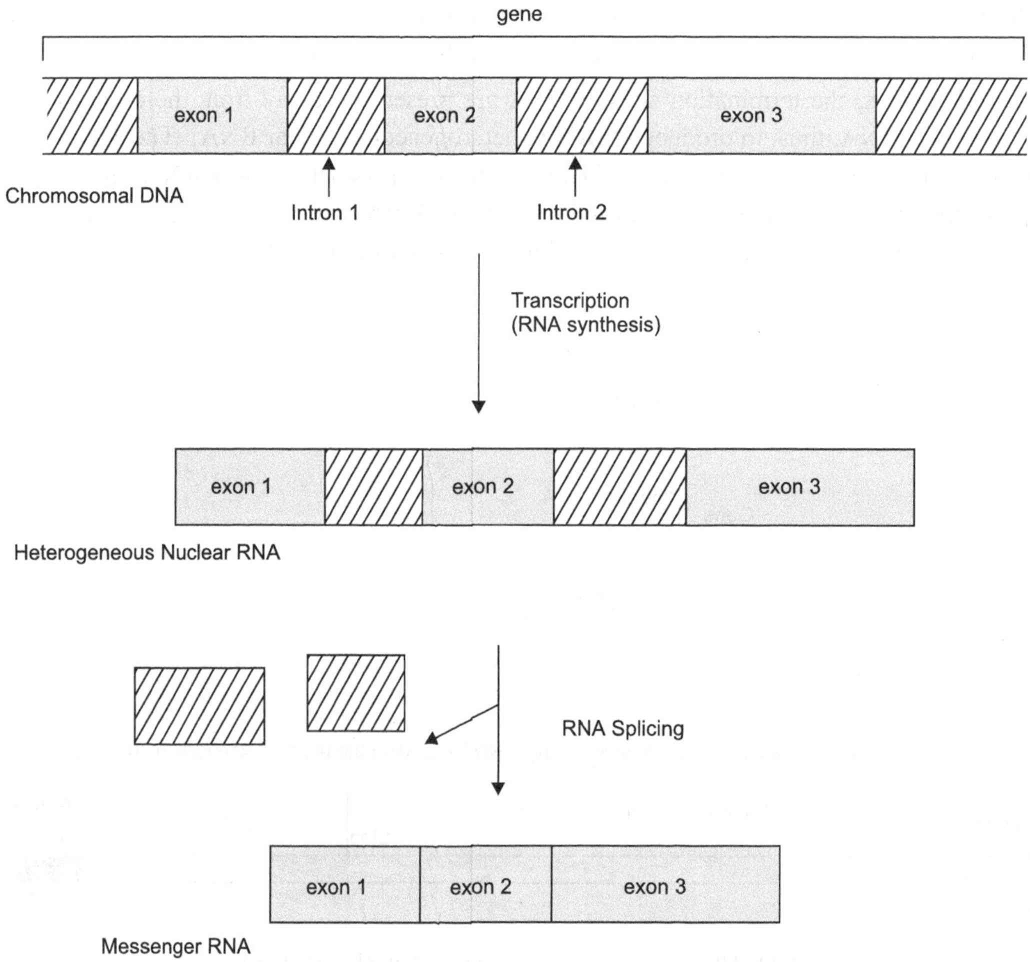

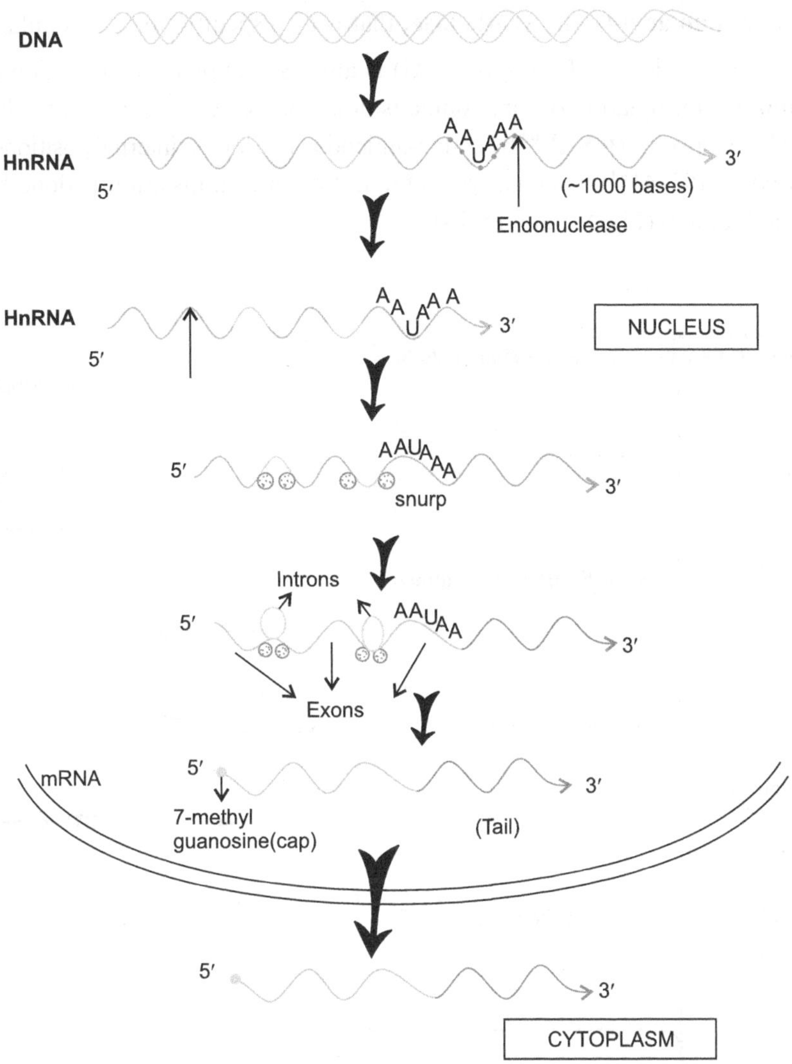

Fig. 8.12 Transcription and processing of mRNA in eukaryotes.

Translation

The process of synthesis of proteins from mRNA (translation of language of nucleic acids into the language of proteins) is called translation. There are **20** different types of **amino acids,** which constitute various proteins, and these amino acids themselves cannot recognize their respective codons in the mRNA. Different amino acids are carried by their specific tRNA molecules at the site of protein synthesis (mRNA). There are about **55** types of **tRNA** molecules available in the cytoplasm, so that one amino acid may have more than one tRNAs.

Various steps for the translation process in **prokaryotes** are:

1. The binding of an amino acid at the **3' end** of its specific **tRNA** is facilitated by an enzyme, **amino acyl-tRNA synthetase,** that is also specific for a particular amino acid.
2. The initiation codon **AUG** (rarely GUG) is always occupied by the amino acid, **formyl methionine** (formylation of methionine is done by the enzyme transformylase) which is carried by its specific **tRNA^{fmet}** by codon-anticodon pairing (in internal positions, methionine is carried by tRNA^{met}). (In eukaryotes the initiation of translation is done by the amino acid, methionine) (Fig. 8.13 and 8.14)

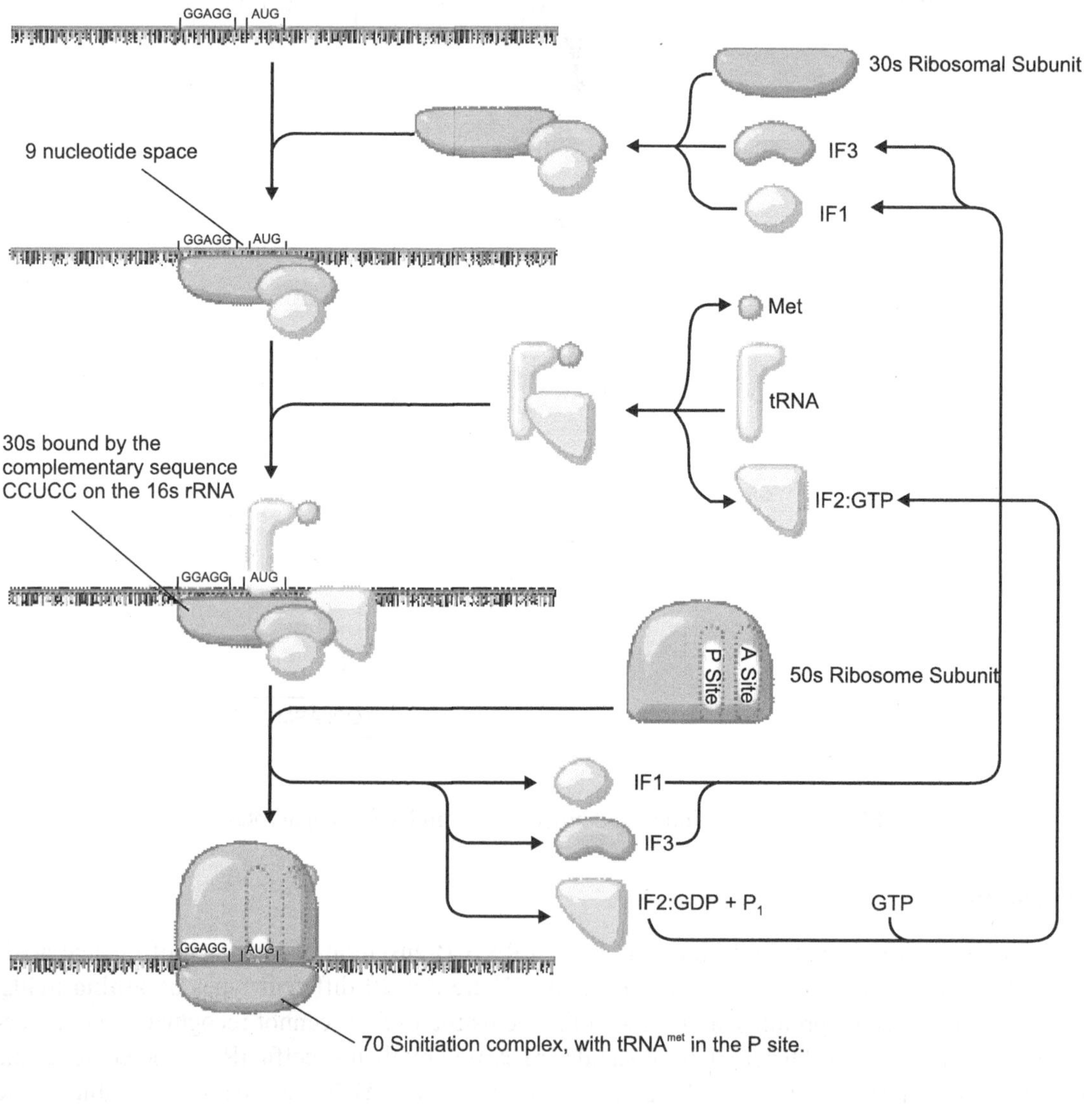

Fig. 8.13 Initiation of translation in prokaryotes.

3. Initiation factor 3 **(IF3)** helps in the dissociation of 70S ribosome into 30S and 50S subunits. Now, the IF3 attached with **30S** subunit, binds at the **5′ end** of the **mRNA.** (In eukaryotes the initiation codon lies far away from the 5′ end of mRNA and the smaller subunit (40S) of ribosome moves along the mRNA to find the initiation codon).

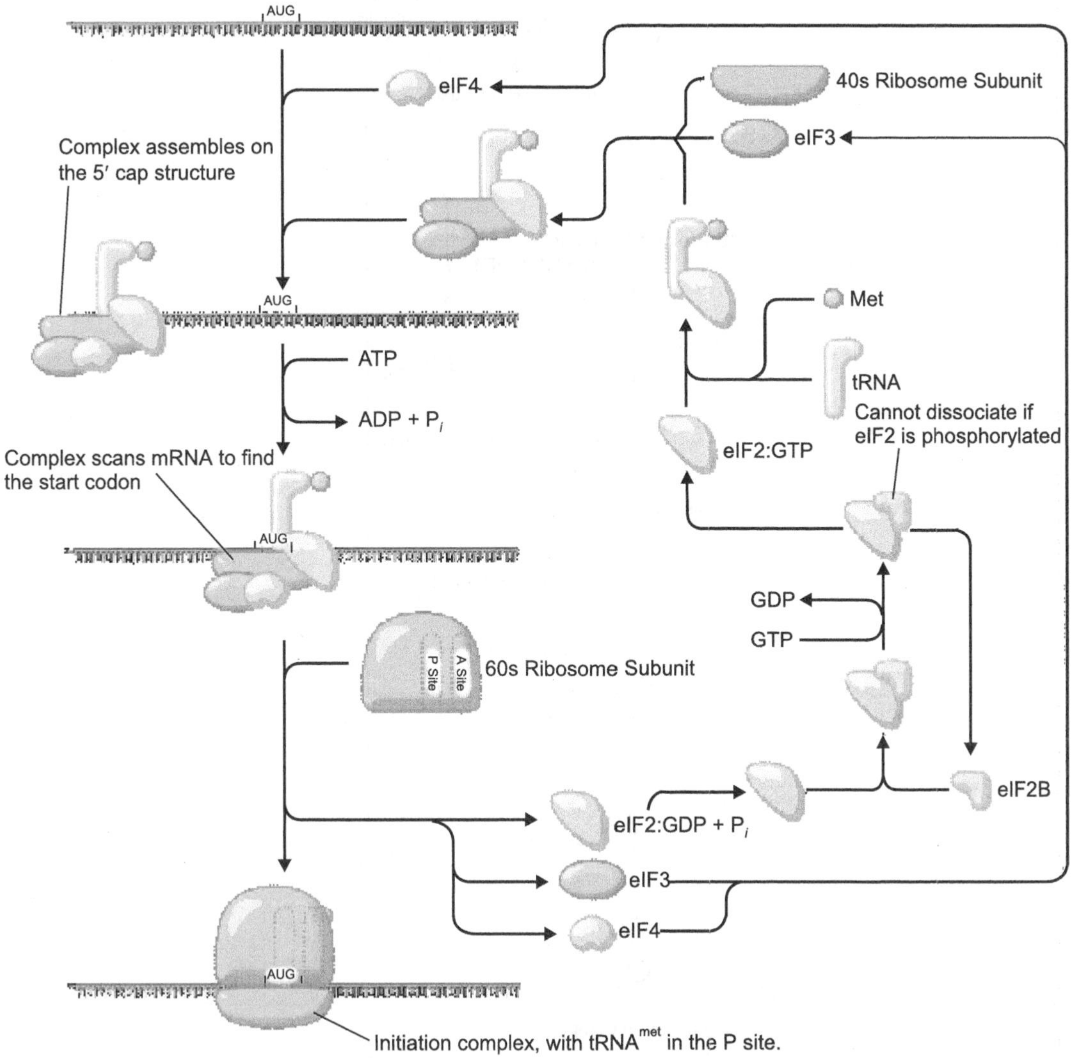

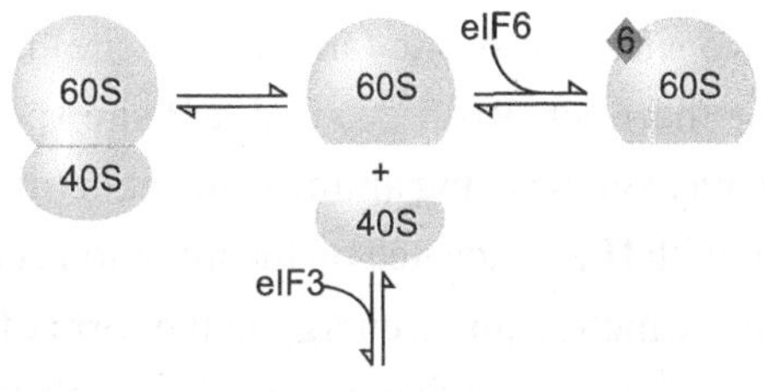

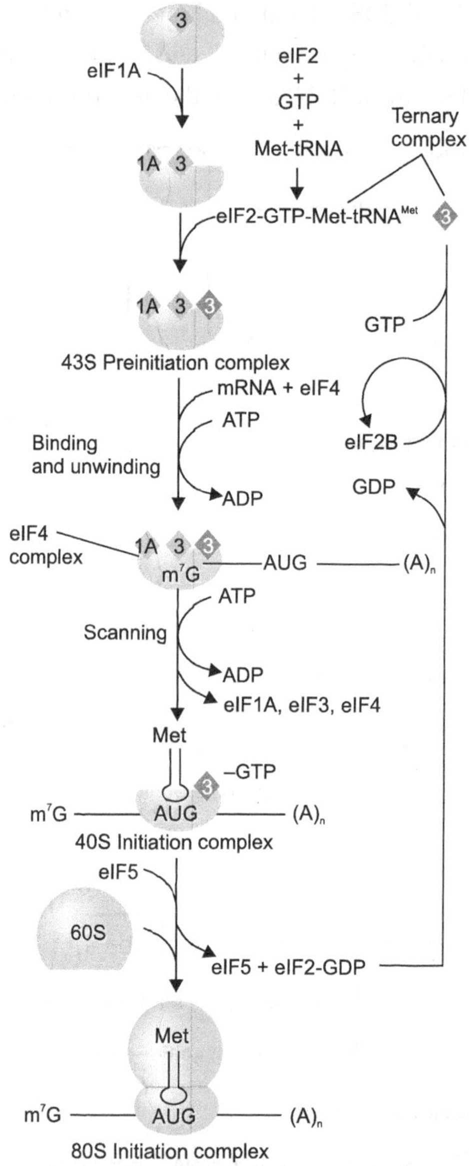

Fig. 8.14 Initiation of translation in eukaryotes.

4. The ribosomes have two sites, the A-site (amino acyl site) for the location of new amino acyl-tRNA (except for the fmet-tRNA^{fmet}) and a P-site (peptidyl site) for locating the tRNA attached with newly synthesizing amino acid chain (di and polypeptide). The **fmet-tRNA^{fmet}** after binding with **IF2,** is located on the initiation codon **AUG** at the **P-site** of the 30S subunit of ribosome which requires energy in the form of **GTP.** The role of IF1 is not known. (In eukaryotes more initiation factors *i.e.*, eIF1 to eIF6, are involved in initiation).

5. After initiation of the translation process, **elongation** process for the formation of dipeptide and ultimately the polypeptide chain is started by the attachment of **AA_2-tRNA$^{AA}{}_2$** (amino acyl tRNA or second amino acid attached with its specific tRNA) at its respective codon at the empty **A-site** with the help of the elongation factors **EFTu** (temperature unstable elongation factor) and **EFTs** (temperature stable elongation factor). This process also requires **GTP.** (In eukaryotes the elongation factors eEF1and eEF2 are involved in the elongation process).
6. Now, the enzyme **peptidyl transferase** helps in the formation of **peptide bond** at the **A-site** between the —COOH group of AA_2-tRNA^{AA2} (peptidyl tRNA for second and subsequent rounds of elongations) and $-NH_2$ group of fmet (thereafter, between 2nd and subsequent amino acids) (Fig. 8.15).
7. Thereafter, the naked **tRNA** at the **P-site** is **removed** with the help of the elongation factor EFG and the energy GTP as the ribosome moves by 3 nucleotides along the mRNA in 5′→3′ direction. Now, the P-site carries the fmet-AA_2-tRNA^{AA2} (tRNA attached with a dipeptide).

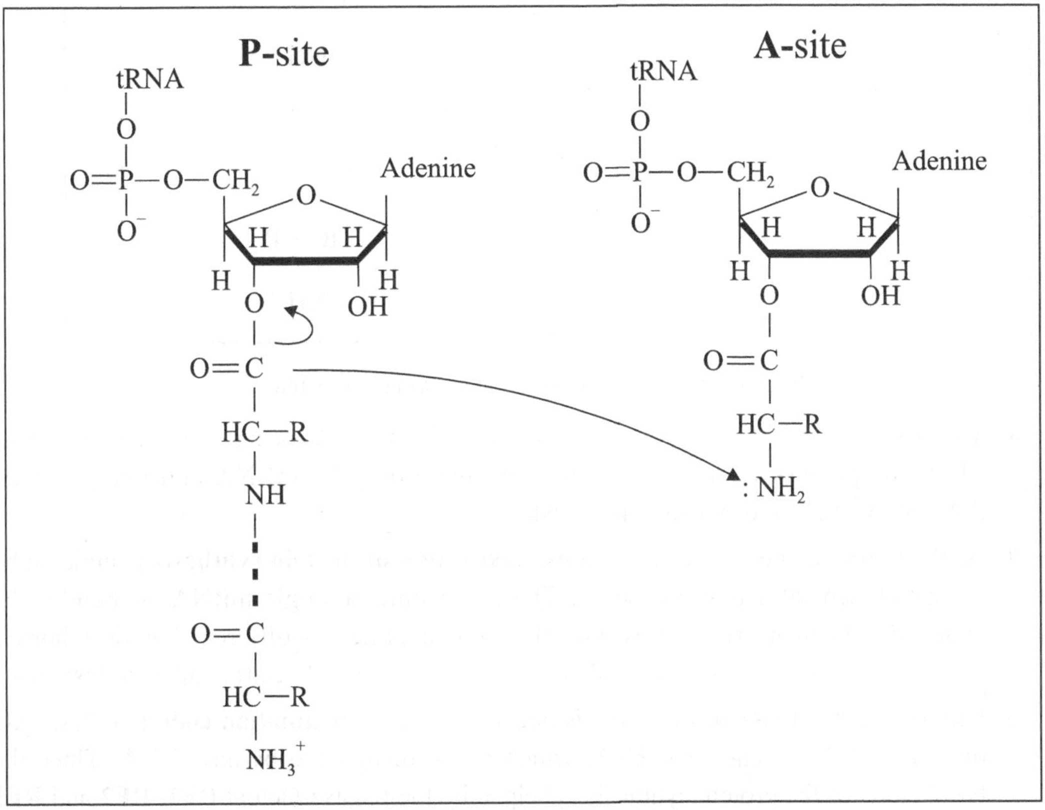

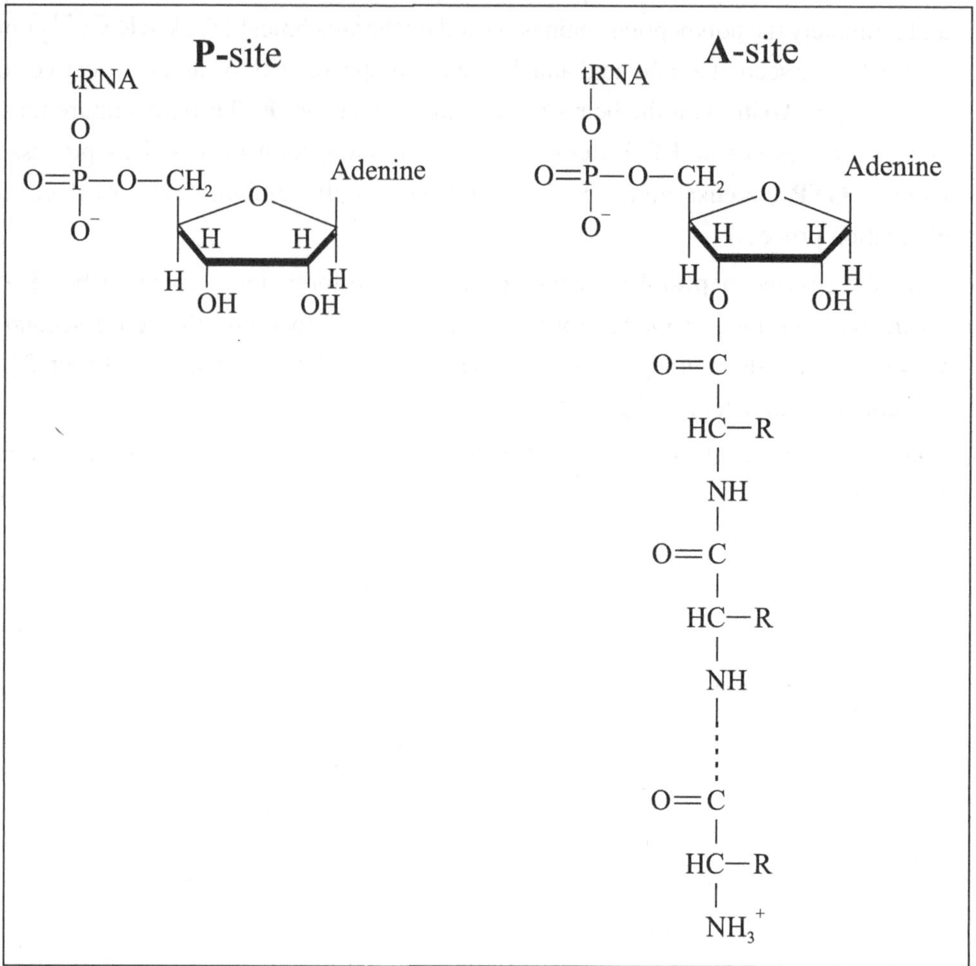

Fig. 8.15 Peptide bond formation in growing polypeptide.

8. The **AA_3-tRNA^{AA3}** (3rd amino acid attached with its tRNA) occupies the empty **A-site.** Thus this process of movement of ribosome along the mRNA and entry of new AA_ntRNA^{AAn}, at the A-site is repeated.
9. As the ribosome moves by two codons, **next round of protein synthesis** is initiated by the attachment of a new ribosome. Thus, at a time, a single mRNA is found to be attached with many ribosomes with their polypeptides of different length, (shortest polypeptide at the 5' end of the mRNA and longest at the 3' end) , called **polysomes.**
10. Ultimately, the **A-site** of ribosome is occupied by the **termination codon** (UAA,UAG or UGA) at the 3′ end of mRNA, which is not recognized by any tRNA. Thus, the termination of the protein synthesis is helped by the release factors **RF1, RF2** and **RF3** (in eukaryotes eRF1), which release the newly synthesized polypeptide chain from the P-site (Fig. 8.16).

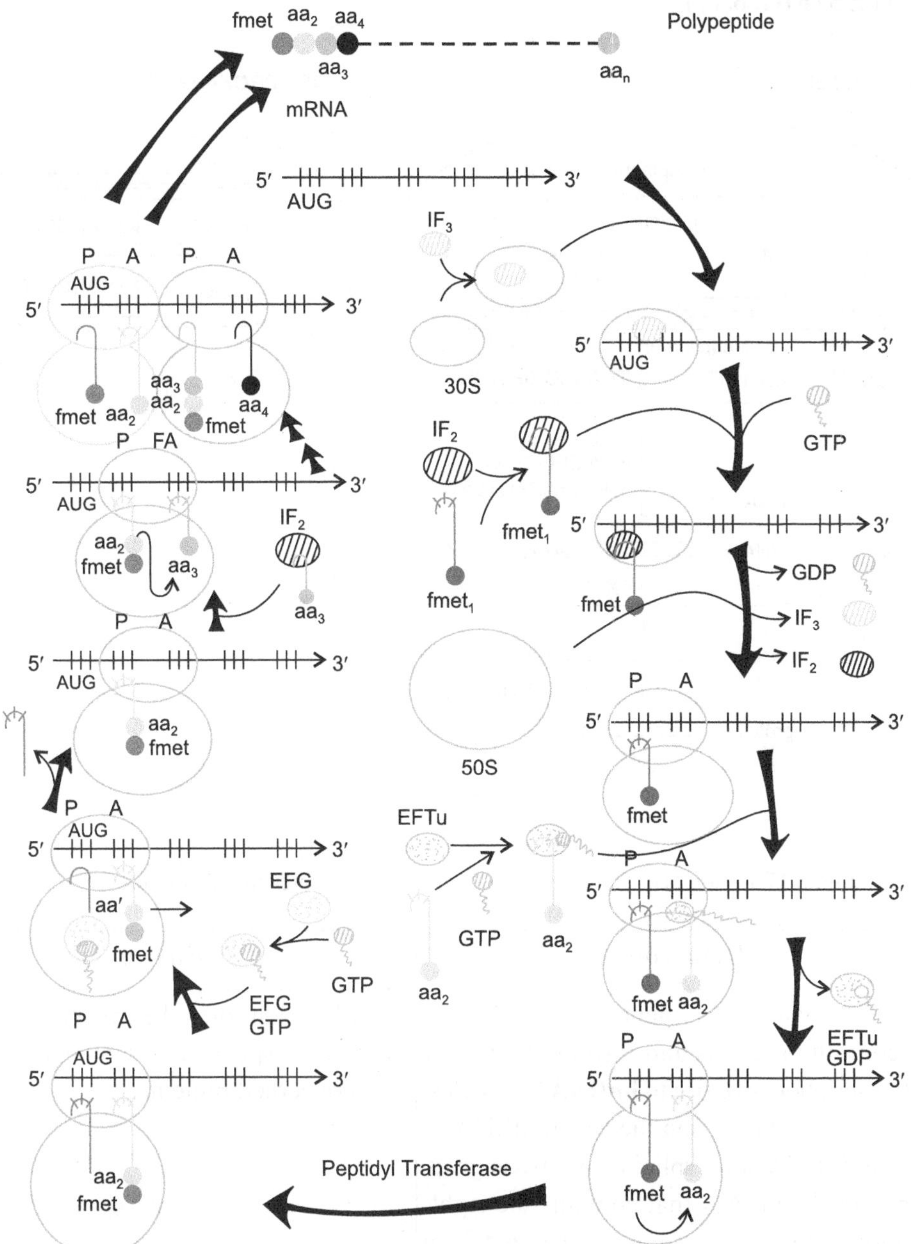

Fig. 8.16 Translation process of protein synthesis in prokaryotes.

11. The newly synthesized polypeptide is **modified** by the removal of formyl group of the first amino acid, formyl-methionine. Moreover, a few amino acids from the N-terminal (–NH_2 terminal) or from the C-terminal (– COOH terminal) or from both the termini may be removed. The primary structure of the polypeptide gets folded variously to produce three-dimensional **secondary** and **tertiary** structures.

DIFFERENCES IN PROTEIN SYNTHESIS IN PROKARYOTES AND EUKARYOTES (FIG. 8.17)

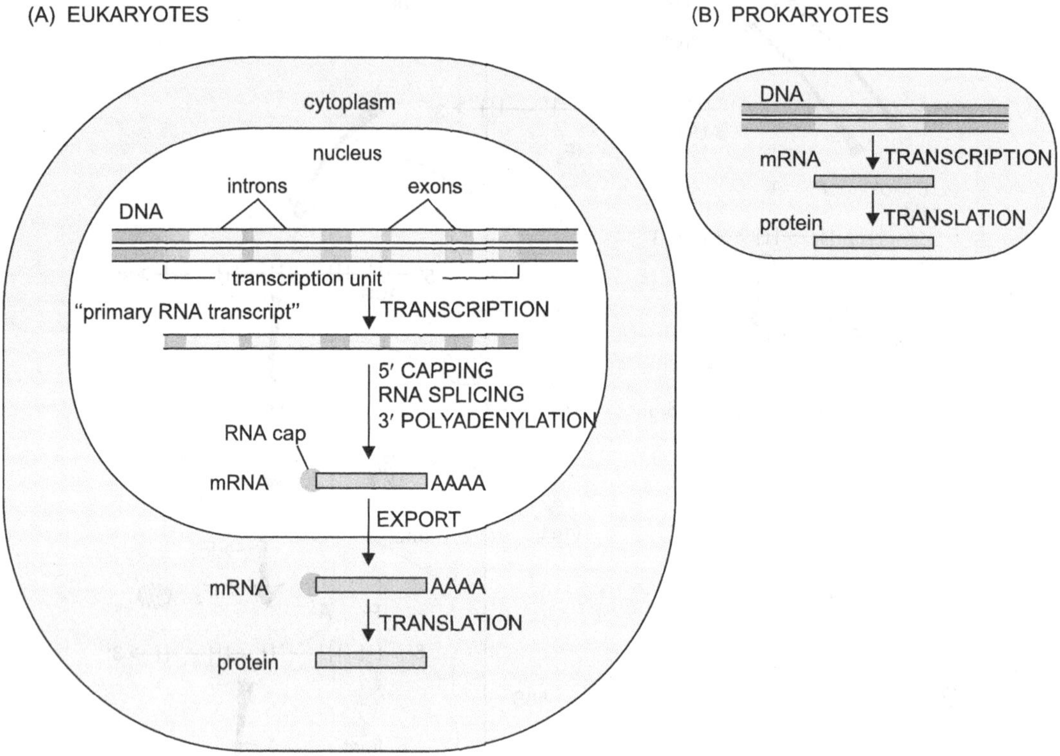

Fig. 8.17 Differences in protein synthesis in prokaryotes and eukaryotes

S. No.	*Eukaryotes*	*Prokaryotes*
1.	In eukaryotes many genes are split genes (containing exons and introns) that after transcription give rise to **hnRNA** molecules, which are longer than the actual mRNA. This hnRNA, after splicing of introns gives rise to the mRNA that contains the right information of the protein synthesized from it (exons).	In all prokaryotes the mRNA synthesized corresponds exactly the length of the concerned protein.
2.	During processing, a **7mG cap** at 5′ end and **poly A tail** at the 3′ end is added to give final mRNA.	The mRNA does not contain such cap and tail.

3.	The **initiation codon** is located a few nucleotides **away from the 5′ end** of mRNA, and during translation process the ribosome after attaching at the 5' end, moves along the mRNA to find the initiation codon.	Here, the **initiation codon** in mRNA lies exactly at its **5′ end.**
4.	Since, the chromosomes containing DNA are separated from the cytoplasm by the nuclear membrane, **transcription** occurs in **nucleus** and **translation** in **cytoplasm.** Thus, transcription and translation processes occur separately at different time periods.	The transcription and translation processes occur almost **simultaneously** in the **cytoplasm** due to the absence of a nuclear membrane.
5.	Ribosomes involved in translation process in thecytoplasm are of **80S** type, whereas, in organelles, mitochondria and chloroplasts, they are generally of 70S type.	**70S** type ribosomes are involved in the translation process.

QUESTIONS

1. Write short notes on:
   (*i*) **Two dimensional structure of tRNA**
   (*ii*) Extended anticodon hypothesis
   (*iii*) **Central dogma**
   (*iv*) Reverse transcription
   (*v*) **RNA polymerases**
   (*vi*) Promoter sites
   (*vii*) Modification of released polyopeptide
   (*viii*) Initiation of protein synthesis (translation)
   (*ix*) **Ribosomes**
   (*x*) **Differences in protein synthesis of prokaryotes and eukaryotes.**
2. Discuss the termination of mRNA in prokaryotes.
3. **Write a note on termination and maturation of mRNA in eukaryotes.**
4. **Give an account of the mechanism of protein synthesis (translation) in prokaryotes.**
5. Discuss the protein synthesis (translation) in eukaryotes.

Chapter 9

GENETIC CODE

A correlation exists between sequences of nucleotides in DNA (or mRNA) and amino acids in the proteins synthesized, and this relationship is called **'Genetic Code'** (Fig. 9.1).

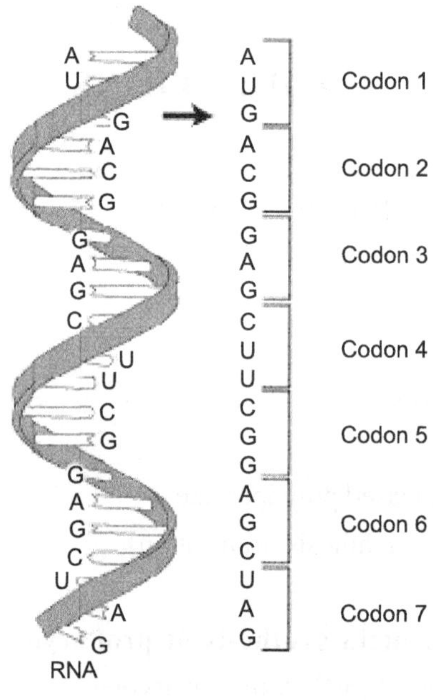

Fig. 9.1 The genetic code.

Since, DNA is made up of only four types of nucleotides (A, T, C and G) whereas proteins contain 20 types of amino acids, therefore, one-letter (4 × 1 = 4 possible codons) or two-letter (4 × 4 = 16) codes would not be enough to unambiguously encode 20 amino acids.

U	C	A	G

Singlet genetic code.

		U	C	A	G	First letter
Second letter	**U**	UU	CU	AU	GU	
	C	UC	CC	AC	GC	
	A	UA	CA	AA	GA	
	G	UG	CG	AG	GG	

Doublet genetic code.

The direction is 5′ to 3′.

		2nd base					
		U	C	A	G		
1st base	U	UUU Phenylalanine	UCU Serine	UAU Tyrosine	UGU Cysteine	U	**3rd base**
		UUC Phenylalanine	UCC Serine	UAC Tyrosine	UGC Cysteine	C	
		UUA Leucine	UCA Serine	UAA Ochre (*Stop*)	UGA Opal (*Stop*)	A	
		UUG Leucine	UCG Serine	UAG Amber (*Stop*)	UGG Tryptophan	G	
	C	CUU Leucine	CCU Proline	CAU Histidine	CGU Arginine	U	
		CUC Leucine	CCC Proline	CAC Histidine	CGC Arginine	C	
		CUA Leucine	CCA Proline	CAA Glutamine	CGA Arginine	A	
		CUG Leucine	CCG Proline	CAG Glutamine	CGG Arginine	G	
	A	AUU Isoleucine	ACU Threonine	AAU Asparagine	AGU Serine	U	
		AUC Isoleucine	ACC Threonine	AAC Asparagine	AGC Serine	C	
		AUA Isoleucine	ACA Threonine	AAA Lysine	AGA Arginine	A	
		AUG Methionine, *Start*	ACG Threonine	AAG Lysine	AGG Arginine	G	
	G	GUU Valine	GCU Alanine	GAU Aspartic acid	GGU Glycine	U	
		GUC Valine	GCC Alanine	GAC Aspartic acid	GGC Glycine	C	
		GUA Valine	GCA Alanine	GAA Glutamic acid	GGA Glycine	A	
		GUG Valine	GCG Alanine	GAG Glutamic acid	GGG Glycine	G	

Triplet genetic code.

Fig. 9.2 Singlet, doublet and triplet genetic codes.

Therefore, a **triplet code,** based on three nucleotides can give rise to $4 \times 4 \times 4 = \mathbf{64}$ **codons** (a sequence of 3 nucleotides on mRNA is called a codon) that can code for 20 amino acids (Fig. 9.2).

The genetic code was discovered in 1960s through significant contributions of **M.W. Nirenberg, J.H. Matthei, Hargobind Khorana,** etc. For their work, Hargobind Khorana, M.W. Nirenberg and Robert Holley got Nobel Prize in 1968.

Hargobind Khorana (Fig. 9.3) used synthetic RNA homopolymers (*e.g.,* poly U gave poly phenylalanine peptide, poly C gave poly proline peptide and poly A gave poly lysine peptide) and copolymers (*e.g.*, poly UA, AC, etc.) to synthesize proteins. In this way, he could be able to assign a number of amino acids to their specific triplet codons. Thus, with homopolymers he found that:

UUU codon codes for **phenylalanine,**

CCC codon codes for **proline,**

AAA codon codes for **lysine.**

In this experiment, since, Poly G attained secondary structure, therefore, it did not give any polypeptide.

Fig 9.3 Dr. Hargobind Khorana.

M.W. Nirenberg and **P. Leder** (1964) used synthetic RNAs or codons for binding with radioactively labeled amino acids (one at a time) and passing the mixture (ribosome, codon$_n$, AA$_n$-tRNA$_n$) through a nitrocellulose membrane. After autoradiography (exposure of photographic film by putting it over nitrocellulose membrane in dark), they assigned most of the codons to their amino acids whenever the radioactivity was detected over nitrocellulose membrane due to the formation of complex, **ribosome-codon$_n$-AA$_n$-tRNA$_n$** and its retention over nitrocellulose membrane due to its larger size (when the labeled amino acid corresponded to its specific codon). Radioactivity could not be detected over nitrocellulose membrane in those situations when the labeled amino acids did not match the added codons and the smaller sized radioactive AA$_n$-tRNA$_n$ passed through the nitrocellulose membrane. This experiment was performed for all the 20 types of labeled amino acids (one at a time) and the genetic code dictionary was deciphered (Fig. 9.4 and 9.5).

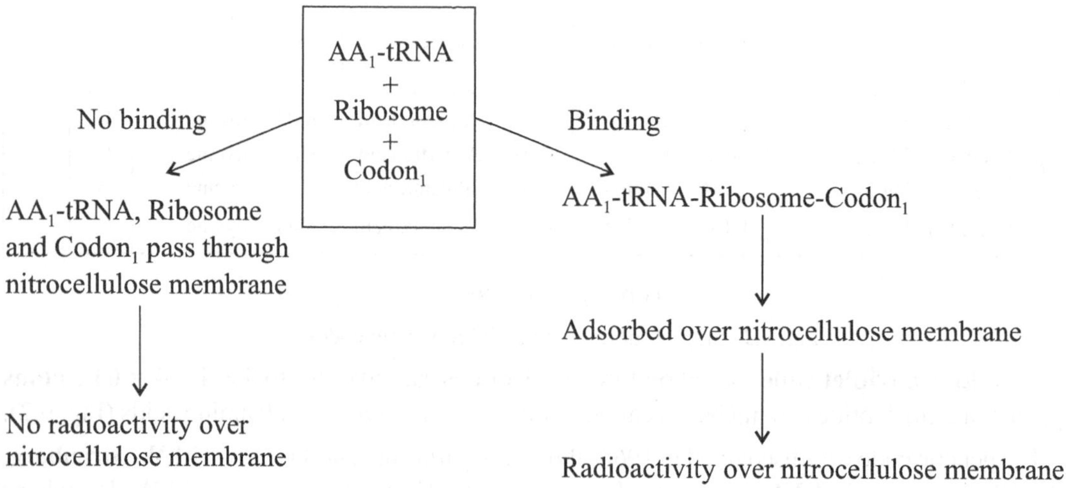

Fig. 9.4 Binding technique for the codon assignment.

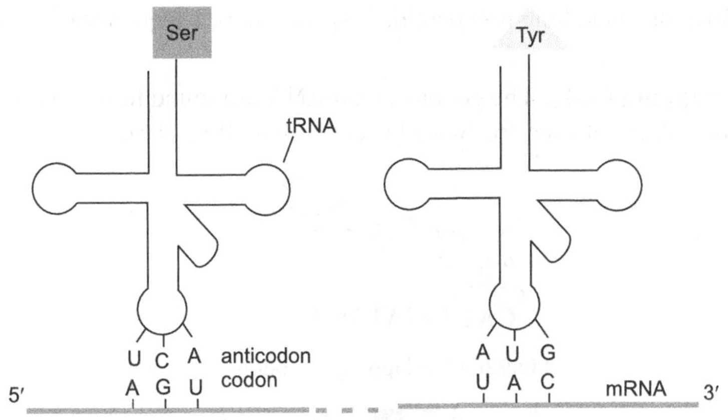

Fig. 9.5 Codon-anticodon binding for amino acids, serine (left) and tryptophan (right).

Out of these 64 codons on mRNA, **61 codons** specify 20 types of amino acids in which two are **initiation codons** (AUG and GUG) and three are **termination codons** (UAA, UAG and UGA). Whenever, any of the three termination codons is encountered on mRNA at the A-site of ribosome, synthesis of protein chain terminates, as no tRNA (with an amino acid) can recognize these termination codons and the polypeptide chain separates from the ribosome.

Characteristics of Genetic Code

1. **Triplet Code:** Each codon is made up of three successive nucleotides in mRNA that specifies an amino acid in the polypeptide.

$$\overbrace{\textbf{CAT}}^{aa_1}\quad \overbrace{\textbf{GAT}}^{aa_2}\quad \overbrace{\textbf{CAT}}^{aa_3}$$

2. **Universal Code:** The genetic code is applicable universally, *e.g.*, it is same from lower organisms, like viruses to higher organisms, like human being.

   Exception: A different and more primitive genetic code is found in the mitochondria of some eukaryotes.

3. **Comma-less Code:** The sequences of codons in mRNAs are always continuous and pauses are not found in between the two adjacent codons. Thus, after one amino acid by a codon, the second amino acid is automatically coded by the next 3 nucleotides (codon) and no letters (nucleotides) are wasted for telling that one amino acid has been coded and now the second should be coded.

$$\overbrace{\textbf{CAT}}^{aa_1}\quad \overbrace{\textbf{GAT}}^{aa_2}\quad \overbrace{\textbf{CAT}}^{aa_3}$$

Comma-free genetic code

$$\overbrace{\textbf{CATG}}^{aa_1}\quad \overbrace{\textbf{GAT}}^{aa_2}\quad \overbrace{\textbf{ACAT}}^{aa_3}$$

Genetic code with comma

Thus, a 30 amino acid long polypeptide is specified by a sequence of 90 nucleotides long mRNA.

4. **Non-overlapping Code:** The codons in a mRNA are found to be non-overlapping and a base in mRNA is not used for two adjacent codons (Fig. 9.6).

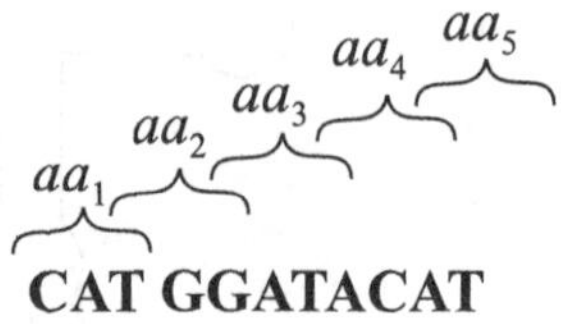

Fig.9.6 Overlapping of codons

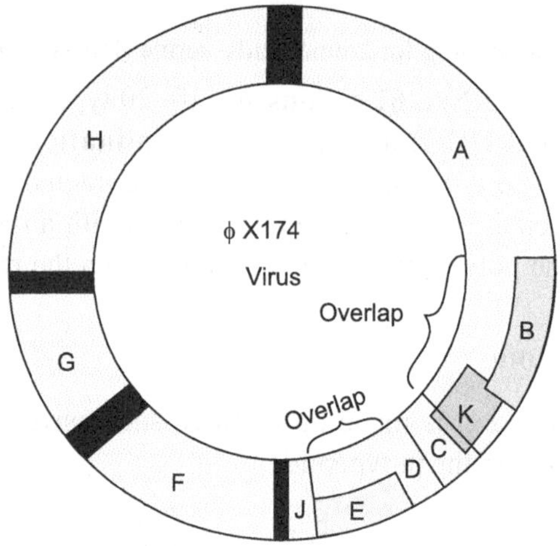

Fig. 9.7 Overlapping genes in virus ΦX174.

In the virus ΦX174, that has single-stranded genomic DNA, though the genes are overlapping, but this overlapping occurs at different time and space (Fig. 9.7). Thus, in this virus inspite of overlapping genes, the genetic code is non-overlapping.

5. **Unambiguous Code:** It means, one codon always represents its specific one amino acid only and has no double meaning, *e.g.*, the codon UUU always codes for the same amino acid, phenylalanine and not for any other amino acid.

   Exception: The codon **GUG** is normally meant for the amino acid **valine,** but sometimes it is also used as the initiation codon (at first position or 5′ end in mRNA) coding for the amino acid **methionine** (or formyl methionine).

6. **Degenerate Code:** In a triplet genetic code, out of 64 possible codons, all codons have some meaning. In a cell of an organism, there are 20 types of amino acids and 61 types of codons coding for various amino acids (and 55 types of tRNAs). Thus, one amino acid can have more than one codons (called **synonymous codons**) that can be recognized by the same tRNA molecule. In these synonymous codons, the first and second nucleotides remain the same, whereas, the third nucleotide can be different. Thus, the pairing of the anticodon in the same tRNA with the synonymous codons is allowed due to wobbling (unstability) in 3′ base of synonymous codons (in mRNA) and 5′ base of anticodon (in tRNA).

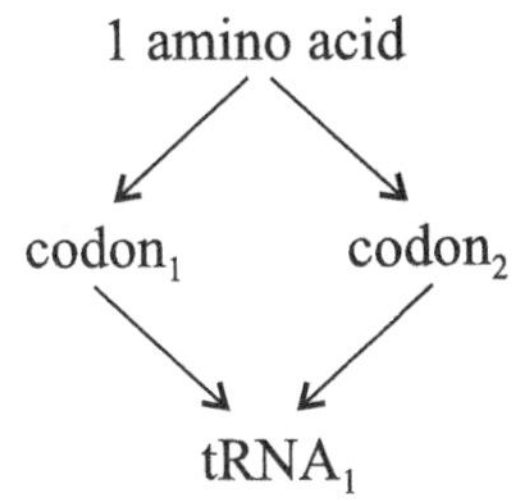

For example, the two synonymous codons for phenylalanine differ in the 3′ base, whereas, the other two 5′ bases remain the same (3rd base degeneracy). Thus, often the 3′ base of the synonymous codons is not important and this reduced specifity is called 3rd base degeneracy.

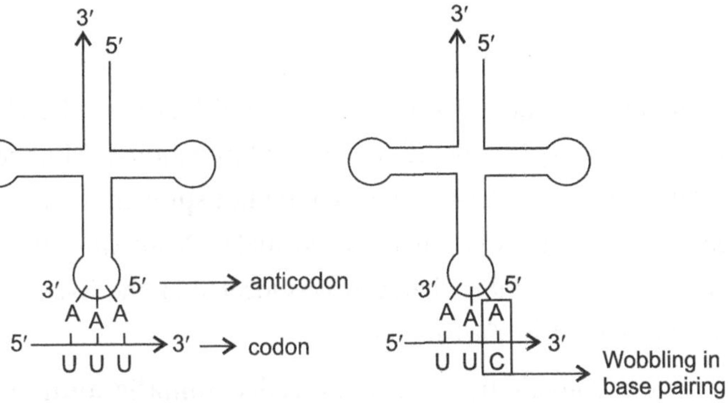

Fig. 9.8 Synonymous codons for phenylalanine.

In a cell of an organism, there are 55 types of tRNAs, 20 types of amino acids and 61 types of codons coding for various amino acids. Thus, one amino acid can have more than one codons (called **synonymous codons**) that can be recognized by the same tRNA molecule. Only Tryptoplan and methionine have only one codons, all other amino acids are specified by 2-6 synonymous codons.

QUESTIONS

1. **What is genetic code? Describe its general properties.**
2. Describe an experiment which helped in deciphering the genetic code.

   or

   Discuss an experiment conducted by Hargobind Khorana and workers (*et. al.*) for codon assignments.
3. Write short notes on frame-shift mutations and their role in understanding the genetic code.

   or

   Give some evidences which support that genetic code does exist in living systems.
4. **Write short notes on:**
   (*i*) Synonym codons
   (*ii*) Wobble hypothesis
   (*iii*) Initiation and termination codons.

Chapter 10 MUTATIONS

Mutations are the abrupt changes in genetic material that alter the chemical structure of genes at molecular level (gene mutations or point mutations). Mutations occur more frequently in certain regions of DNA that are called **hot spots** of mutations. The first mutant was isolated by **Morgan** (1910), that was a white-eyed male *Drosophila melanogaster* (fruit fly, where the normal eye colour is red) and this character was located on the X-chromosome. Thereafter, about 500 mutants were discovered in *Drosophila*.

Mutations may occur in somatic tissues and are called **somatic mutations,** which are non-heritable; whereas **germ line mutations** occurring in gametes are heritable. Mutations may be **lethal** or **nutritional** (auxotrophs requiring additional supplements in the minimal nutrient medium) or may cause some **morphological, biochemical** or **regulatory changes** or may impart **resistance** to antibiotics, pesticides, etc., or **silent mutations** that do not cause any change or **back mutations** where the second mutation reverts mutational character to the normal wild type feature.

Mutations that can occur **spontaneously** or may be **induced** by certain mutagens that may be caused mainly by any of the following mechanisms (See Fig. 10.1).

1. Base Substitutions. In base substitution, a normal base in the genetic material is replaced by another base, and they are the most common causes of mutations. Base substitutions can be of two types:

(*i*) ***Transitions:*** Where, a purine base is replaced by another purine or a pyrimidine base by another pyrimidine, *e.g.,*

A G (A purine is replaced by G purine)
| → |
T C (T pyrimidine is replaced by C pyrimidine)

G A (G purine is replaced by A purine)
| → |
C T (C pyrimidine is replaced by T pyrimidine)

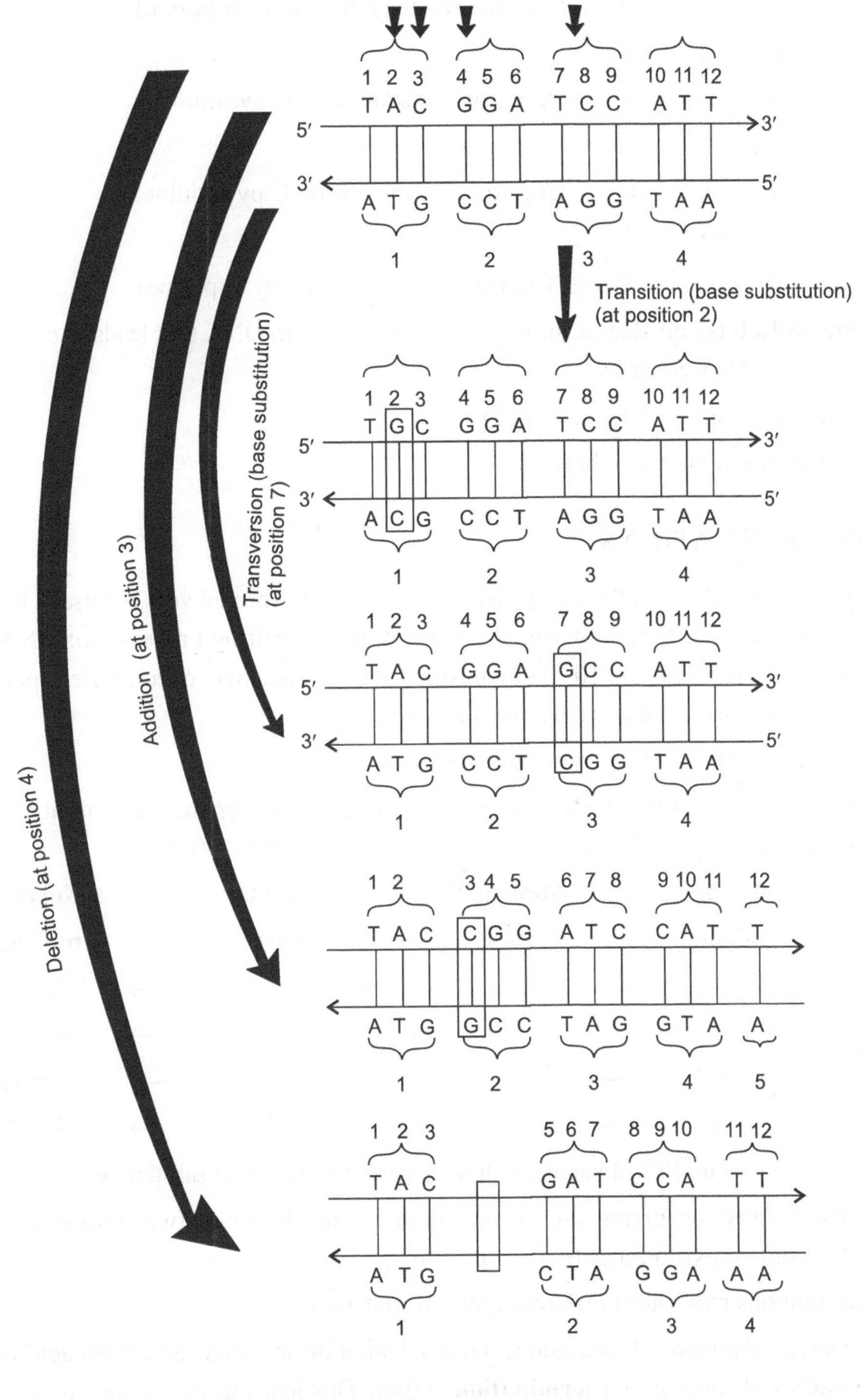

Fig.10.1 Different types of mutations.

(*ii*) ***Transversions***: Where, a purine is replaced by a pyrimidine or *vice-versa*, *e.g.*,

T		G	(T pyrimidine is replaced by G purine)
|	→	|	
A		C	(A purine is replaced by C pyrimidine)

A		C	(A purine is replaced by C pyrimidine)
|	→	|	
T		G	(T pyrimidine is replaced by G purine)

2. Frame Shift. It occurs due to shifting of reading frame in DNA that leads to change in the pattern of codons in mRNA either by:

(*i*) ***Addition*** of one or a few bases in DNA or

(*ii*) ***Deletion*** of one or a few bases.

SPONTANEOUS MUTATIONS

The replication of DNA in all the organisms is generally performed very precisely, but, sometimes due to rare mistakes performed during the replication or repairing processes of DNA, generally at a very low frequency (10^{-7} to 10^{-12}/organism), the mutations may occur, called spontaneous mutations. The reasons of these mutations may be many:

1. Base Substitutions. They can be of two types:

(*i*) ***Transitions***: occurring due to tautomeric changes (tautomerism is the ability of a molecule to exist in more than one forms) in normal bases (Fig. 10.2).

Wrong pairing		**Abnormal base**		**Normal base**		**Normal pairing**
C = A°	⟵	A°	⟵	A	⟶	A = T
C ≡ G°	⟵	G°	⟵	G	⟶	G ≡ C
A = C°	⟵	C°	⟵	C	⟶	C ≡ G
A = T°	⟵	T°	⟵	T	⟶	T = A

Fig. 10.2 Tautomeric shifts in bases to cause base substitutions.

(*ii*) ***Transversions***: occurring due to mistaken pairing between two purines (*e.g.*, A-G pairing) or two pyrimidines (*e.g.*, C-T pairing).

Base substitutions may cause different types of mutations:

(*a*) *Non-sense Mutations.* Here, due to base substitution in DNA, an amino acid codon in the mRNA changes into a **termination codon.** This leads to the premature termination of polypeptide synthesis, the length of the polypeptide depending upon the position of the mutant termination codon (Fig. 10.3 and 10.4).

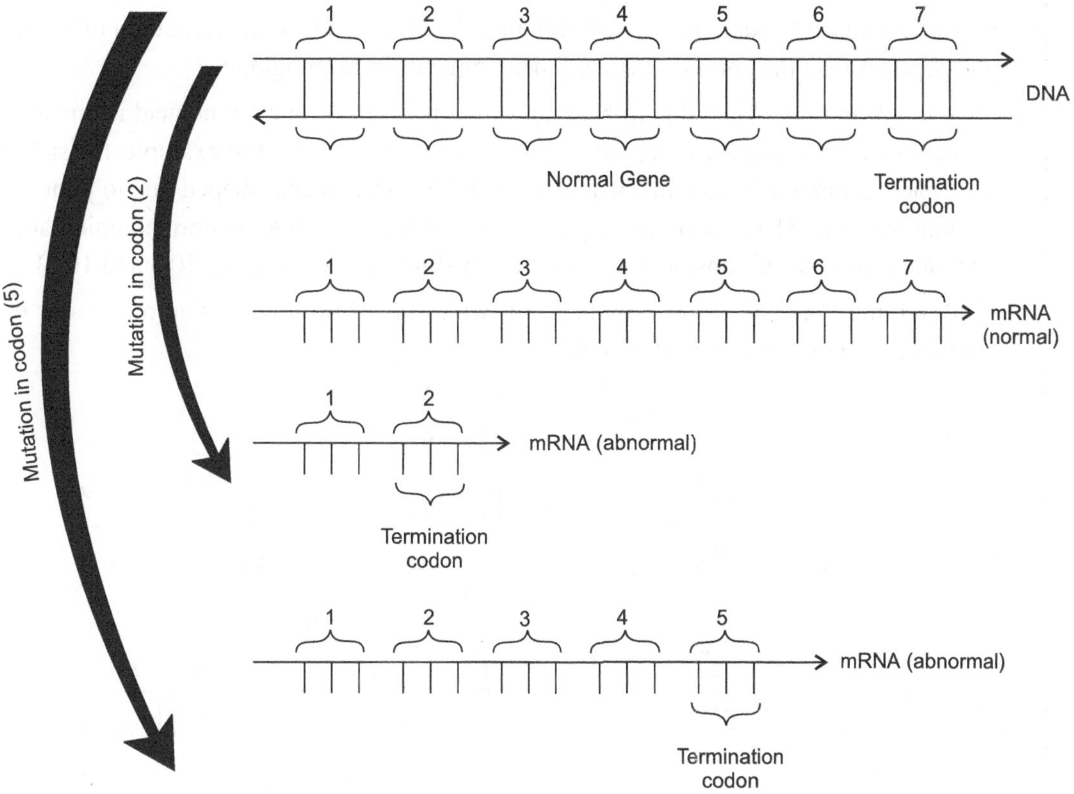

Fig. 10.3 Nonsense mutation.

	1	2	3	4	5	6	7	8
DNA:	TAC	TGA	GTG	GCT	CGC	GCT	TCG	ACT
	ATG	ACT	CAC	CGA	GCG	CGA	AGC	TGA
				↓				
mRNA	AUG	ACU	CAC	CGA	GCG	CGA	AGC	UGA
				↓				
Normal protein :	Met	Thr	His	Arg	Ala	Arg	Ser	Stop
DNA :	TAC	TGA	GTG	GCT	CGC	GCT	TCG	ACT
	ATG	ACT	CAC	TGA	GCG	CGA	AGC	TGA
				↓				
mRNA:	AUG	ACU	CAC	UGA	CGC	CGU	AGC	UGA
				↓				
Abnormal Protein:	Met	Thr	His	Stop				

Fig. 10.4 Nonsense mutation in codon 4 to produce smaller polypeptide.

(*b*) *Reverse of Non-sense Mutation.* Here, the normal termination codon in mRNA changes into an amino acid codon due to mutation in DNA. This leads to the formation of longer polypeptides than the normal size, *e.g.*, many mutants in haemoglobin.

(*c*) *Missense Mutation.* When due to base substitution in DNA, one amino acid is replaced by another amino acid, the mutation is called missense mutation. For example, the sickle cell anaemia disease in humans, where the RBC becomes sickle shaped due to poor O_2 carrying capicity. Here, in β-globin polypeptide of haemoglobin the normal amino acid glutamic acid at the 6th position is replaced by **valine** amino acid (Fig. 10.5 and 10.6).

(The tetramer haemoglobin is made up of two kinds of proteins— two molecules of α-globin and two molecules of β-globin)

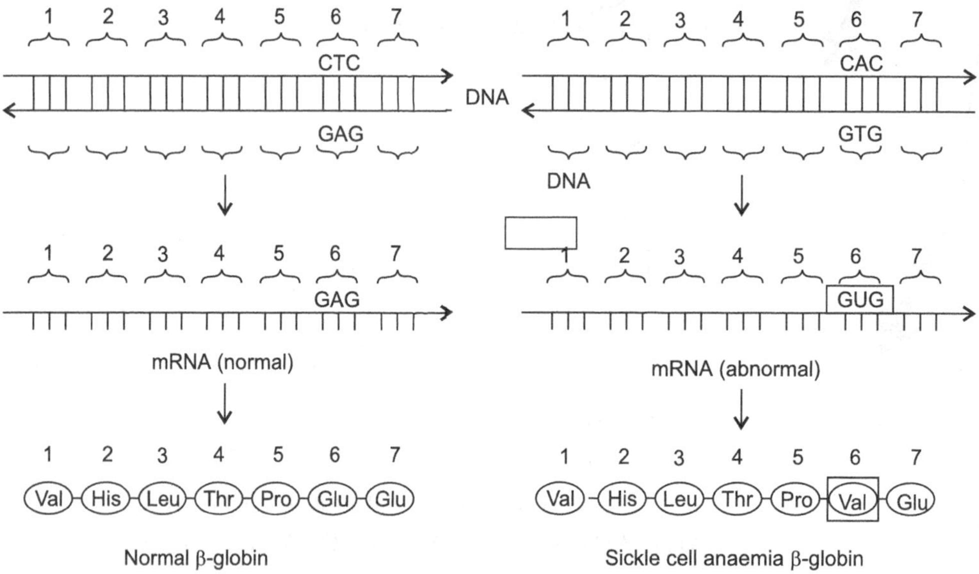

Fig. 10.5 Missense mutation in β-globin gene of haemoglobin.

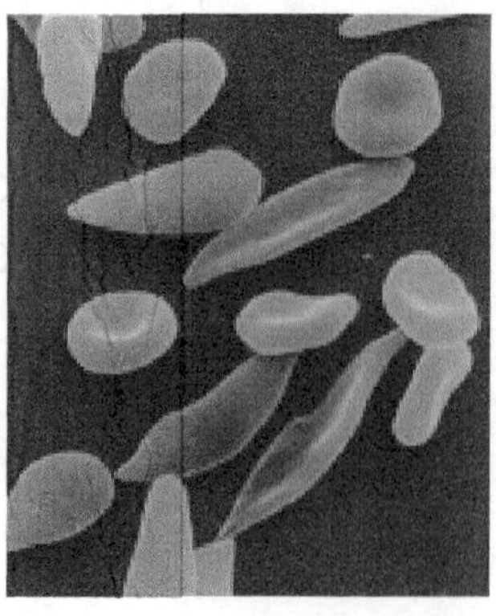

Fig. 10.6 Sickle shaped red blood cells of a sickle cell anaemia patient.

2. Frame Shift Mutations

In this case, addition or deletion of bases in DNA may occur due to loop formation in the template DNA strand at the time of DNA replication, so that one or a few bases present in the loop may not be copied (deletions) or copied twice (additions). This may also occur due to breakage and reunion of DNA, leading to addition or deletion of some bases (Fig. 10.7).

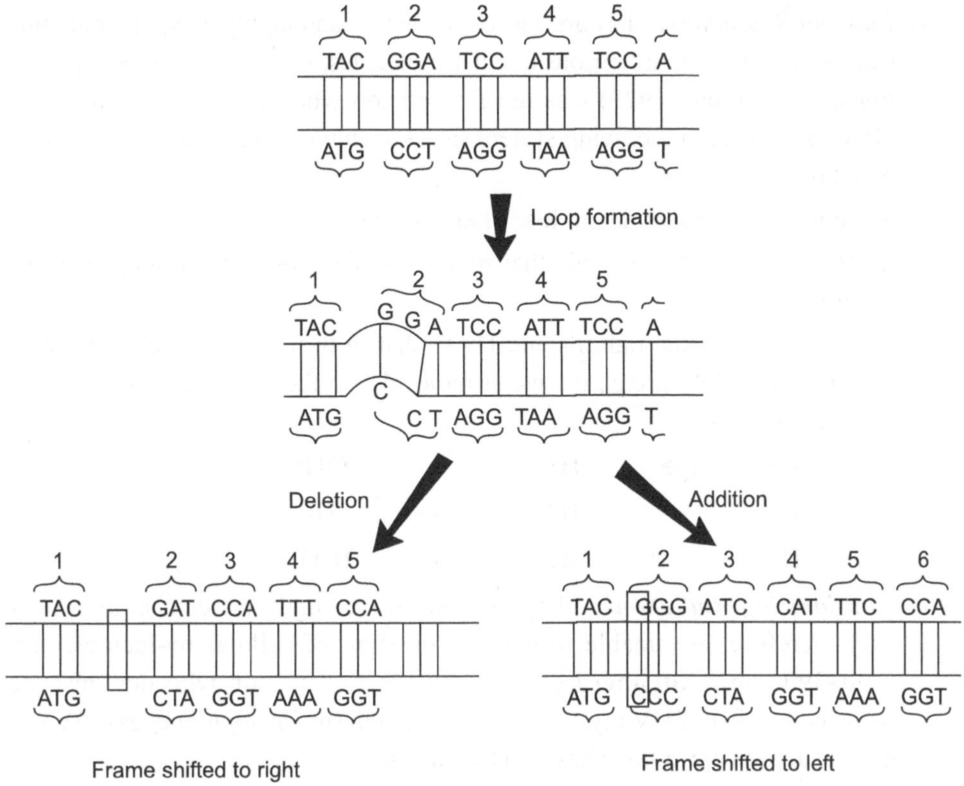

Fig. 10.7 Frame-shift mutation due to loop formation in DNA.

Additions may also occur due to unequal crossing over between homologous chromosomes at the time of meiotic Prophase (Fig. 10.8).

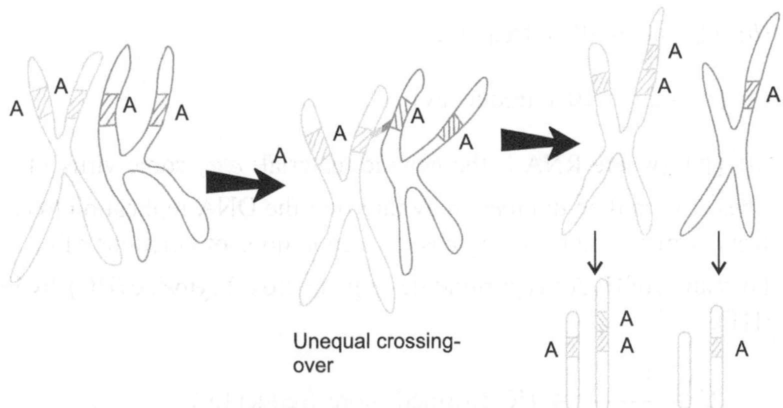

Fig. 10.8 Frame-shift mutation due to unequal crossing-over in homologous chromosomes during meiosis.

INDUCED MUTATIONS

The mutations in the genetic material can be induced artificially by some physical or chemical mutagens that increase the rate of mutation.

1. **Physical Mutagens.** They can be ionizing or non ionizing radiations.

(*i*) ***Ionizing Radiations*:** They are short-wavelength and highly energetic radiations that can cause ionization of various cellular molecules, *e.g.*, X-rays, γ-rays, β-rays, neutrons, etc. **Muller** (1927) was the first person who demonstrated the mutagenic effect of X-rays in the living system (in *Drosophila* or the fruit fly). X-rays induce mutations by:

1. Ionization of bases due to loss of electrons.
2. Breakage of DNA strands that may cause deletions or rearrangement of bases after reunion.
3. Formation of free radicals and H_2O_2, that in turn may induce breaks and other changes in DNA, *e.g.*, the warer molecules in the cells may lead to free radical formation as:

$$H_2O \rightleftharpoons \mathbf{H^\circ} + \mathbf{OH^\circ}$$
$$H^\circ + H^\circ \rightarrow H_2$$
$$OH^\circ + OH^\circ \rightarrow \mathbf{H_2O_2}$$

(*ii*) ***Non-ionizing Radiations*:** They are long-wavelength, lesser energetic radiations and, therefore, are unable to cause ionization of cellular molecules, *e.g.*, UV (190–390 nm). **Altenberg** (1930) was the first person who demonstrated the mutagenic effect of UV rays in the living system (in *Drosophila* eggs). UV radiation usually acts on pyrimidine bases in two ways:

1. Most frequently intrastrand **pyrimidine dimmer** formation (Fig. 10.9):

(*a*) $\overset{\leftrightarrow}{TT}$ (about 50% frequency)

(*b*) $\overset{\leftrightarrow}{TC}$ (about 40% frequency)

(*c*) $\overset{\leftrightarrow}{CC}$ (about 10% frequency)

(*d*) $\overset{\leftrightarrow}{UU}$ (where RNA is the genetic material, *e.g.,* some viruses)

These pyrimidine dimmers may interrupt the DNA replication to cause lethal effects or may lead to wrong base incorporations or deletions of bases.

2. Formation of Hydoxyl pyrimidines, *e.g.*, hydroxyl cytosine (HC), hydroxyl thymine (HT):

$$C \xrightarrow[H_2O]{UV} HC \text{ (formed more frequently)}$$

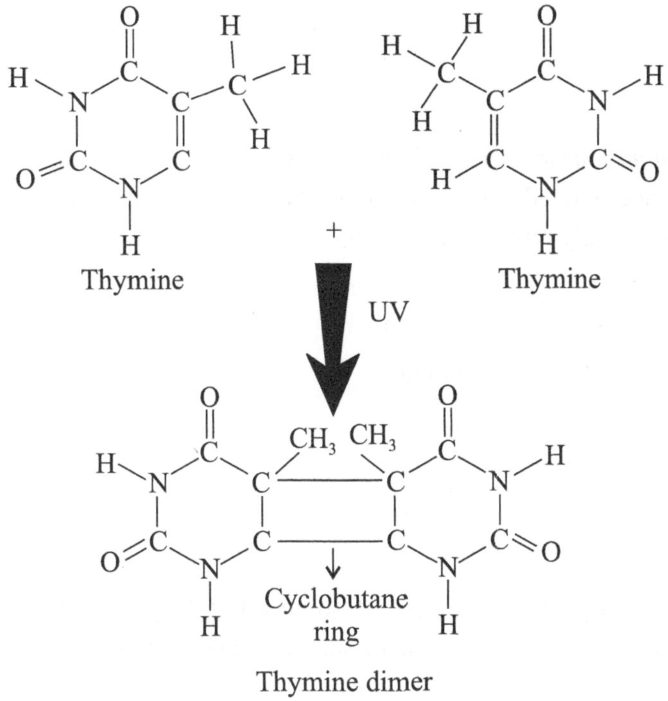

Fig. 10.9 Thymine dimer formation by exposure to UV radiation.

$$T \xrightarrow[H_2O]{UV} HT \text{ (formed less frequently)}$$

The hydroxyl pyrimidines base pair wrongly to cause transitions.

3. UV also induces free radicals and peroxides formation to cause mutations.

2. Chemical Mutagens: There are many types of chemicals that are mutagenic.

(*i*) ***Base Analogues***: They are almost similar in structure to the normal bases found in DNA and may get incorporated in DNA at their places. Moreover, their mutagenic effect is due to the fact they form tautomers more readily than the normal bases, therefore, during DNA replication they readily pair with wrong bases and cause base substitutions. Some of the important base analogues are:

(*a*) *5-Bromouracil:* It is analogue of the base thymine (keto base) and after getting incorporated into DNA, it more frequently forms the enol tautomer (that base pairs wrongly with guanine) due to higher electronegativity of –Br than the $-CH_3$ group. Thus, it causes transitions.

T = A	T° = G
5-Bu = A	5-Bu° = G
(keto)	(enol)

e.g.,

G ↔ A
| |
C T

(*b*) *2-Amino purine:* It is analogue of the base adenine (amino base) and after getting incorporated into DNA, it more frequently forms the imino tautomer that base pairs wrongly with cytosine. Thus, it also causes transitions.

A = T	A° = C
2-Ap = T	2-Ap° = C
(amino)	(imino)

e.g.,

G ↔ A
| |
C T

(*ii*) ***Alkylating Agents*: ** Examples of some alkylating agents are methyl methane sulphonate (MMS), ethyl methane sulphonate (EMS), etc. They cause mutations by affecting the DNA in different ways:

1. EM, most frequently after adding an ethyl group in guanine base, forms the 7-ethyl guanine (7-EG) that due to wrong base pairing causes **transition.**

G = C 7-EG = T

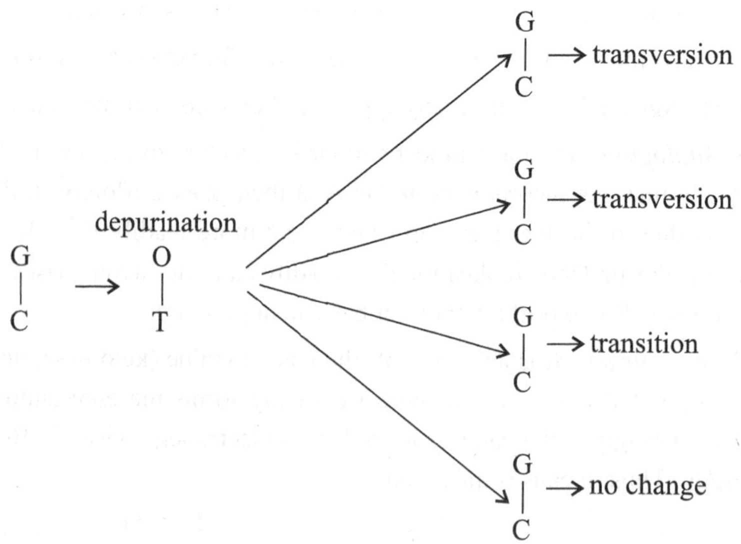

2. Since, 7-ethyl guanine is weakly attached to the deoxy-ribose sugar in DNA, it may get detached **(depurination)** from the sugar. This gap in DNA may interfere with the DNA replication or any of the four bases (T, C, A or G) may get incorporated in front this vacant site to cause transitions, transversion or no change at all, depending upon the incorporated base.
3. Alkylating agents may also cause mutations by inducing alkylation of PO_4^{3-} group in the DNA backbone to cause **breakage** of DNA strand.

(*iii*) ***Dyes*:** Dyes, being larger in size, get intercalated either in front of or in between the bases in DNA to cause additions or delitions that lead to frame shift mutations, *e.g.*, proflavin, acridine orange, etc. (Fig. 10.10 and 10.11).

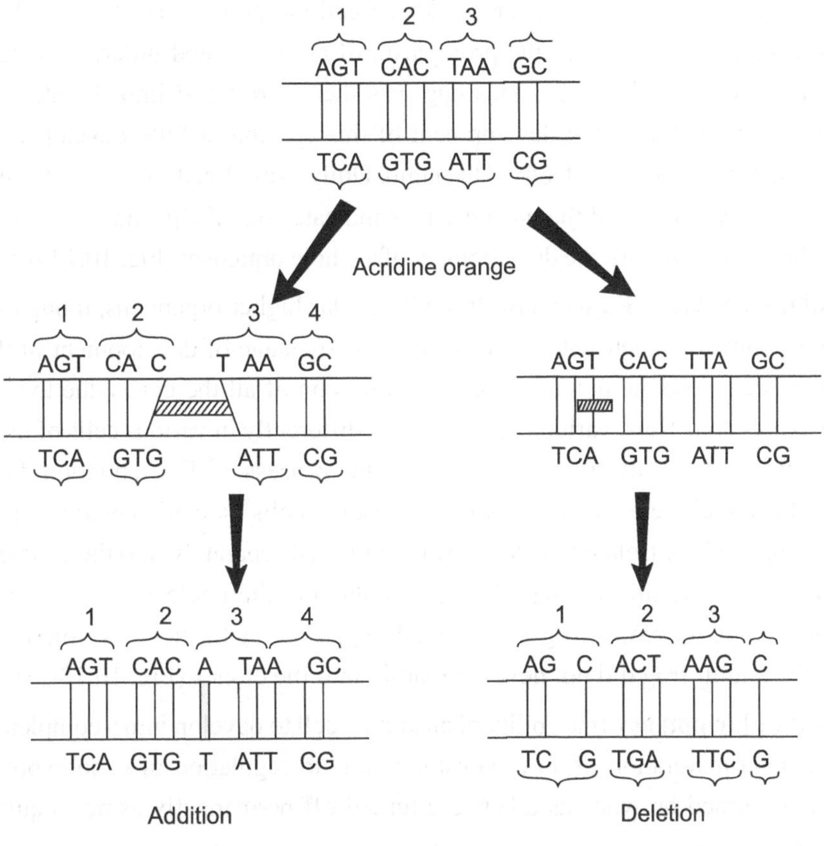

Fig. 10.10 Frame shift mutation by acridine orange.

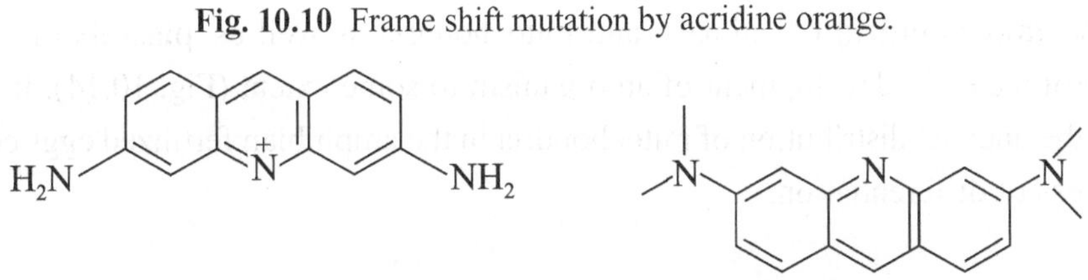

Fig. 10.11 Chemical structures of mutagenic dyes proflavin (left) and acridine orange (right).

DIFFERENTIATION AND DEVELOPMENTAL MUTATIONS

Unicellular organisms have little or no differentiation, whereas, the multicellular higher plants and animals have a variety of cells which originate from a single cell (zygote) by a highly ordered and coordinated process of **differentiation and development.** The information for the differentiation and development of an organism from the zygote is contained in the **DNA** that is expressed in a regulated manner to produce different body parts. Many factors control the process of differentiation and development, these are:

1. **Nucleus.** Nucleus is the main factor controlling differentiation and development of an organism that can be explained by the examples of clones developed by various scientists. **Dolly,** the sheep, was a clone developed by **Ian Wilmut** in 1997. Ian Wilmut isolated the nucleus from the cultured udder (milkbag) cell (somatic cell) of a female sheep and transferred this nucleus into the protoplasm of an enucleated unfertilized egg obtained from a second female sheep. This egg was then introduced into the uterus of a third female sheep and after the development of this egg into a fully developed foetus, this female sheep gave birth to Dolly. This lamb, Dolly, was clonal (genetically similar) to the first female who donated the nucleus. This indicates that the primarily, the nucleus contains the information for the development of a whole organism (Fig. 10.12 see Appendix).
2. **Regulation of Gene Expression.** In multicellular higher organisms, though all the cells have the same complete set of genes, but, at early stage of development of the embryo, different cells lose the potential for the expression of all the genes due to regulation of gene expression. Thus, various genes express differently in various cells of the embryo to give different kinds of proteins that control the process of differentiation. For example, when the nuclei were isolated from the intestinal cells (somatic cells) of the **larva** and **adult** frog, and the isolated nuclei were transformed separately into the enucleated protoplasm of the unfertilized eggs, the eggs containing the nucleus from the larva readily developed into the clonal frogs; whereas, the eggs having the nucleus from the intestinal cell of the adult frog did not develop readily into the clones (Fig. 10.13 see Appendix).

   Thus, the **pluripotency** (the ability of an animal cell to develop into a complete organism) is lost upon maturation of an organism, as due to regulation of gene expression some genes are turned on whereas others are turned off permanently as per requirement.
3. **Cytoplasmic Factors.** The **extranuclear genes** (organeller genes), that are contained in the **mitochondria** of animals, and mitochondria as well as **plastids** of plants, also control the early development of an organism to some extent (Fig. 10.14). It was found that the unequal distribution of mitochondria in the amphibian fertilized eggs controls the process of differentiation.

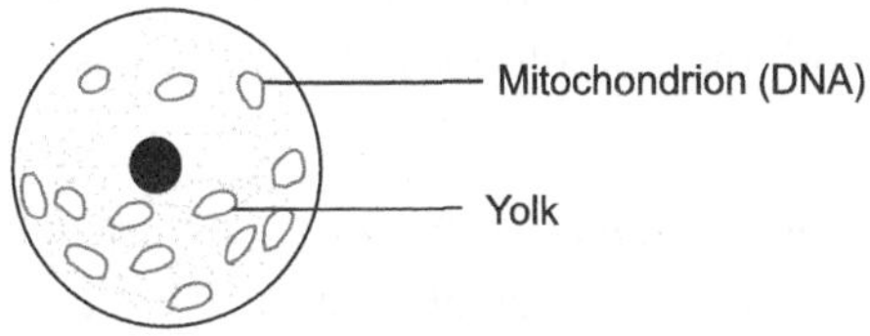

Fig. 10.14 Unequal distribution of mitochondria and yolk in amphibian egg.

4. **Maternal Effect.** In the **fertilized egg** of an organism, though, the nucleus contains one set of chromosomes from the paternal parent and the other set from the maternal parent, the **cytoplasm** is always derived from the **maternal parent.** This cytoplasm, in addition to containing organeller genes, also contain some maternal **nuclear products** (mRNA or proteins) that regulate the early embryonic development that is known as maternal effect (Fig. 10.15 see Appendix).
5. **Environmental Factors.** The environment also affects the development of an organism, for example, in *Drosophila* the normal body colour is brown, but when the normal larvae are fed upon the silver salt containing food, they develop into yellow flies. Some mutant flies of *Drosophila* are also reported that have yellow body colour. There two types of yellow flies (yellow mutant and the yellow flies reared on silver salt containing food), which are genotypically different but phenotypically same, are called **phenocopies** (Fig. 10.16).

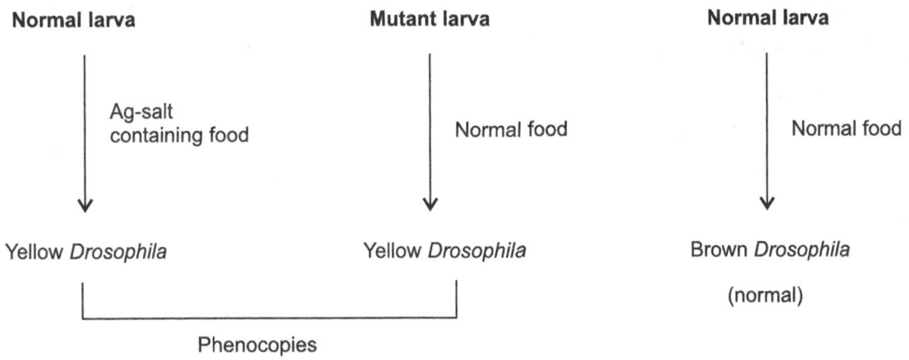

Fig. 10.16 Effect of food on the body colour of *Drosophila* flies.

Developmental Mutants

The differentiation and developmental processes of the organisms have not been fully understood. The **homoeotic genes** in organisms determine the body plan and mutation in these genes may lead to the transformation of one body part into other. Many **developmental mutants** of a number of organisms have been studied that have helped in understanding the process of differentiation and development to some extent.

For example, in *Drosophila* the *antp* gene of the *antennapedia* gene complex after mutation in homozygous condition (both the homologous chromosomes affected) leads to the conversion of antenna into legs.

Similarly, in a plant, *Antirrhinum,* the homoetic mutations lead to the transformation of one flower part into other, *e.g.*, in *apetala* mutants the petals get transformed into bracts or stamens (male flower part) or carpels (female flower part).

QUESTIONS

1. **What do you understand by spontaneous and induced mutations?**

   or

   Write in brief the possible source and mechanism of spontaneous mutations in nature.

   or

   Discuss different kinds of radiations and chemical mutagens utilized for induction of mutations.

   or

   How transitions and transversions can be induced by chemical and physical mutagens?
2. How can base substitutions be induced by a base analogue?
3. **Discuss the mechanism of the mutagenic action of an alkylating agent.**
4. How do ultraviolet radiations bring about changes in genetic material? How can the effect of UV be modified due to exposure to light?
5. Write short notes on:
   (*i*) **Frame shift mutations**
   (*ii*) Acridine dyes
   (*iii*) Somatic mutation

Chapter 11 REGULATION OF GENE EXPRESSION

In living system, different genes give rise to different types of RNAs and proteins, but all the proteins are not needed all the time and in all the types of tissues. Thus, throughout the lifecycle of a multicellular organism, though, all cells contain same set of genes, but in different types of tissues different genes are expressed to produce various types of RNAs and proteins, depending upon the need, *e.g.*, digestive enzymes are specifically produced in alimentary canal, growth hormone in pituitary, insulin in pancreas, testosterone hormone in testis, progesterone and estrogen hormones in ovary, etc. Similarly, in all the organisms (multicellular as well as unicellular), expression of genes also varies with the environment. Thus, gene expression is highly **regulated** and energy is not wasted in the expression of unwanted genes, as protein synthesis requires energy in the form of ATP, GTP, etc. (Fig. 11.1).

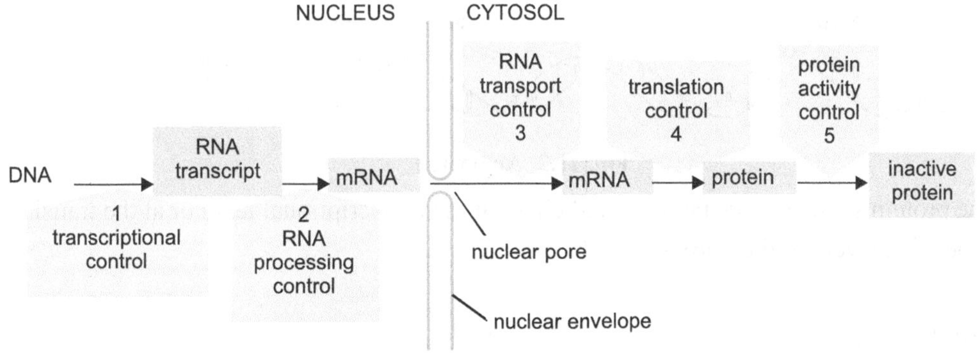

Fig.11.1 Various levels of gene regulation.

In addition to the regulated genes, products of some essential genes are required all the time so that they are not regulated at all and remain continuously expressed all the time, *e.g.*, respiratory enzymes of glycolysis and Krebs cycle, ribosomal proteins, tRNAs, rRNAs, etc. Such genes are called **housekeeping genes.**

In case of regulated genes, in different types of cells, some mechanisms allow the expression of the desired genes. Such regulatory mechanisms have most extensively been studied in prokaryotic systems, particularly in *Escherichia coli*. In prokaryotes, since, the DNA is not bounded by a nuclear envelope and, moreover, the DNA is not found to be associated with any histone and non-histone proteins (naked DNA), the regulatory mechanisms are quite different from those found in eukaryotes.

GENE REGULATION IN PROKARYOTES

In prokaryotic cells, only one type of **RNA polymerase** synthesizes all the three types of RNAs (mRNA, tRNA and mRNA) and, since, the DNA is not bounded by any nuclear membrane, **transcription** (formation of mRNA) and **translation** (formation of proteins) processes occur **simultaneously** in the same cytoplasm. (In eukaryotes, transcription or formation of mRNA occurs in nucleus, whereas, translation or protein synthesis in cytoplasm).

Moreover, all the proteins or enzymes needed for a metabolic pathway are generally synthesized simultaneously by the production of a single **polycistronic mRNA** (one cistron gives rise to one protein), as these genes lie adjacent to each other and are transcribed continually. This polycistronic DNA along with its regulatory sites, promoter and operator, is known as an **operon** (Fig. 11.2). **Promoter** (*p*) is the site where RNA polymerase binds for starting the transcription, whereas, **operator** (*o*) is a small part of promoter that is recognized and bound by the regulatory protein synthesized by a **regulator gene** (*i*), which is not a part of the operon and lies at a distance (up-stream) from the operon. Depending upon the type of metabolic activity, the regulation of gene expression can be of inducible type or of the repressible type. The synthesis of enzymes required for most of the catabolic pathways, *e.g.*, utilization of lactose sugar, are the **inducible systems,** where the presence of lactose, for example, induces the expression of necessary genes. On the other hand, the presence of a metabolite, *e.g.*, tryptophan amino acid supplemented in nutrient medium, may inhibit the synthesis of all the enzymes necessary for its production (anabolic pathways). Such systems are known as **repressible systems**.

Jacob and Monod (1961) were the first to give the models for explaining the induction or repression of enzyme synthesis in *E. coli*.

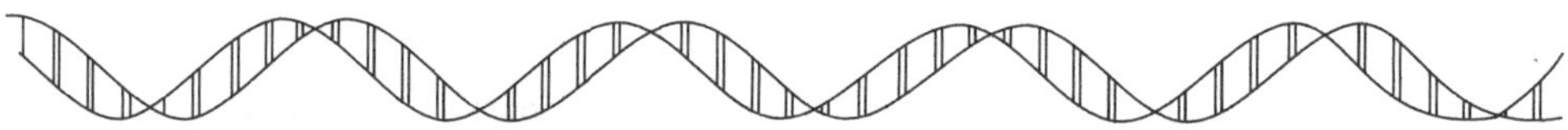

Fig. 11.2 An operon.

The protein synthesis can be regulated either at the transcriptional level or at the translational level. Some of the well-studied operons in *E. coli* are:

1. *Lac* Operon

In *E. coli,* the *lac* operon, which is an **inducible system**, consists of promoter, a operator and three structural genes. These structural genes, after transcription and translation, give rise to the three enzymes, β-galactosidase (product of *z* structural gene), β-galactoside permease (product of ***y*** structural gene) and β-galactoside transacetylase (product of ***a*** structural gene), which are necessary for the metabolism (catabolism) of lactose sugar (a β-galactoside) if present in external nutrient medium (minimal nutrient medium, that is required for the growth of organism, supplemented with lactose) (Fig. 11.3).

β-galactosidase catalyses the reaction:

Lactose → Glucose + Galactose

β-galactoside permease helps in the transport of lactose into the cell from external medium.

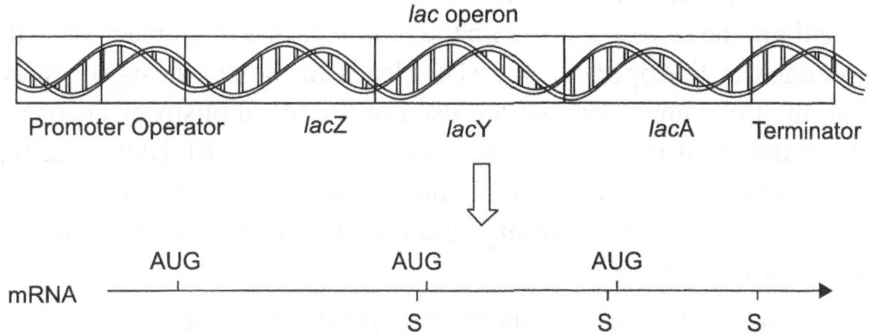

Fig. 11.3 Structure of *lac* operon in *E. coli*.

β-galactoside transacetylase transfers the acetyl group from acetyl CoA to β-galactoside sugars. It also helps in the utilization of non-metabolizable analogues of β-galactoside, which results in the detoxification and excretion of these analogues.

(*i*) **When lactose is absent in nutrient medium.** The **regulator gene *i*** synthesizes a **repressor** protein that recognizes and binds at the operator site in the absence of lactose in external medium and, thus, inhibits the synthesis of polycistronic mRNA by checking the movement of RNA polymerase at the promoter site (Fig. 11.4 and 11.6).

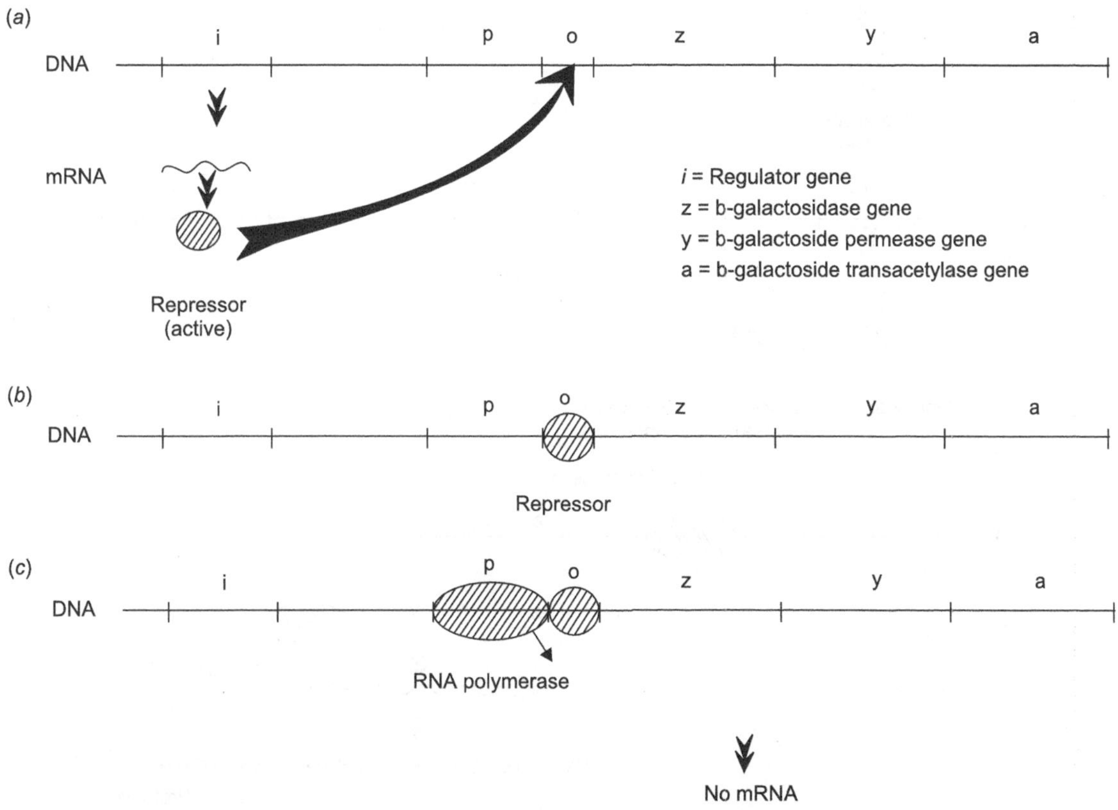

Fig. 11.4 *Lac*-operon in *E. coli* when lactose is absent in external medium.

(*ii*) **When lactose is present in nutrient medium.** When lactose is added in the minimal nutrient medium, the lactose binds with the repressor making it inactive so that it becomes unable to attach at the operator site. This leads to the synthesis of mRNA and, after, translation, the three enzymes that are used in the metabolism of lactose. Thus, lactose acts as the **inducer** of gene expression (Fig. 11.5 and 11.6). [After the repressed state, when lactose is added in the external medium, it manages to enter the cell with the help of β-galactoside permease, as a minimal amount of this enzyme is always present to start the metabolic pathway]

[Actually, lac operon normally remains in repressed state and whenever lactose is added in minimal nutrient medium, it becomes induced.]

Instead of lactose, allolactose that is the metabolic product of galactose, is the actual inducer.

Lactose → Glucose + Galactose
↓
Allolactose + Galactobiose]

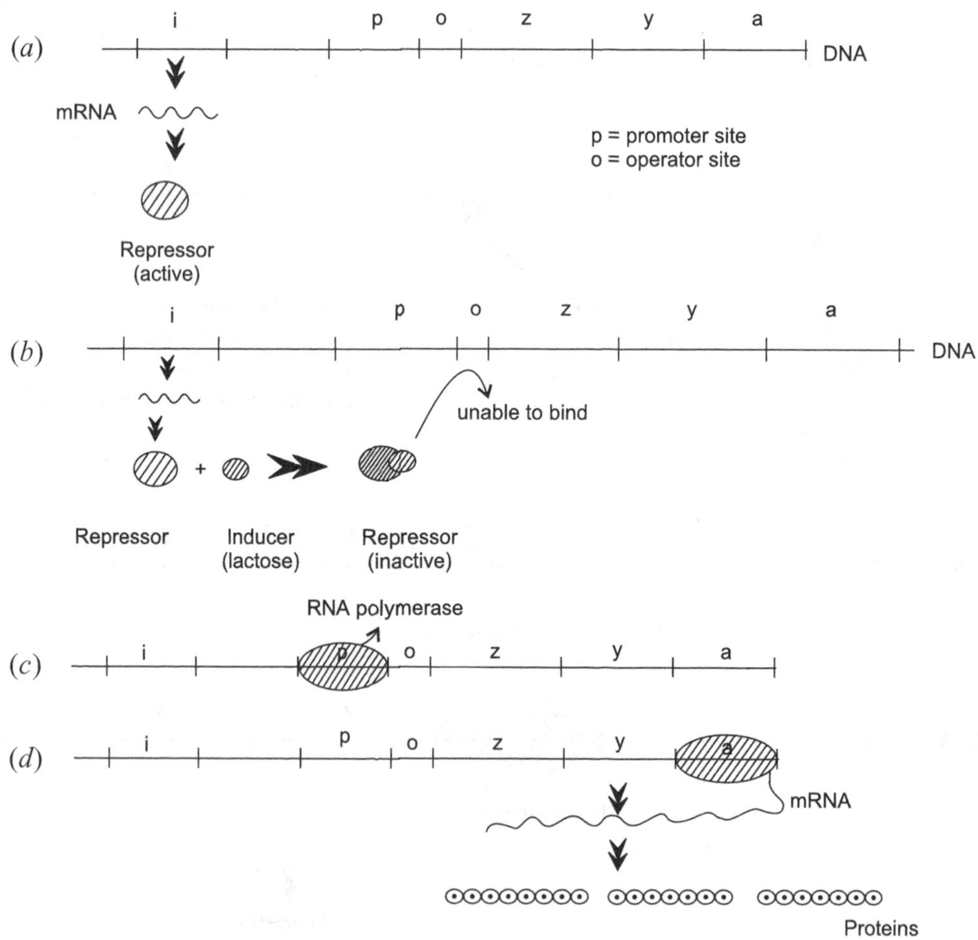

Fig. 11.5 Functioning of *lac* operon in the presence of lactose in *E. coli*.

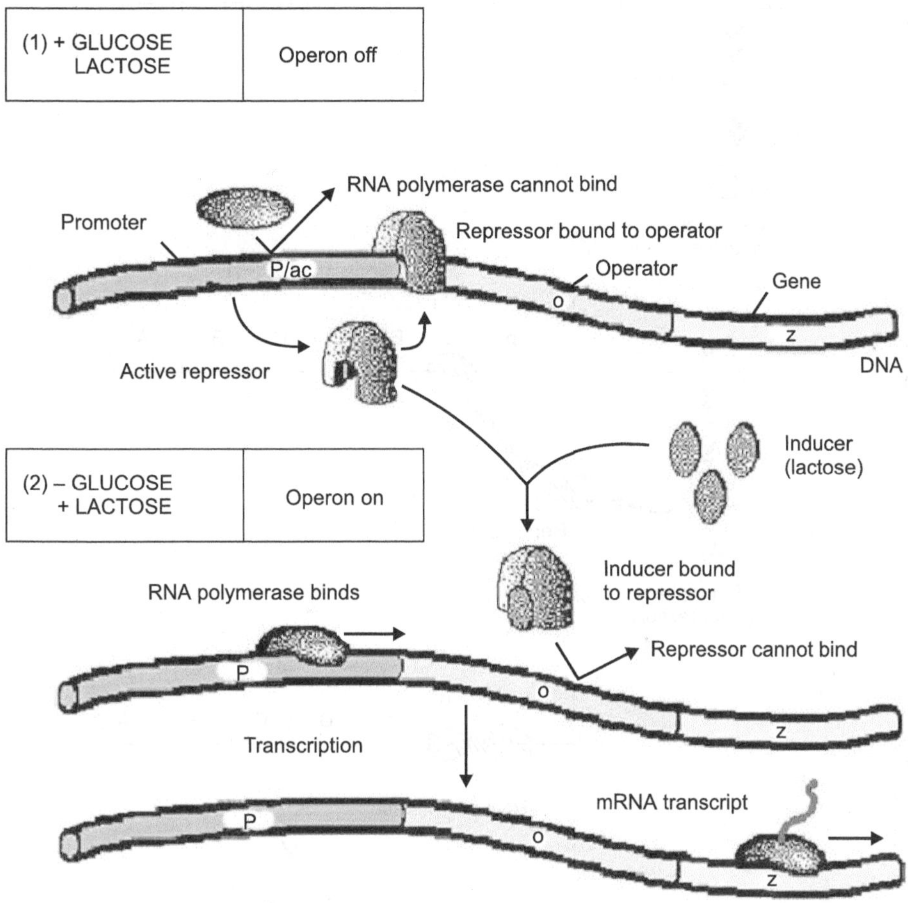

Fig. 11.6 *Lac*-operon in *E.coli.*

2. Tryptophan Operon

In *E. coli,* the *trp* operon, which is a **repressible system,** consists of promoter, operator and five structural genes, trp E, trp D, trp C, trp B and trpA, that after transcription and translation give rise to the respective enzymes necessary for the synthesis of tryptophan amino acid, if it is not present in external nutrient medium.

(*i*) **When tryptophan is present in nutrient medium.** The **regulator gene** ***i*** synthesizes an inactive **repressor** protein that after binding with the corepressor (tryptophan) present in external medium recognizes and binds at the operator site, thus, inhibits the synthesis of polycistronic mRNA by checking the movement of RNA polymerase at the promoter site (Fig.11.7).

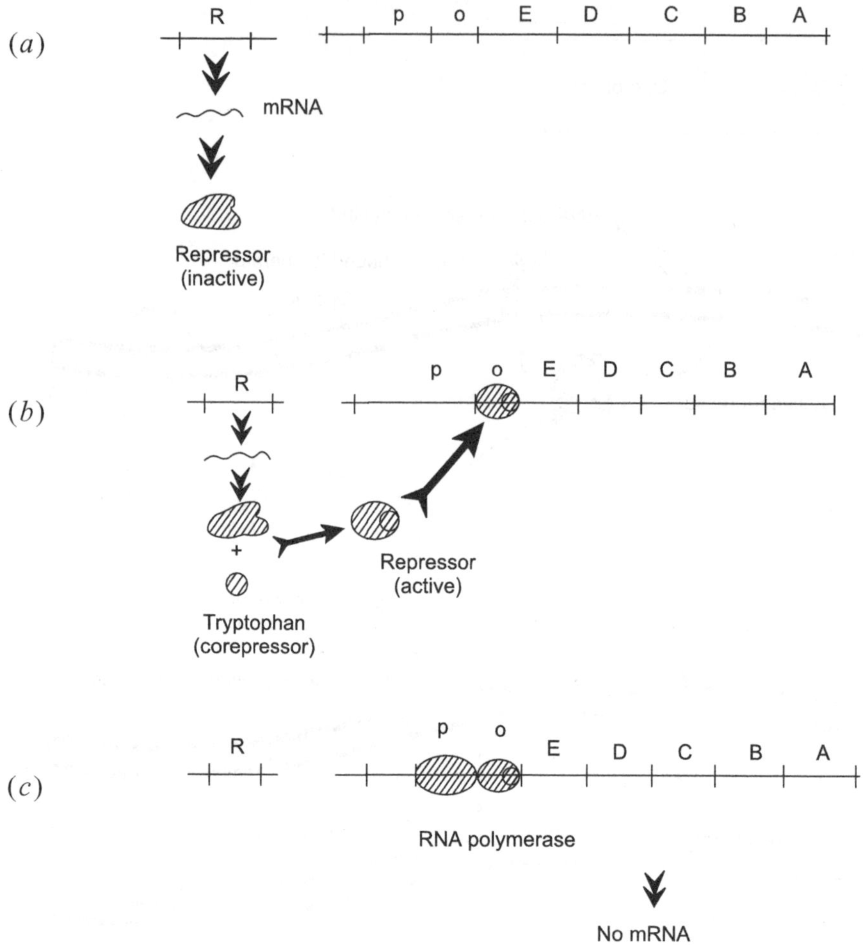

Fig. 11.7 *Trp* operon when tryptophan is present in external medium.

(*ii*) **When tryptophan is absent in nutrient medium.** When tryptophan is absent in the external medium, the inactive repressor is unable to bind at the operator site, thus, leading to the synthesis of mRNA and the corresponding enzymes that are used in the synthesis of tryptophan (Fig. 11.8).

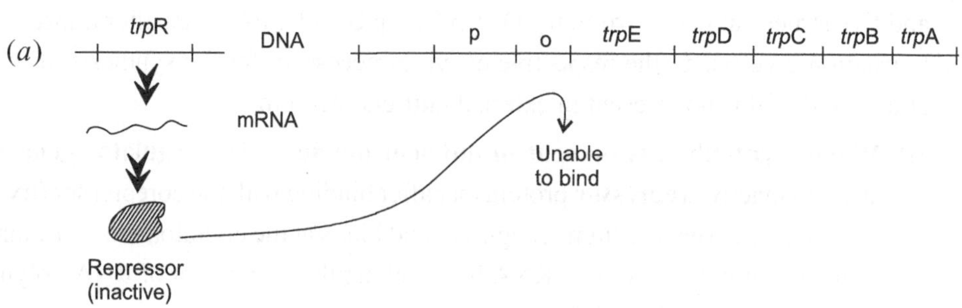

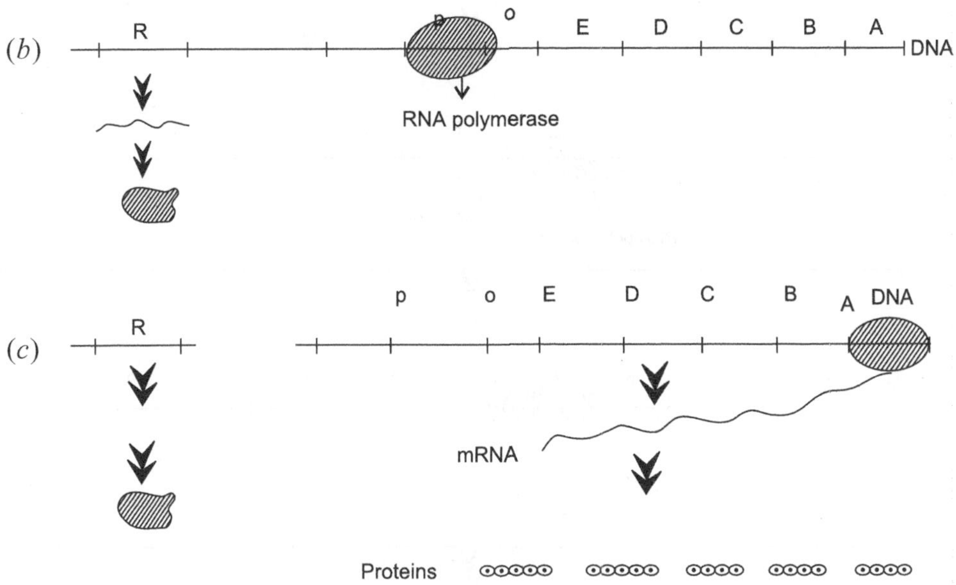

Fig. 11.8 *Trp* operon, when tryptophan is absent in external medium, in *E. coli.*

[Normally, the *trp* operon remains in expressed form, and whenever tryptophan is added in the minimal nutrient medium, it is repressed.]

Post-transcriptional Regulation of Tryptophan Operon

(*i*) **When tryptophan is present in nutrient medium.** One more type of control system of the *trp* operon occurs **post-transcriptionally** (after the start of transcription process) that is based on the synthesis of a **leader polypeptide** when tryptophan is present in the external medium. A 162 base pair long DNA found at the 5′ end of the gene *trp* E (part of gene E), having two successive tryptophan codons, is responsible for the synthesis of leader polypeptide. This regulation is based on the coupling of transcription and translation processes in prokaryotes. After the initiation of transcription, ribosomes one by one get attached to the 5′ end of the newly synthesizing mRNA and start the synthesis of the respective proteins. In the presence of tryptophan, the two tryptophan codons are translated due to the availability of tryptophan in external medium, thus, forming the complete leader polypeptide. This leader polypeptide induces the formation of hairpin loop in the newly synthesizing mRNA, causing the detachment of RNA polymerase from the coding strand of DNA. Thus, termination of further transcription occur (Fig. 11.9).

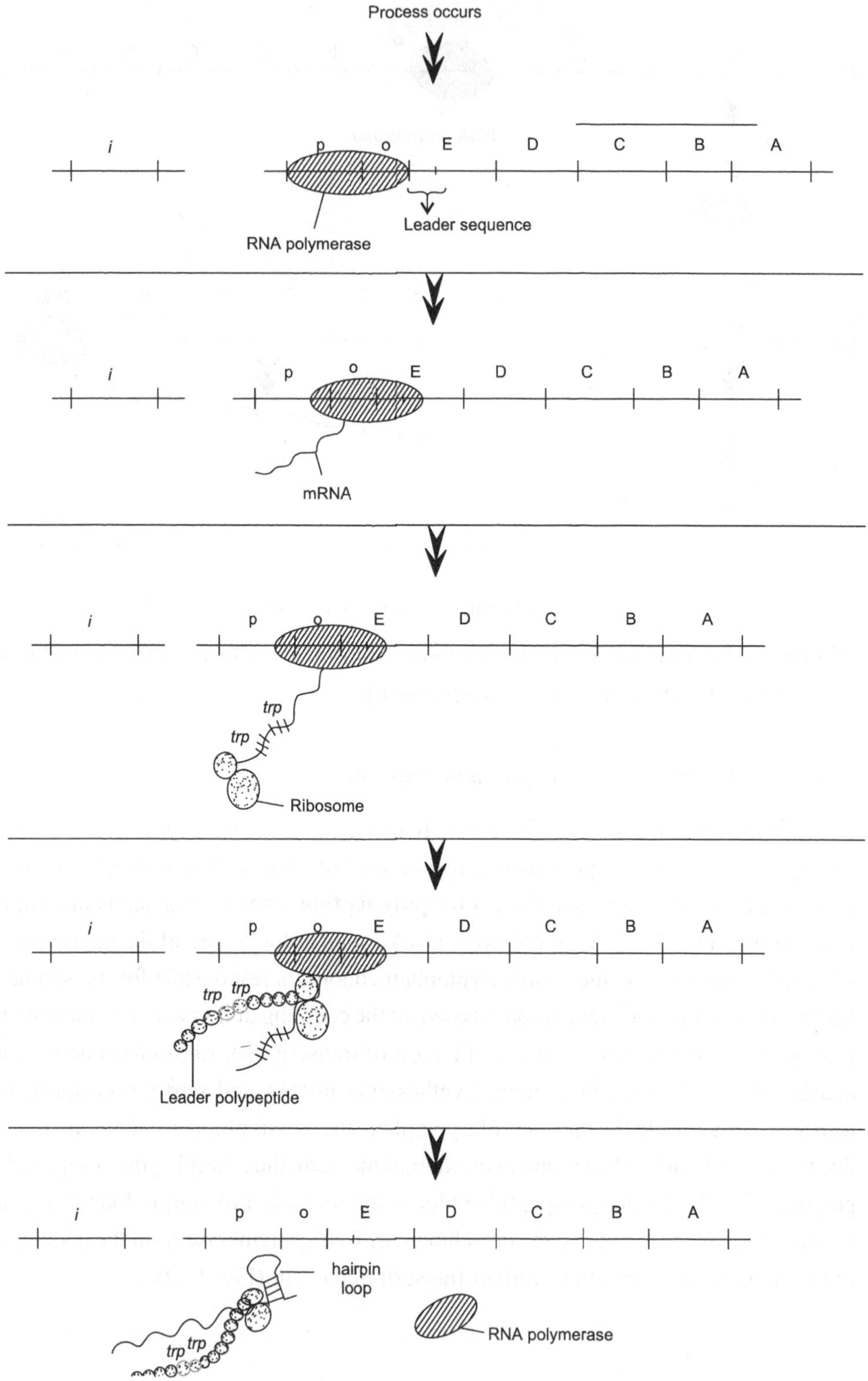

Fig. 11.9 Post-transcriptional control of *trp* operon in the presence of tryptophan.

(*ii*) **When tryptophan is absent in nutrient medium.** In the absence of tryptophan, the ribosome gets stalled at the two tryptophan codons and no leader polypeptide is synthesized. Thus, no hairpin loop is formed in mRNA and the RNA polymerase remains attached with the DNA and completes the transcription process to form a polycistronic mRNA, made up of five cistrons. The translation of this mRNA synthesizes all the enzymes required for the biosynthesis of tryptophan (Fig. 11.10).

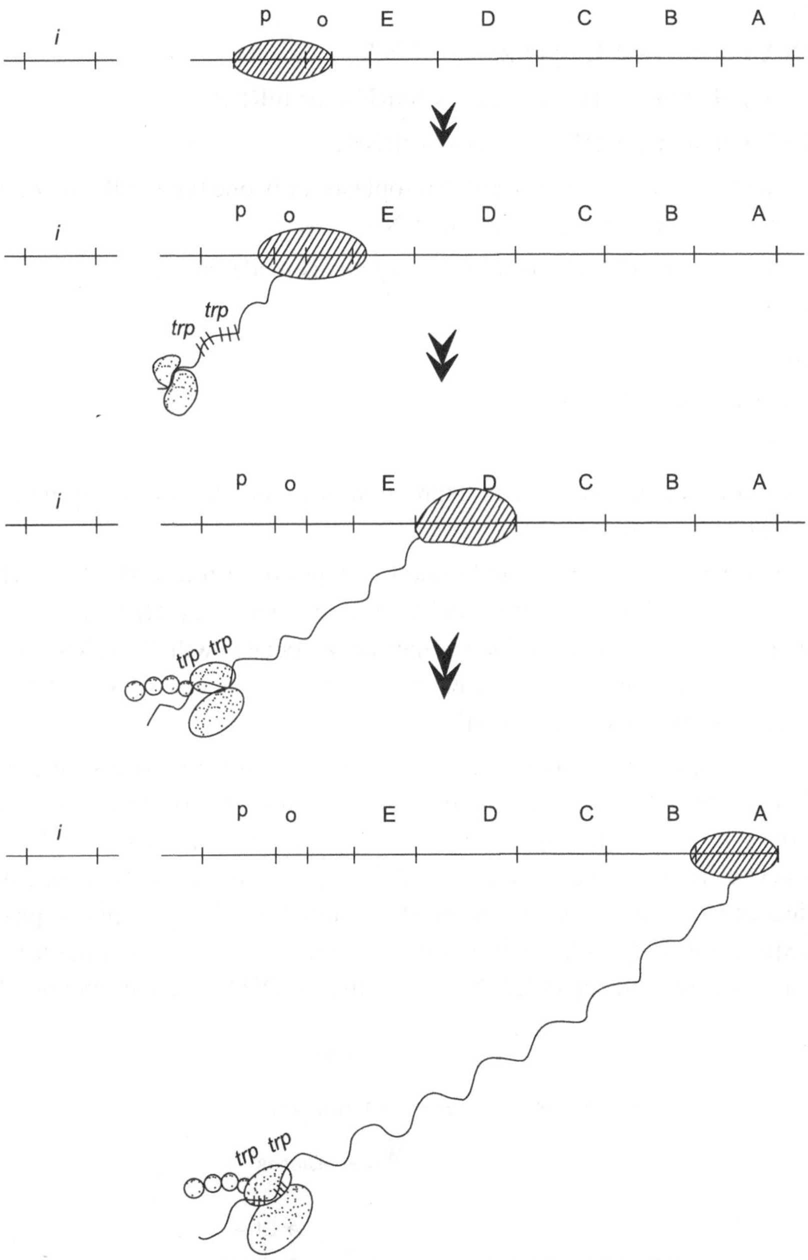

Fig. 11.10 Post-transcriptional control of *trp* operon in the absence of tryptophan.

GENE REGULATION IN EUKARYOTES

In eukaryotes, only 2-15% of the genes are expressed at a particular time and in comparison to the prokaryotes, eukaryotes have more complex gene regulation systems, like :

1. **Availability of Many Types of RNA Polymerases.** In these organisms, a gross level of gene expression regulation is found by having three types of RNA polymerases in the cell :

   (*a*) **RNA polymerase I,** synthesizes rRNA.

   (*b*) **RNA polymerase II,** synthesizes hnRNA or mRNA.

   (*c*) **RNA polymerase III,** synthesizes tRNA.

In the cell organelles, mitochondria and chloroplasts, only one type of RNA polymerase, like in prokaryotes, synthesizes all the three types of RNAs.

2. **Chromatin Proteins.** Chromatin in eukaryotes is made up of:

   (*a*) DNA

   (*b*) Histones

   (*c*) Non-histones

   (*d*) RNA

   More specific regulation of gene expression is excercised by the chromatin proteins, histones and non-histones.

   Histones are highly conserved and same five types of histones, H1, H2A, H2B, H3 and H4, are found in all the primitive and advanced eukaryotes. Histones repress the gene activity non-specifically and as long as they are associated with the DNA and they repress the activity of the genes. Thus, for RNA synthesis in nucleus, these histones have to be separated from the DNA temporarily.

   Non-histones are highly variable and they vary not only from one species to other species, but also from tissue to tissue in the same organism. Unlike histones, which are same in all the types of tissues, non-histones, being variable, give rise to different types of mRNAs in different tissues. **Gilmour** and **Paul (1973)** performed the **chromatin reconstitution** experiments in mammals and showed that non-histones play a **positive** role in gene expression. They isolated the chromatin from different types of tissues, *e.g.*, thymus and bone marrow, and separated their constituents, DNA, histones and non-histones.

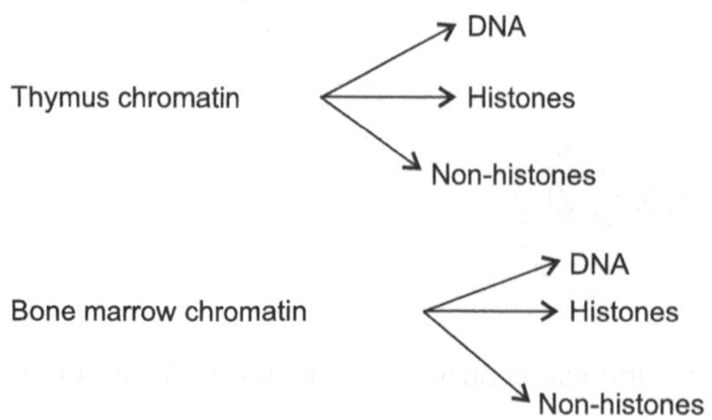

Thereafter, they reconstituted all the possible types of chromatins by recombining the various chromatin components in different combinations. The type of mRNA produced from these artificially reconstituted chromatin depended upon the non-histones and not upton the DNA and histones. Thus, irrespective of the DNAs and histones, the reconstituted chromatins that contained the thymus type of non-histones gave rise to the thymus-type of mRNAs, whereas those containing bone marrow type non-histones gave bone marrow type of mRNAs (Fig. 11.11 see Appendix).

3. **Regulation at the Transcriptional Level.** For regulation of gene expression at the mRNA synthesis level in eukaryotes, **Britten** and **Davidson** (1969) proposed a **'Gene-battery model'** or **'Britten-Davidson model',** which is based on the presence of a large proportion of repetitive DNA in eukaryotes. (Those DNA sequences that are found in single copy are known as unique sequences, whereas the DNA sequences that are repeated many times or found in multiple copies are called repetitive or redundant DNA). These repetitive DNA sequences may play some regulatory role in gene expression.

The 'Britten-Davidson model' has:

1. A **sensor site,** which regulates the activity of integrator gene. The integrator gene is transcribed only when the sensor site is activated by some hormones and proteins.
2. An **integrator gene** that is equivalent to the regulator gene *i* of operons in prokaryotes. Under the control of sensor site, it produces an activator RNA (or protein) that activates the receptor site.
3. A **receptor site,** which is equivalent to operator site in bacterial operons, is found adjacent to the producer gene.
4. **Structural gene** is similar to the structural genes of operons and gives rise to the mRNA (Fig. 11.12).

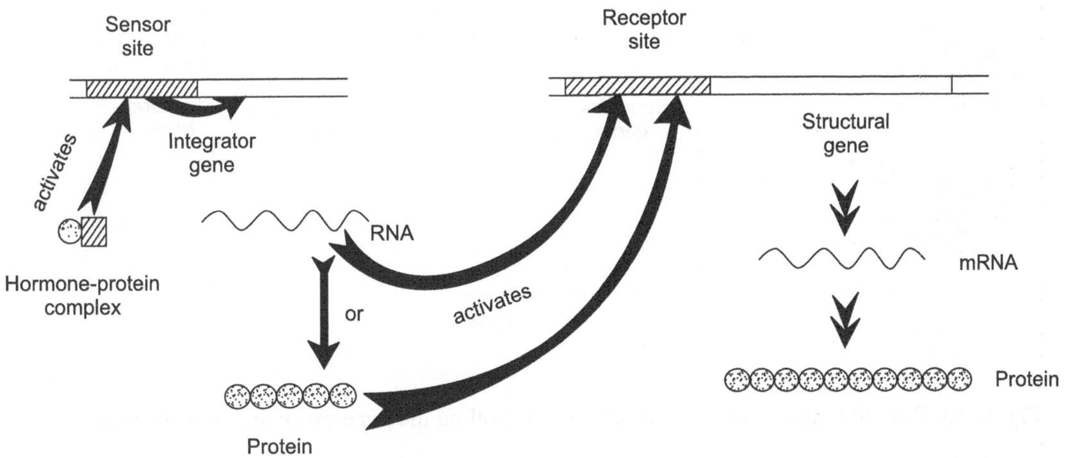

Fig. 11.12 Britten and Davidson's model of regulation of gene expression.

It has been proposed that the presence of multiple copies of receptor sites and sensor sites may control the expression of genes. Thus, according to this model, a **battery** is made up of:

1 Sensor

↓ controlling

1 or many integrators

↓ controlling

1 or many receptors

↓ controlling

1 or many structural genes

In this way, all the structural genes controlled by a **sensor** make a battery.

The repetitive **receptor sites,** thus, may control the gene expression as (Fig. 11.13):

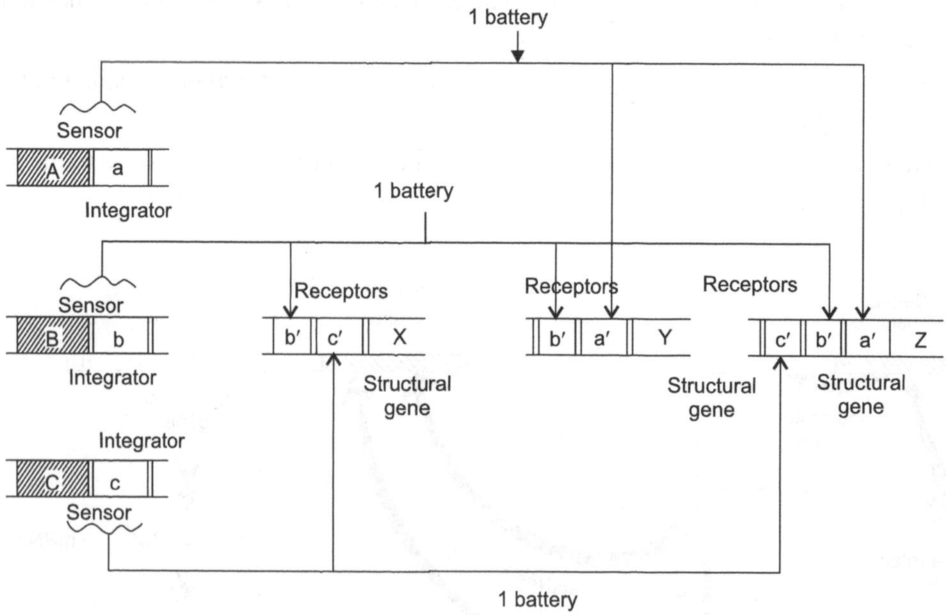

Fig. 11.13 Role of repititive of receptor sites in controlling the gene expression in eukaryotes.

Similarly, the repititive **integrator genes** may also help in the regulation of gene expression (Fig. 11.14).

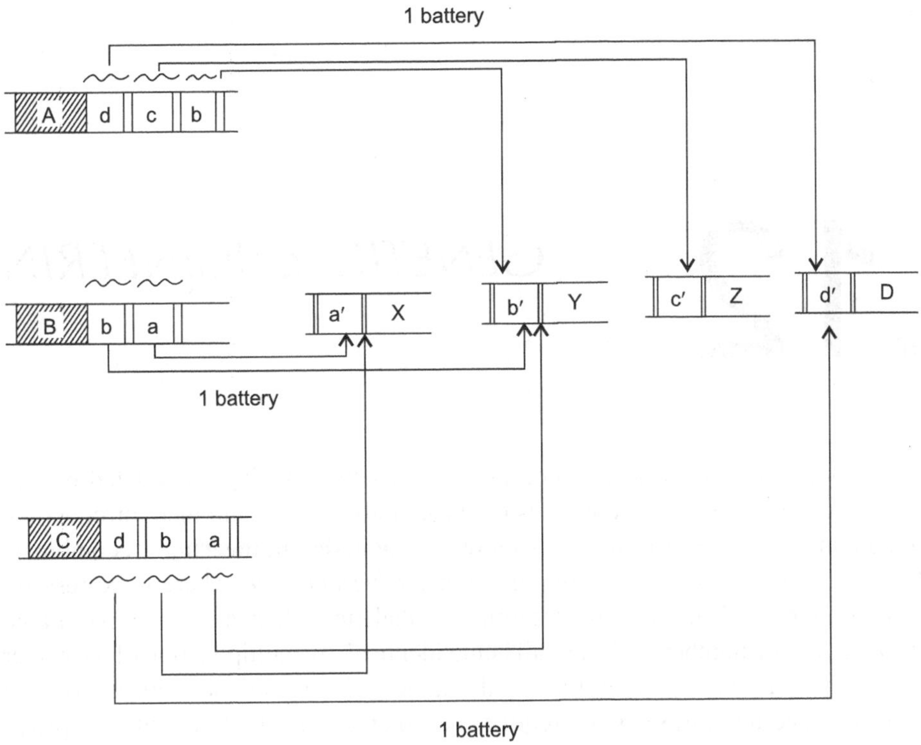

Fig. 11.14 Role of repititive integrator genes in controlling the gene expression in eukaryotes.

4. **Regulation at the Translational Level.** In many amphibians, the unfertilized egg contains some mRNAs that remain untranslated until it is fertilized by a sperm, *e.g.*, in *Xenopus* translation of mRNA for fibronectin protein occurs after 6–7 hours of fertilization. Thus, fertilization, here, controls the gene expression at translation level.

QUESTIONS

1. Describe *lac* operon in *E. coli.*

2. Discuss a repressible system of gene expression regulation in prokaryotes.

3. With the help of diagrams, explain the role of leader sequence in controlling the gene expression in *E. coli.*

4. How histones and non-histones are involved in regulation of gene expression in eukaryotes ?

5. Explain gene-battery model of gene regulation.

6. Discuss the role of repetitive DNA in gene regulation in eukaryotes.

Chapter 12 GENETIC ENGINEERING

The technology of production of recombinant DNA by inserting an isolated or artificially synthesized fragment of DNA at a desired place into a vector and then introduction of this recombinant DNA into a host cell is known as **genetic engineering** (recombinant DNA technology), *e.g.*, transfer of human insulin gene into *Escherichia coli* where it expressed to produce the insulin protein that is used for the treatment of diabetes. Thus, a single copy of a gene can be cloned into an indefinite number of copies, all being identical, by multiplication of the bacterium or phage containing that particular gene in the usual manner. This technique is also known as **gene cloning.** Genetic engineering in **eukaryotes** requires new methods and tools as, unlike in prokaryotes, the chromosomes are found in the nucleus and, moreover, many genes are split genes (exons interrupted by introns) that cannot be correctly expressed when cloned in prokaryotes (prokaryotes do not have machinery for the processing of HnRNA, and when cloned as such in prokaryotes, the entire HnRNA is translated into respective protein).

Now a days, **polymerase chain reaction** (PCR) discovered by **Carry Mullis** (1985), involving *Taq* DNA polymerase (that is obtained from a thermophilic bacterium *Thermus aquaticus* found growing in hot springs) is used to obtain millions of copies of DNA segment of choice. In this technique of **DNA amplification,** the two strands of the desired DNA fragment act as templates that are separated **(denatured)** by increasing the temperature for 90–98°C in an automatic PCR machine. After adding the two **primers** (artificially synthesized short segments of single-stranded DNA) that are complementary to the 3'ends of the two templates, the temperature is lowered to 40–60°C to allow the **base pairing** of the primers with their respective complementary sequences in the template DNAs. Now, the addition of ***Taq* DNA polymerase,** dNTPs (all the four nucleotides) and other requirements lead to the synthesis of the complementary sequences using the primers to produce **double-stranded DNA** molecules. This cycle is repeated as per requirement (20-30 times) where the temperature is alternately increased (for denaturation of DNA) and decreased (for renaturation of templates with primer and polymerization) to get millions or billions of copies of the desired DNA fragment, that can be purified by the gel-electrophoresis. Whenever the temperature is increased for the denaturation of DNA, *Taq* DNA polymerase remains active and is not destroyed, thus, allowing automation of the process; whereas, the ordinary DNA polymerase obtained from the mesophilic microorganisms gets denatured at higher temperatures of 90–98°C required for denaturation (Fig. 12.1 and 12.2). The PCR technique is not only used for the DNA amplification (in gene amplification, DNA fingerprinting, etc.), but, is also used for **RAPD** (Randomly Amplified Polymorphic DNA that is used as molecular marker in genome mapping), **site-directed mutagenisis** (by inducing mutation at the desired location in the primers), etc.

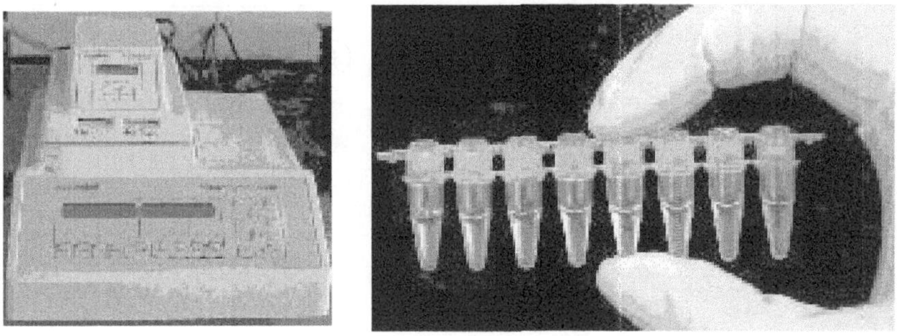

Fig. 12.1 PCR machine (left) and PCR tubues (right).

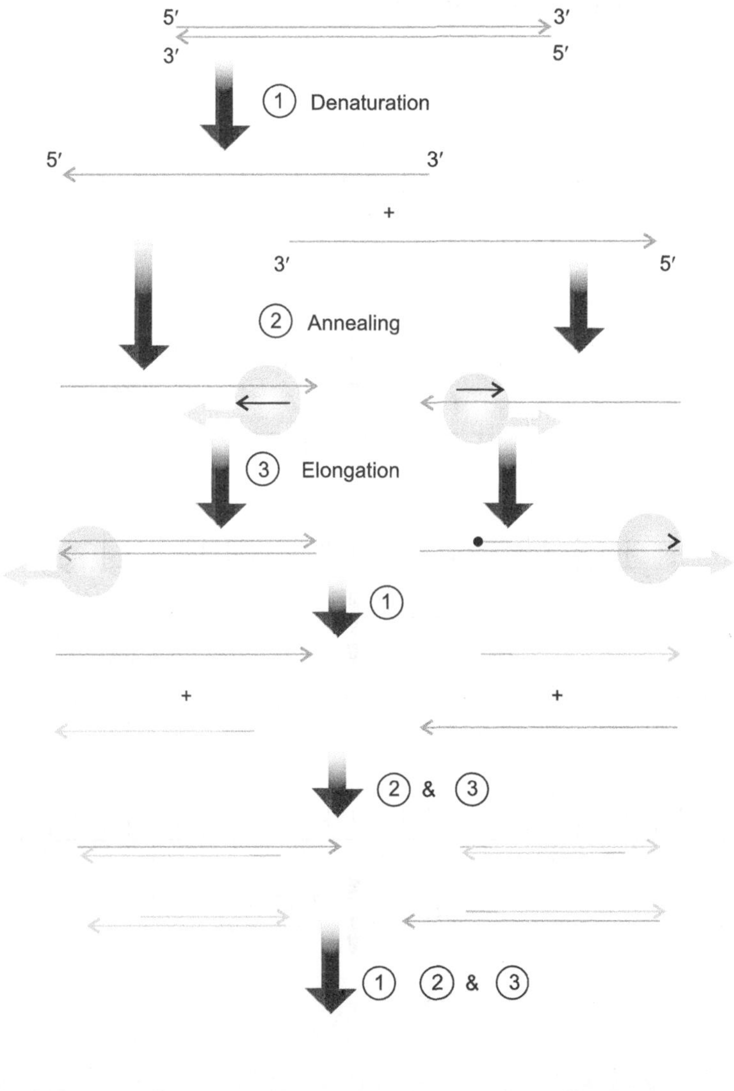

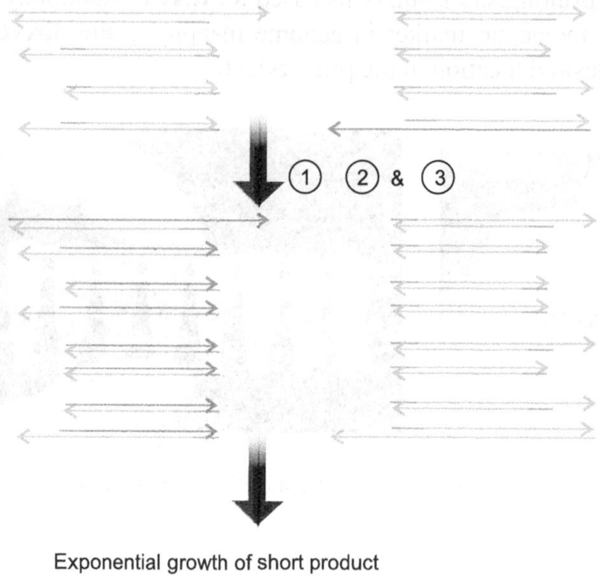

1
2
&
3
Exponential growth of short product

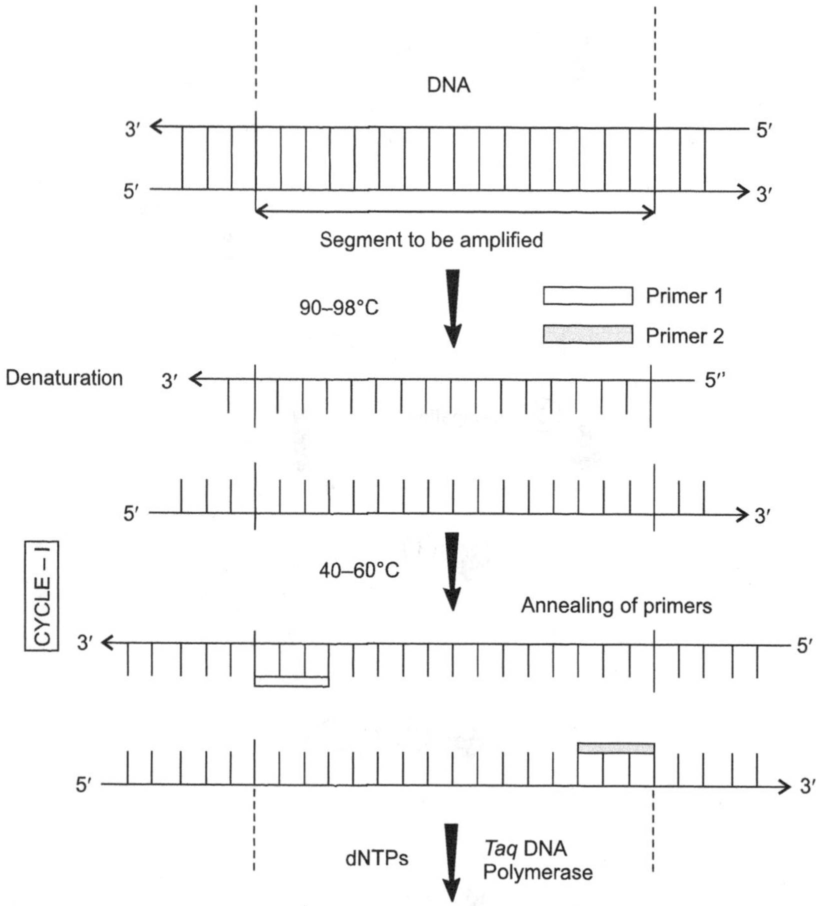

DNA
3′
5′
5′
3′
Segment to be amplified
90–98°C
Primer 1
Primer 2
Denaturation
3′
5″
5′
3′
CYCLE – I
40–60°C
Annealing of primers
3′
5′
5′
3′
dNTPs
Taq DNA
Polymerase

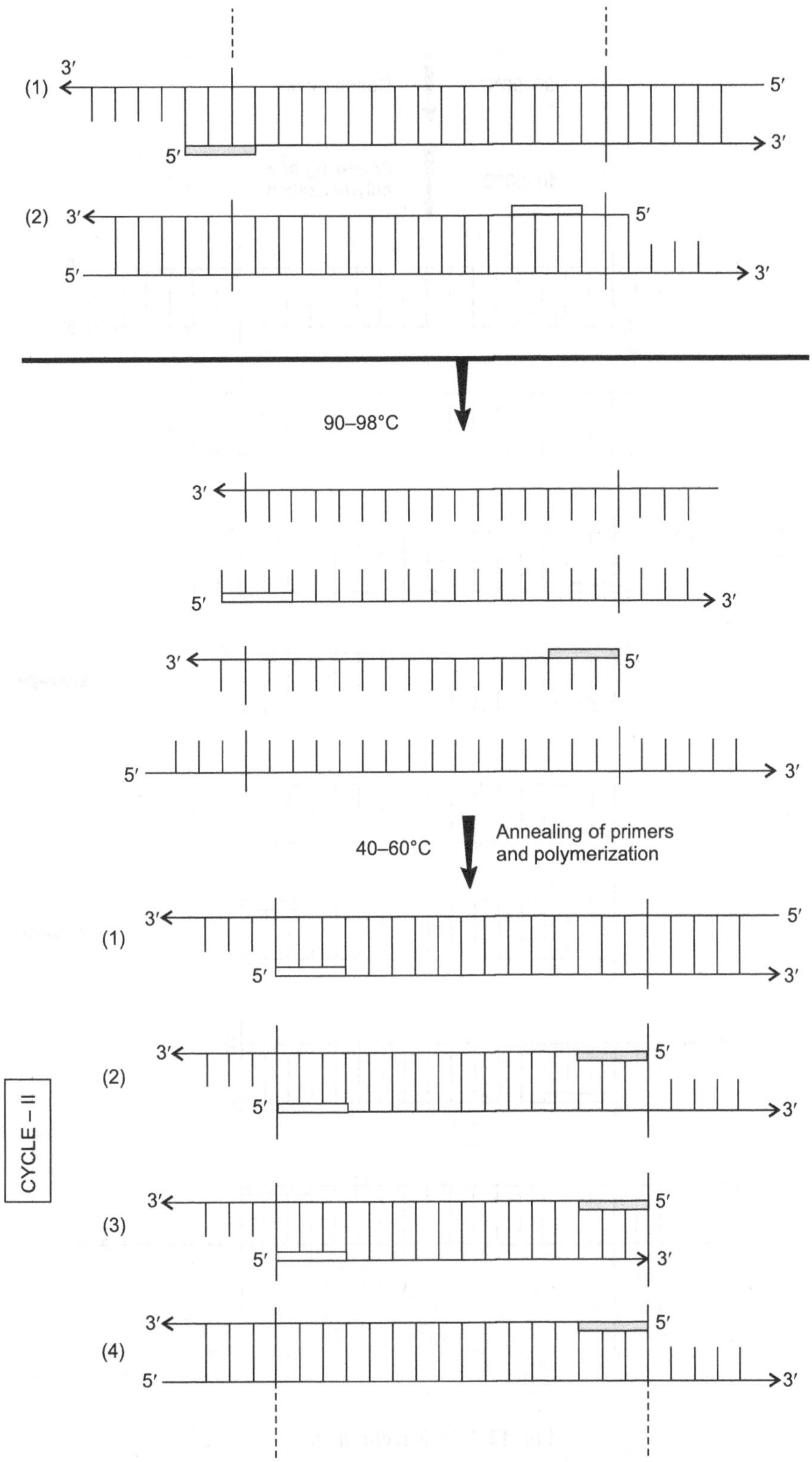
(1)
(2)
90–98°C
40–60°C
Annealing of primers
and polymerization
CYCLE – II
(1)
(2)
(3)
(4)

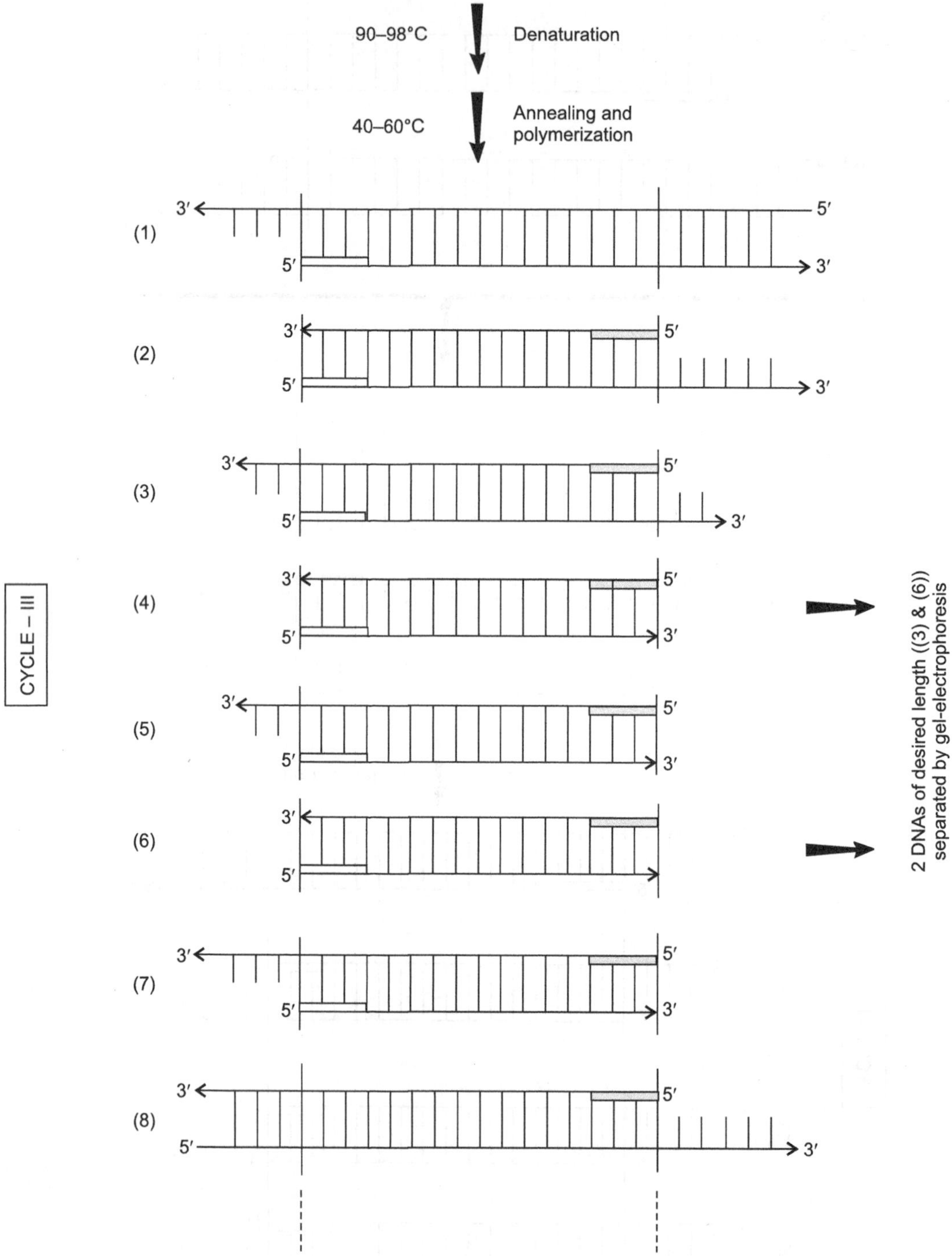

Fig. 12.2 PCR technique.

TOOLS FOR GENETIC ENGINEERING

1. **Restriction Enzymes:** For isolation of a DNA fragment and production of recombinant DNA, the DNAs are desired to be cut at specific sites by the restriction enzymes. These enzymes are named by 3 or 4 letter abbreviations that identify their origin, and the Roman numerals, I, II, III, etc., indicating the isolation of several types of enzymes from the same microorganism. The restriction enzymes obtained from various microorganisms **recognize** and **cut** the DNAs at **specific sites** that are short (4–8 base pairs long palindromic sequences) double-stranded nucleotide sequences. Different restriction enzymes recognize different specific sequences. For some restriction enzymes the recognition and cutting sites on the DNA are located at different positions, whereas, for others the recognition and cutting sites are the same (used in recombinant DNA technology) (Table 12.1). Among the restriction enzymes that are used in genetic engineering, some cut the double stranded DNA molecule at different locations **(staggered cuts)** to produce short complementary sticky ends that may reanneal (*e.g. EcoRI* obtained from *Escherichia coli*) (Fig. 12.3).

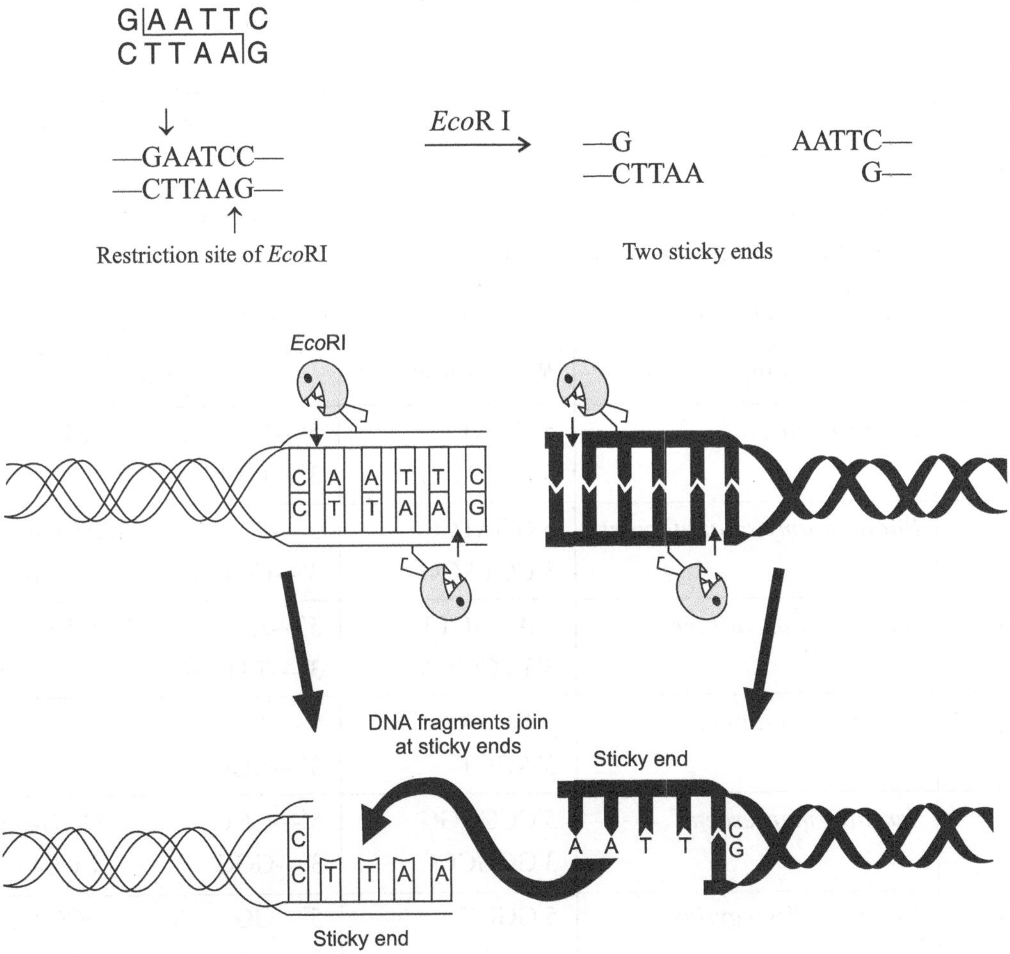

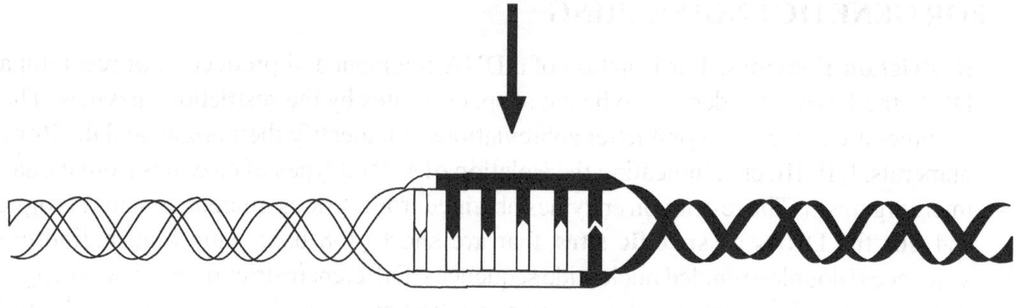

Recombinant DNA

Fig. 12.3 Restriction enzyme action of *Eco*RI.

whereas, others cut the two strands of DNA at the same locations to produce blunt ends (*e.g. Hae*III obtained from *Haemophilus*) (Fig. 12.4).

CCC | GGG
GGG | CCC

Restriction site of *Sma*I

↓ —GGCC— —CCGG— ↑   —*Hae*III→   —GG GG— / —CC CC—

Restriction site   Two blunt ends

Fig. 12.4 Restriction site of *Sma*I and restriction enzyme action of *Hae*III.

Table 12.1: Recognition sites and cut ends produced by some restriction enzymes.

Enzyme	*Source*	*Recognition Site*	*Cut*
*Eco*RI	*Escherichia coli*	5′GAATTC 3′CTTAAG	5′---G AATTC---3′ 3′---CTTAA G---5′
*Bam*HI	*Bacillus amyloliquefaciens*	5′GGATCC 3′CCTAGG	5′---G GATCC---3′ 3′---CCTAG G---5′
*Hind*III	*Haemophilus influenzae*	5′AAGCTT 3′TTCGAA	5′---A AGCTT---3′ 3′---TTCGA A---5′
*Taq*I	*Thermus aquaticus*	5′TCGA 3′AGCT	5′---T CGA---3′ 3′---AGC T---5′
*Sam*I	*Serratia marcescens*	5′CCCGGG 3′GGGCCC	5′---CCC GGG---3′ 3′---GGG CCC---5′
*Hae*III	*Haemophilus egytius*	5′GGCC 3′CCGG	5′---GG CC---3′ 3′---CC GG---5′

*Pst*I	*Providencia stuartii*	5′CTGCAG 3′GACGTC	5′---CTGCA G---3′ 3′---G ACGTC---5′
*Sal*I	*Streptomyces albae*	5′GTCGAC 3′CAGCTG	5′---G TCGAC---3′ 3′---CAGCT G---5′

2. **Vectors:** Vectors are the **carrier** DNA molecules (or microorganisms) for introduction of foreign DNA into the host cell. These vectors multiply in their usual manner even after carrying the foreign DNA, so that the inserted foreign DNA also replicates along with the parent vector DNA. The vectors should possess a site (*e.g.* restriction site) for foreign DNA insertion without disrupting any essential function, a selection marker (*e.g.* antibiotic resistance) for identification of the recombinant DNA containing the foreign DNA, etc. Some important vectors used in recombinant DNA technology are:

(*i*) **Plasmids:** Plasmids are the extrachromosomal, self-replicating, double-stranded **circular DNA** molecules (1–1000 in number) found in many prokaryotic cells and a few eukaryotic cells (*e.g.* yeast). They confer properties like, antibiotic resistance, fertility, toxin production, etc., and are not essential for the survival of the host cells, *e.g.* pBR322 plasmid (Fig. 12.5). All the suitable properties to be used as vector may not be present in a plasmid and they are modified by inserting the DNA segments imparting desired features (*e.g.* antibiotic resistance) or removing undesired parts. They can be easily isolated and transferred to other bacteria and can accommodate 10-15 kbp (kilo base pair) of DNA. The recombinant plasmids are transformed into a suitable host cell (*e.g. E. coli*) for multiplication.

Cosmids are the plasmids containing *cos* site of λ-phage that regulates the packaging of DNA into phage particles and can accommodate 35-45 kbp of foreign DNA.

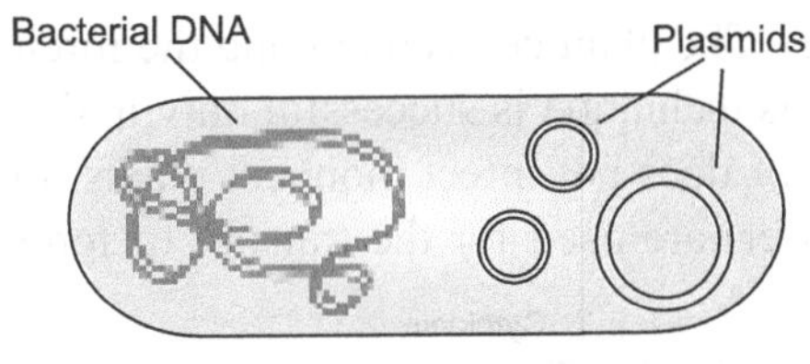

Fig. 12.5 A bacterial cell with plasmids.

(*ii*) **Viruses:** Most commonly, bacteriophages (*e.g.* λ-phage that can accommodate 22 kbp of foreign DNA) are used as vectors that contain linear DNA molecules. The non-essential DNA parts of the phage DNA are removed so that the foreign DNA can be inserted into it to produce the recombinant DNA (the protein coat of the phage can accommodate only a definite length of DNA). This recombinant DNA is allowed to infect the host cell where it replicates to produce a large number of progeny phage particles each containing a foreign DNA fragment. In addition to the bacteriophages, certain plant (*e.g.* CaMV or Cauliflower Mosaic Virus, TMV or Tobacco Mosaic Virus -a RNA virus) and animal (*e.g.* SV-40 or Simian Virus-40) viruses are also used as vectors in plants and animals, respectively. The use of viruses as vectors is restricted as they have narrow host range.

(*iii*) **Artificial Chromosomes:** Certain chromosomes have been prepared artificially to be used as vectors that can accommodate several hundred kbp of foreign DNA (sometimes the whole chromosome), *e.g.* YAC (Yeast Artificial Chromosome) by isolating and cloning telomeres (terminal parts of chromosomes) and centromeres of yeast chromosomes in a plasmid (Fig. 12.6), BAC (Bacterial Artificial Chromosome) etc.

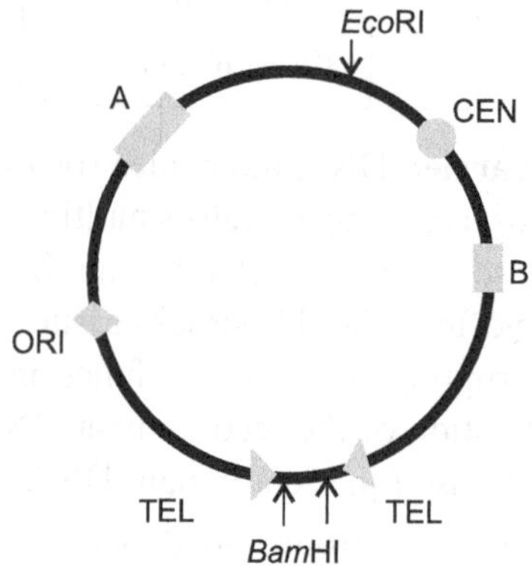

Fig. 12.6 The YAC vector.

(*iv*) **Agrobacterium:** *Agrobacterium tumefaciens* bacterium, that cause crown gall disease on the stem of dicot plants, contains a plasmid called **Ti plasmid** (tumour inducing). This Ti plasmid has a **T-DNA** region into which the foreign DNA can be inserted. This T-DNA is transferred to the host plant cell and gets incorporated into the chromosome of host cells (Fig. 12.7 in Appendix and 12.8). Another species, *A. rhizogenes* causes hairy root disease in dicots and contains a plasmid, **Ri plasmid** (root-inducing plasmid). Ti and Ri plasmids are used as vectors in dicot plants. For cloning purpose, the disease-causing part of the plasmid is removed. Thus, after infection by the *Agrobacterium* carrying foreign DNA to its host plant cell (the cultured cells or the leaf discs, root sections, floral parts, etc.), though the foreign DNA is expressed, but, it does not cause any disease (crown gall/hairy root disease). The plant cells containing the foreign DNA are regenerated into complete plantlets. This technique is successful only in dicot plants and not in monocot plants as *Agrobacterium* does not infect monocot plants. In monocots, other techniques (*e.g.* direct gene transfer) are used for the transfer of foreign DNA.

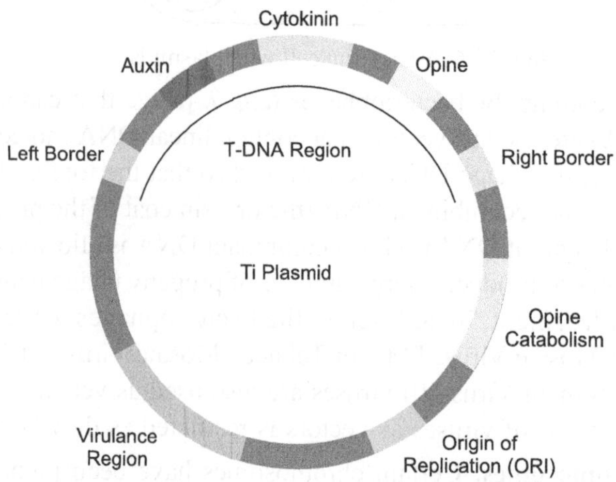

Fig. 12.8 Ti plasmid of *A. tumefaciens*.

3. **Direct DNA Transfer:** Several technologies for direct DNA transfer are used in plants, animals and microbes:

(*i*) **Gene Gun:** Gold or tungsten particles coated with foreign DNA are projected into the target cells with the help of gene gun machine. After penetrating the cell wall or cell membrane, the DNA gets incorporated into the genomic DNA of the cell (Fig. 12.9).

Fig. 12.9 Particle gene gun.

(*ii*) **Polyethylene Glycol (PEG):** Polyethylene glycol (15–25%) stimulates the uptake of DNA by the protoplasts (cells without cell walls).

(*iii*) **Electroporation:** Exposure of protoplasts to electric current for very short duration of time induces temporary pores in the cell membrane through which the DNA enters readily and gets incorporated into the genomic DNA of the cell.

(*iv*) **Microinjection:** This technology involves the injection of DNA using micropipettes (0.5-10 μm tip diameter) into the target cell that is immobilized on a solid support (coverslip or slide).

(*v*) **Macroinjection:** Here, the DNA is injected using needles having diameter greater than the cell diameter. In rye (*Secale cereale*) DNA was successfully injected into stem below the floral meristem (actively dividing tissue) from where it reached the sporogenous cells to develop into the transgenic plants containing the foreign DNA.

(*vi*) **Liposomes:** In this technique the DNA is encapsulated into lipid bags and then they are fused with the protoplasts so that the DNA enters into the cell.

Some Important Techniques Used in Recombinant DNA Technology

1. **Gel Electrophoresis:** In this technique, the **DNA,** after extraction, is cleaved into fragments of different length by treating with a **restriction enzyme.** An **agarose** (or polyacrylamide) **gel slab** (about 10 cm long and 0.5 cm thick) is prepared by pouring hot solution of agarose and a buffer into a mould (tray). When still in molten condition, a comb is placed on one side of the slab, which is removed after solidification to produce **wells** in the agarose gel slab. This slab is then immersed in a buffer solution in a tray and the mixture of the **DNA fragments** are **loaded** into the wells with the help of a micropipitte. Under **electric field** the DNA fragments, being negatively charged, start moving towards the anode. The speed of movement depends upon the molecular weight of the DNA fragments, so that the lighter fragments move faster than the heavier fragments. Thus, different fragments of different lengths appear as various **bands** (Fig. 12.10).

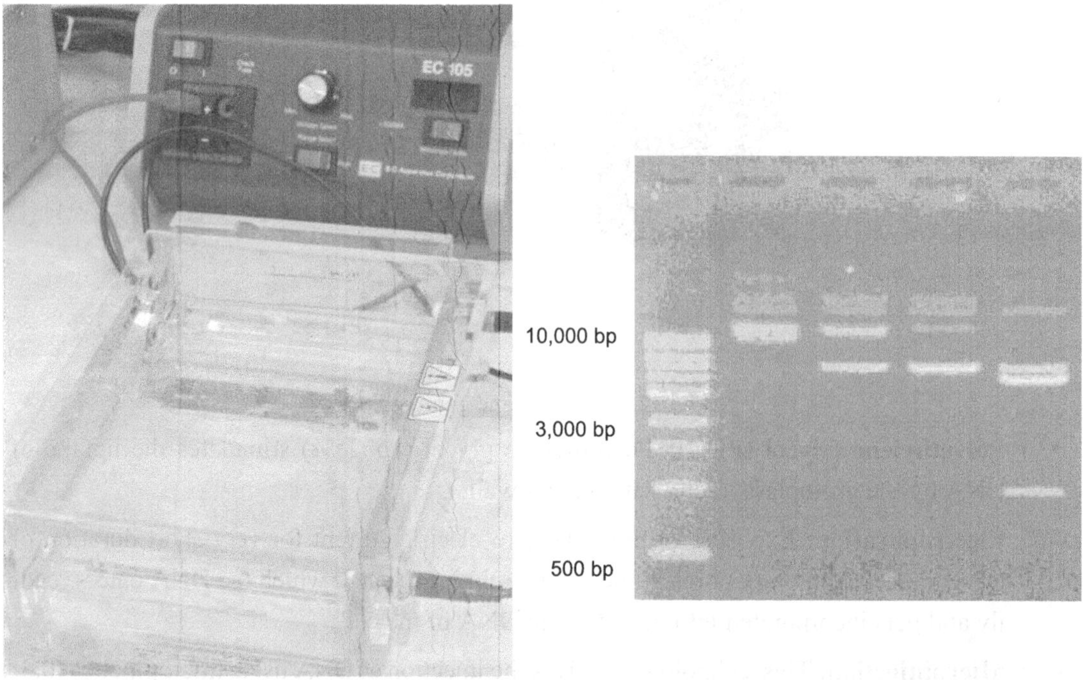

Fig. 12.10 Gel electrophoresis apparatus (left) and agarose gel with DNA bands (right).

2. **Blotting:** A mixture of DNA/RNA/proteins is separated by gelelectrophoresis (as bands) and hybridized with a labeled probe. Then these bonds are transferred to a **nitrocellulose membrane** by putting it on the surface of gel and this technique is known as blotting. The blotting of **DNA** (hybridized with labeled DNA or RNA probe with radioactive ^{32}P) bands is known as **Southern blotting** after the name of **E.M. Southern,** (Fig. 12.11), whereas the blotting of RNA (hybridized with labeled DNA or RNA probe with radioactive ^{32}P) is called **northern blotting** and of protein bands (on polyacrylamide gel bound with labeled antibody probe with radioactive ^{125}I) is called **western blotting.**

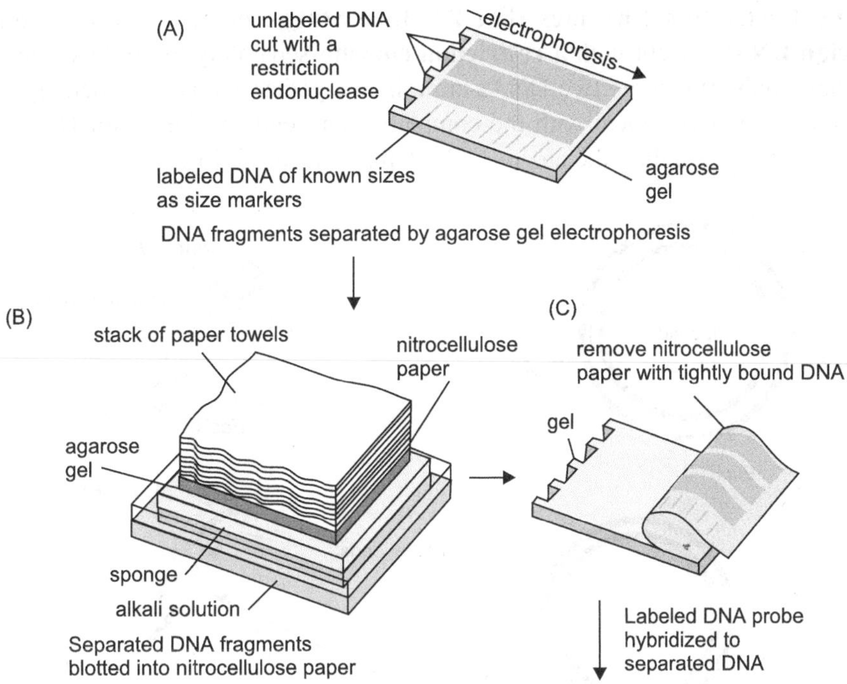

Fig. 12.11. Technique of Southern blotting.

RECOMBINANT DNA TECHNOLOGY

Integration of isolated or artificially synthesized **DNA** into a **vector** to produce **recombinant DNA** is known as **recombinant DNA technology.** The **recombinant DNA** thus produced can be screened by the selection markers (*e.g.* antiobiotic resistance in the vector may be lost by integration of foreign DNA) (Fig.12.12) and separated by gelelectrophoresis from other DNAs (*e.g.* reannealed circular vector and foreign DNAs). This recombinant DNA is then transferred into a host (target) cell to generate useful products (*e.g.* insulin) or to produce transgenic organisms (*e.g.* disease resistant crop plants) to be used for the welfare of human beings. Whenever required, the foreign DNA can be retrieved by treating with the same restriction enzyme. Some of the important techniques for the generation of recombinant DNA, depending upon the requirement, are:

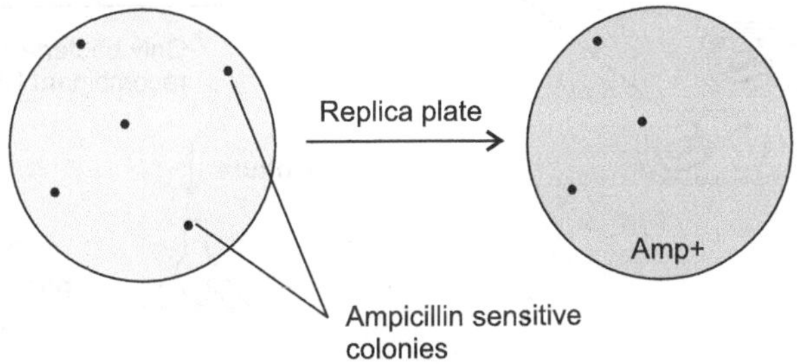

Fig. 12.12. Screening of ampicillin sensitive recombinant clones by replica plating in ampicillin-containing medium.

1. **Using Restriction Enzymes that Produce Staggered Cuts:** The **vector** and the **foreign DNAs** are cut with a **restriction enzyme** separately to produce sticky complementary ends. When the two DNAs are mixed together, the complementary ends of the foreign DNA may anneal with the complementary ends of the vector DNA, which are joined finally by adding the enzyme, **DNA ligase** (Fig. 12.13).

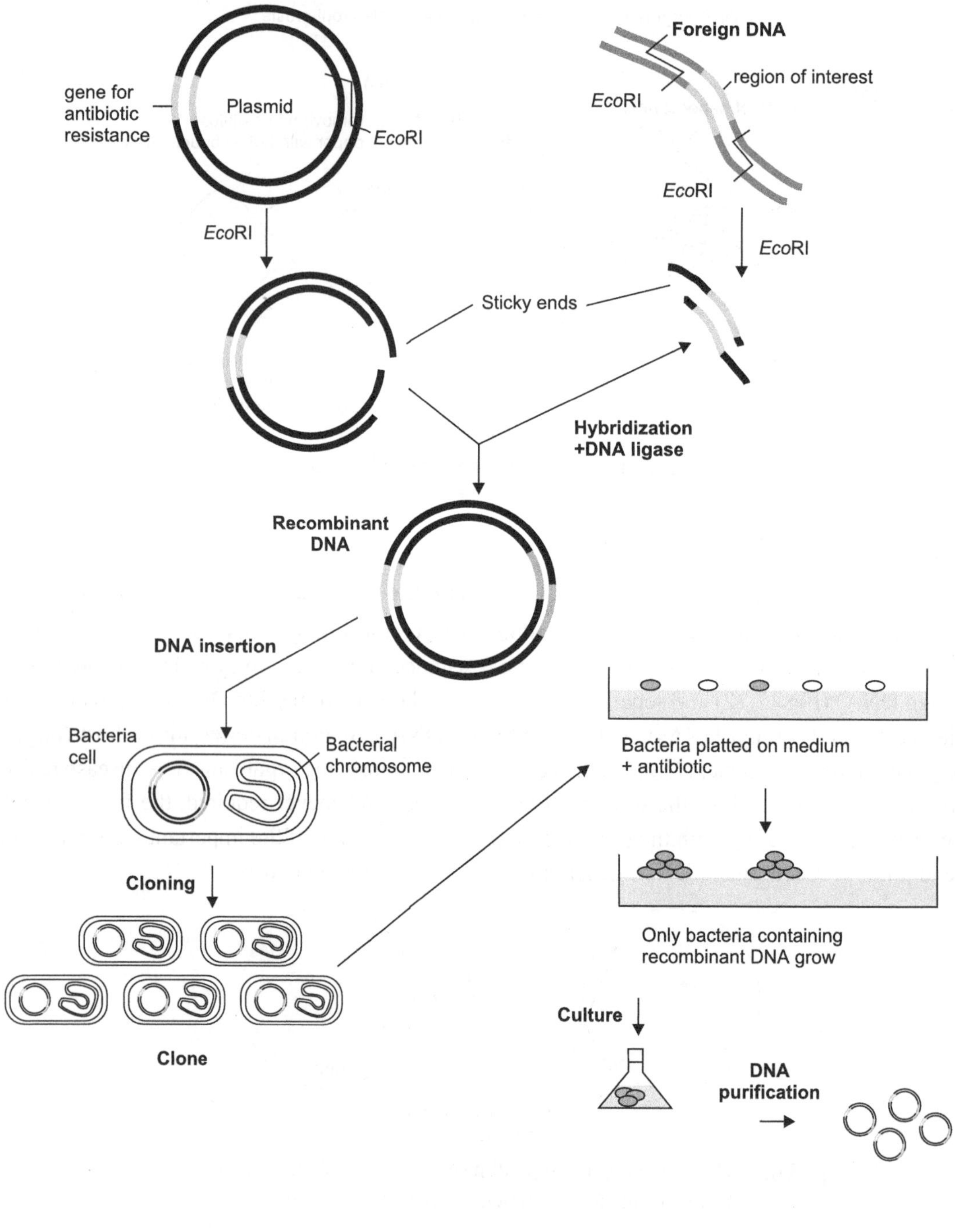

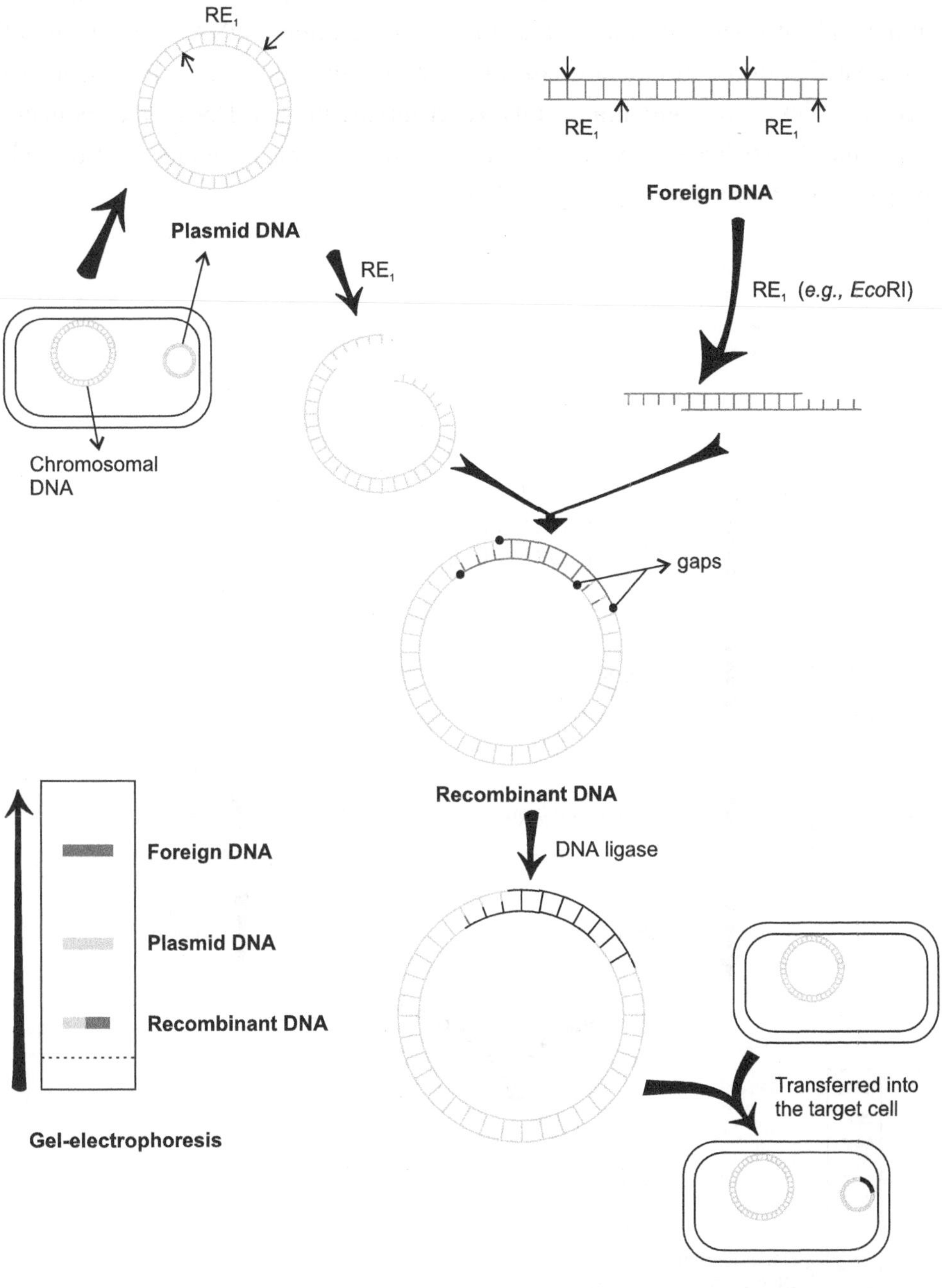

Fig. 12.13 Recombinant DNA technology using *Eco*RI, REI = Restriction Enzyme I.

2. **Blunt End Ligation by T4 DNA Ligase:** In this method, the plasmid DNA is made linear by cutting with a **restriction enzyme** to produce **blunt ends.** When artificially synthesized restriction site (linker DNA) of a restriction enzyme is added separately into

the linear vector DNA and foreign DNA in the presence of **T4 DNA ligase** (which can link two double-stranded DNA segments with blunt ends), the linker DNA gets attached at the ends of vector DNA as well as to that of the foreign DNA. Now, the two DNAs are treated with the **restriction enzyme** that cuts the restriction site (linker DNA) to generate sticky **complementary ends.** After mixing the two DNAs, recombinant DNA is produced after ligation by the DNA ligase enzyme. This technique is also used in the preparation of cDNA library (Fig. 12.14).

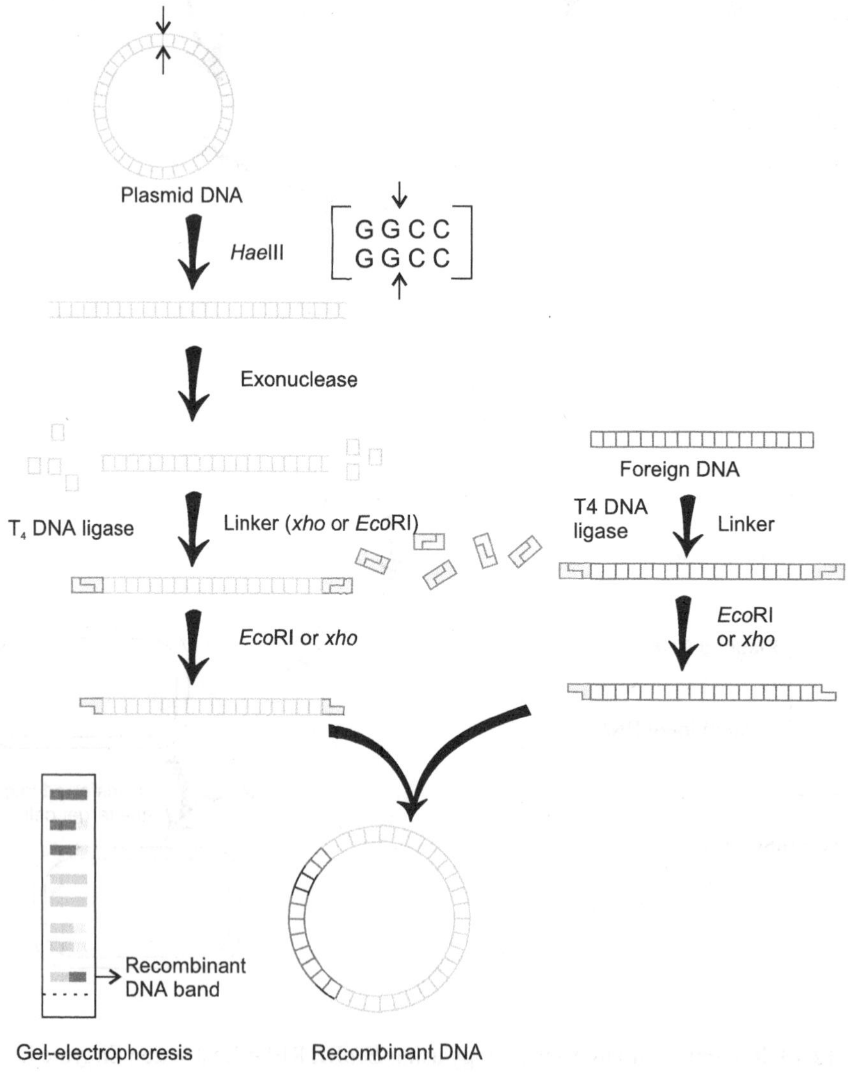

Fig. 12.14 Recombinant DNA formation by blunt-end ligation using T4 DNA ligase.

3. **Adding Poly dG at the 3' End of Vector and Poly dC at the 3' End of Foreign DNA:** The vector DNA after cutting with a **restriction enzyme** that generates **blunt ends,** is added with **dGTP** in the presence of enzyme terminal transferase to produce

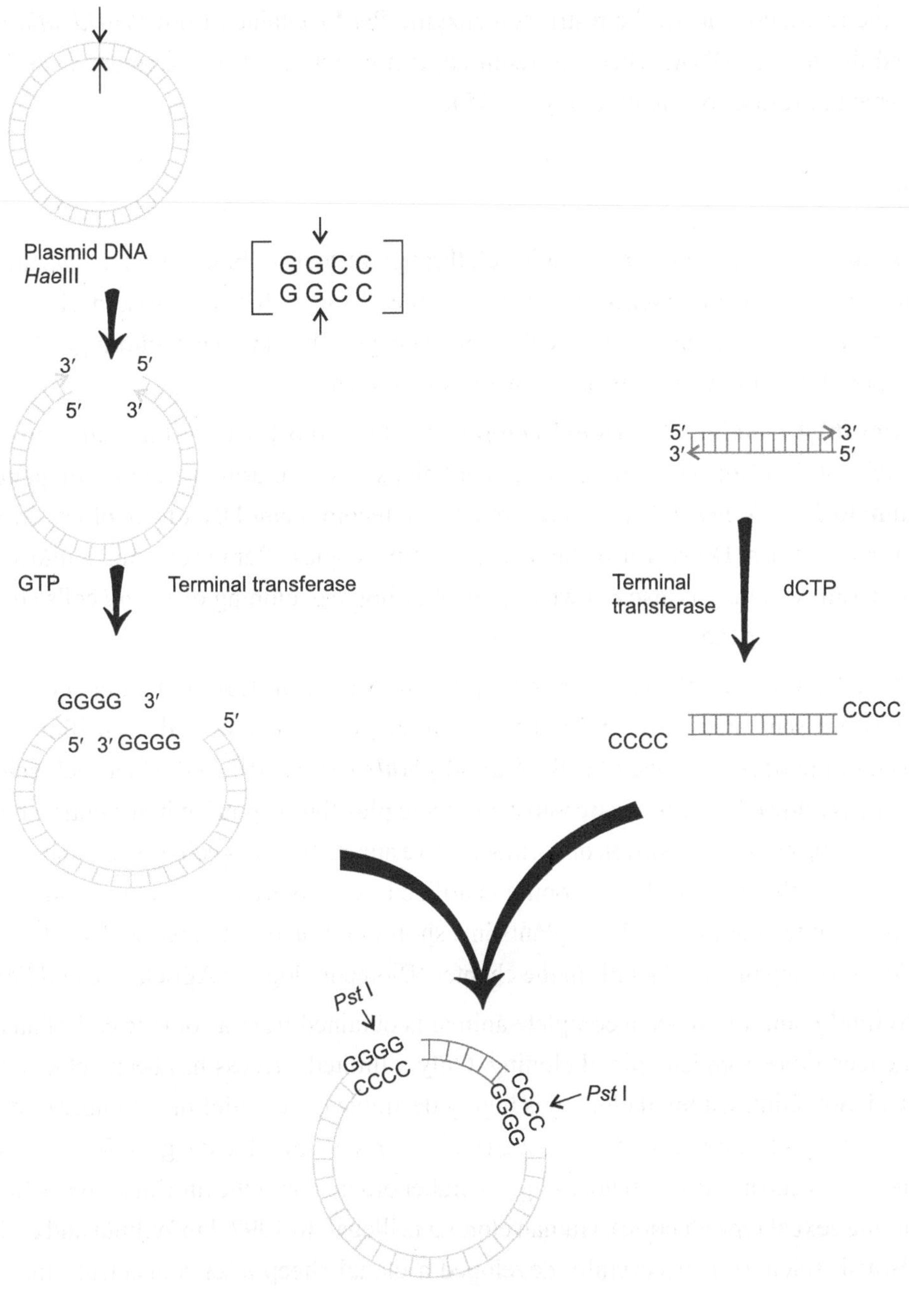

Fig. 12.15 Recombinant DNA formation by adding poly G and poly C tails in plasmid vector and foreign DNA, respectively.

poly dG at the 3' ends of this linear vector DNA (addition of new nucleotides during polymerization always occurs at the 3' end) . Similarly, the foreign DNA is mixed with **dCTP** to generate poly dC at its 3' ends. When the two DNAs are mixed together, recombinant DNA is produced due to complementary dG : dC base pairing. This **dG : dC** is the restriction site of the restriction enzyme **Pst I** (obtained from *Providentia stuartii*) and the foreign DNA, whenever required, can be retrieved by treatment with the same particular restriction enzyme (Fig. 12.15).

CLONING

A clone of cells is derived from a single cell through **mitosis** or binary fission, etc. and all the cells of a clone are genetically **identical** (except for the changes that may arise during and after cloning). The process of obtaining clone of cells from a single cell is known as **cloning.** Cloning has been successfully done in microbes, plants as well as in animals.

1. **Microbial Cloning: Microbial cloning** has been most successfully and most extensively utilized for the welfare of human beings, *e.g.* culturing of clones of **genetically unmodified** *Escherichia coli* cells are used in recombinant DNA technology. Moreover, the recombinant DNA is also obtained in multiple copies after introducing it into a suitable bacterial cell and it is also known as **gene cloning,** *e.g.* cloning of *E. coli* cells containing human insulin gene.
2. **Plant Cloning:** In plants, for cloning purpose cells from leaf or stem or root or flower parts are separated from the tissues and a single cell is cultured in artificial nutrient medium to obtain a clone of cells. **Clonal plants** can be obtained when each **cell** of this clone is allowed to **regenerate** into a complete **plantlet** by putting it in a nutrient medium containing correct proportion of hormones, like auxins (induces cell growth) and cytokinins (induces cell division). Plant cloning is utilized extensively in agriculture and forestry to get unlimited number of clonal plants in a short duration of time, particularly for rare and slow growing plants. (Details in the chapter 'Biotechnology in Agriculture and Forestry')
3. **Animal Cloning:** When a complete animal is obtained from a somatic cell of an animal, the technique is called animal cloning. Only a limited success has been achieved in this field. Sometimes, animal cloning is highly desirable to get indefinite number of desirable similar type of animals by the cloning (which does not involve the genetic changes due to meiosis occurring at a particular stage in higher organisms at the time of gametes formation during sexual reproduction). Human cloning is illegal. In 1997, Ian Wilmut and colleagues (British scientists) successfully developed a clonal sheep after transferring the nucleus from an udder cell of a female sheep into the protoplasm of an enucleated unfertilized egg. This egg was then transplanted into the uterus of another female sheep (surrogate

mother) for the development into a normal lamb, named **Dolly.** It was genetically identical to the sheep that donated the nucleus from the udder cell (Fig. 12.16). (Details in the chapter 'Mutations') .

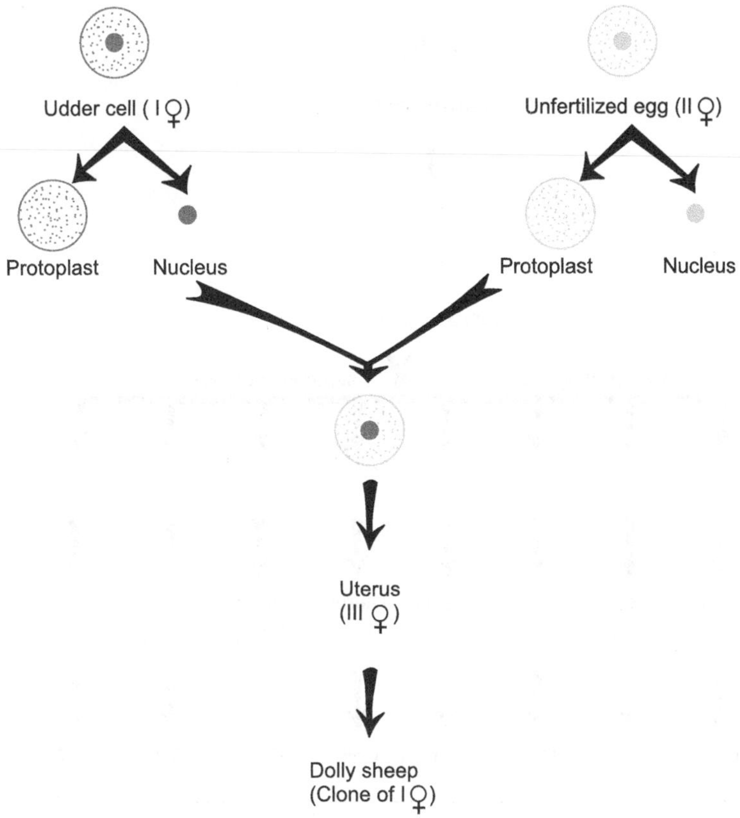

Fig. 12.16 Development of a sheep clone, Dolly.

After Dolly, the Japanese scientists developed eight clonal calves in 1999 by embryo splitting technique, where, at the 8-celled stage of the sexually obtained embryo, the cells were separated and each cell was introduced into the uterus of a separate cow to get eight clonal calves identical to each other (Fig. 12.17).

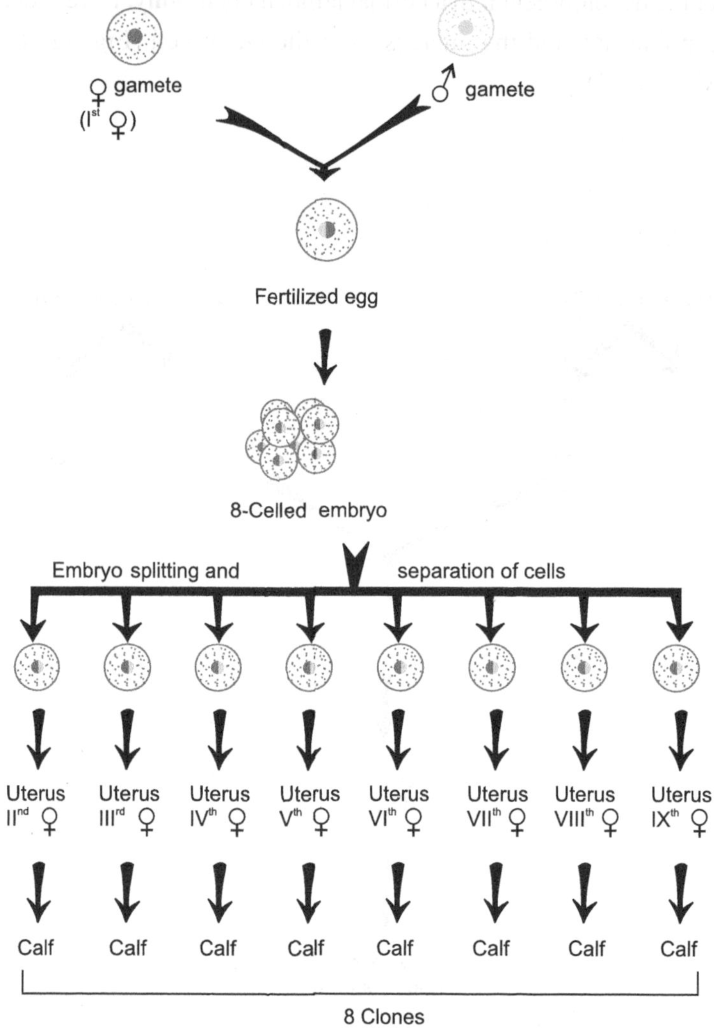

Fig. 12.17 Development of clonal calves by embryo-splitting technology.

American scientists reported the cloning of rhesus monkey, that is closest to human beings, in 2000 by embryo splitting technique. In this technique, the normal zygote produced after the sexual reproduction was allowed to divide mitotically and at 4-celled stage, the four cells of the embryo were separated. Each cell was then introduced into the uterus of a separate surrogate mother for the development of cell into a baby monkey.

DNA LIBRARIES

DNA library is the collection of DNA fragments of entire or partial genome of an organism that are cloned separately while attached to a suitable vector inside a host cell. DNA library is analogous to an academic library where each clone carrying a particular DNA fragment is similar to a book and the number of cells of a single clonal culture in a culture vessel is analogous to the number of books of that particular book. DNA libraries of a variety of organisms have been constructed, *e.g., E. coli, Saccharomyces cerevisiae, Bacillus subtilis, Drosophila* (fruit fly), *Arabidopsis* (a higher plant), etc., which are used for various purposes, like, isolation of a desired gene or DNA fragment, study of the structure of genes, regulation of genes, genome mapping, etc. with the help of specific probes. The DNA library is of the two types:

1. Genomic DNA Library

Genomic DNA library (gene bank) of an organism is the collection of cloned random DNA fragments covering the entire genome that may contain both **introns** (do not contain any information of amino acids and after transcription, though, they are present in Hn RNA, but, during processing they are removed to produce mRNA) and **exons** (sequences containing information required for protein synthesis) in eukaryotes. The genomic DNA library of an organism by shotgun method can be constructed by following the steps:

(*i*) Genomic **DNA** is **extracted** from the cells of an organism.

(*ii*) The extracted DNA is cut into random fragments of reasonable size by a particular **restriction endonuclease** (RE).

(*iii*) A suitable **vector** (*e.g.* a plasmid, a cosmid, etc.) is made linear by cutting with the same restriction enzyme to get the same complementary ends as in the genomic DNA fragments.

(*iv*) The two types of linear DNAs (vector and individual genomic DNA fragments) are mixed together in a culture vessel and the complementary ends of the **recombinant DNA**, thus, obtained are joined together by adding the **DNA ligase** enzyme.

(*v*) These **recombinant DNAs** are separately introduced into a host cell (*e.g. Escherichia coli*) and **cloned** by growing in artificial nutrient medium in a culture vessel.

(*vi*) Thus, the individual transformed cells, each containing a genomic DNA fragment, are inoculated at the surface of solidified nutrient medium in a Petri plate and allowed to grow into separate colonies (clones). Individual colonies can also be transferred into the liquid medium in separate culture vessels. The set of cloned DNA fragments from the entire genome of an organism maintained separately in host cells constitutes the **genomic DNA library.** The number of cloned DNA fragments in a genomic DNA library increases with increase in the size of genome and the human genome comprises more than 8,00,000 cloned fragments (Fig. 12.18).

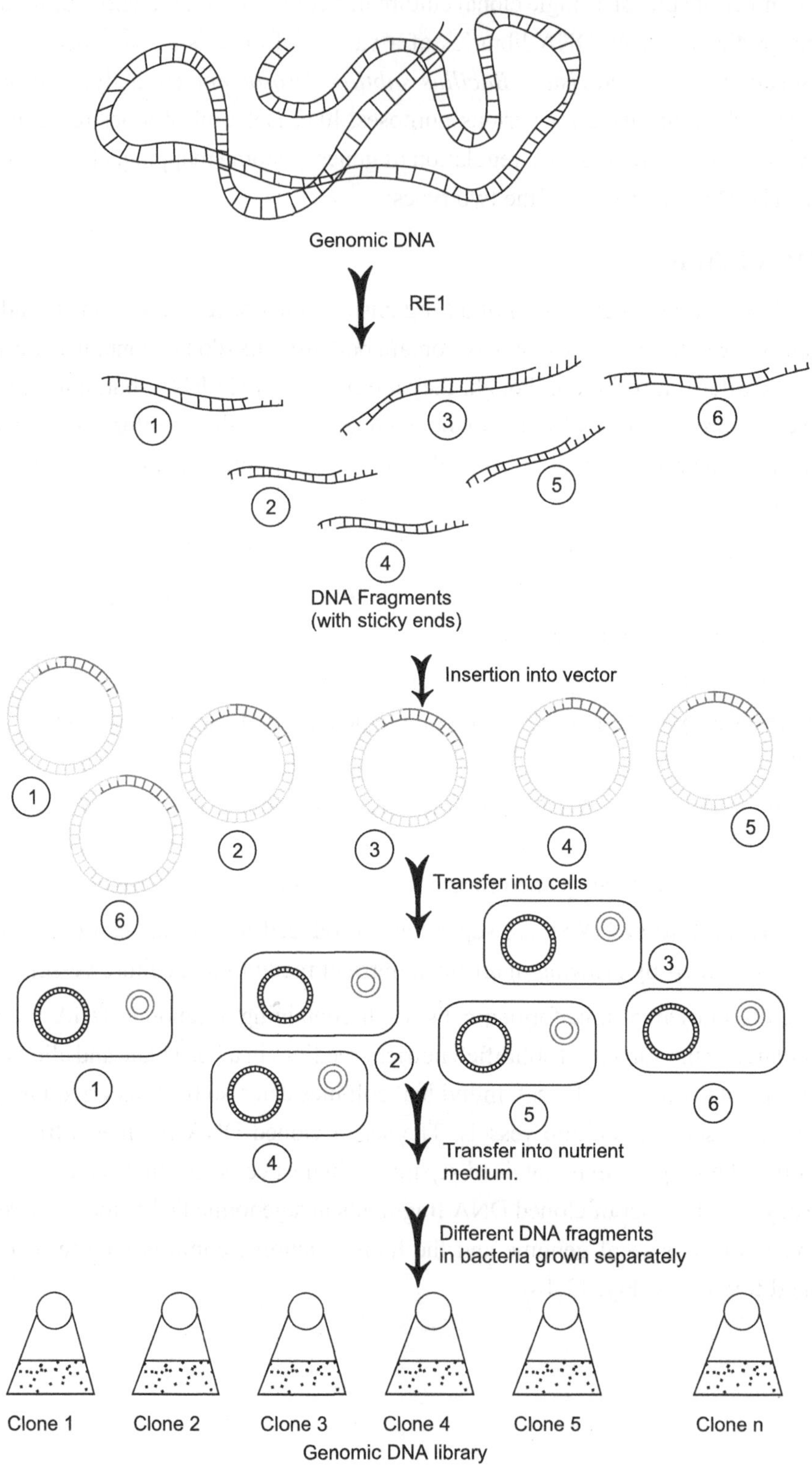

Genomic DNA
RE1
1
2
3
4
5
6
DNA Fragments
(with sticky ends)
Insertion into vector
1
2
3
4
5
6
Transfer into cells
1
2
3
4
5
6
Transfer into nutrient medium.
Different DNA fragments in bacteria grown separately
Clone 1
Clone 2
Clone 3
Clone 4
Clone 5
Clone n
Genomic DNA library

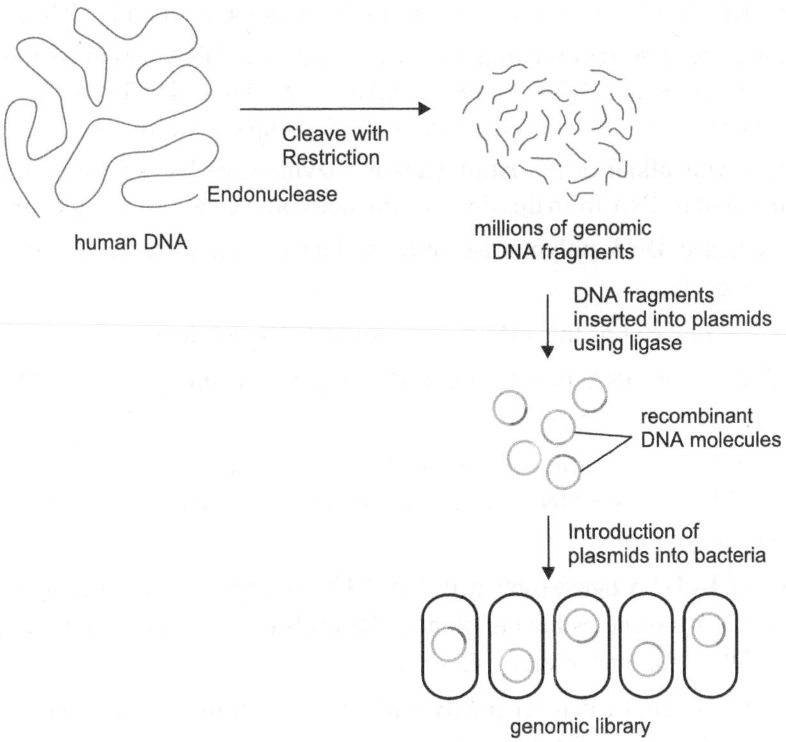

Fig. 12.18 Technique of genomic DNA library formation (RE1 = Restriction enzyme 1).

2. cDNA Library

Many organisms have very large genome and the construction of genomic DNA library in such organisms would require maintenance and screening of a large number of clones. Moreover, the fragments cloned in genomic DNA library may include introns as well that should be removed before the translation process of protein synthesis as they do not contain any information for amino acids. When such bacterial clones are utilized for the commercial production of a particular protein, the transformed host bacteria do not possess the machinery required for the processing of the HnRNA (heterogeneous RNA) transcribed from that particular split gene to produce mRNA. These problems can be overcome by constructing the cDNA libraries.

cDNA is the complementary DNA (copy DNA) of mRNA that lacks introns and other undesirable nucleotide sequences. Thus, cDNA library of an organism contains only the expressed genes (cDNAs) as they are prepared from the mRNA by the following steps:

(*i*) **Extraction** of **mRNAs** from the cells (in unicellular organisms) or different tissues engaged in active synthesis of proteins (in multicellular higher organisms, different genes are expressed in various tissues to give different types of mRNAs and proteins due to regulation of gene expression) to get all the possible kinds of mRNAs of an organism.

(*ii*) Addition of a **primer** that is made up of a short sequence of dT, causes base pairing with the complementary sequence of poly A tail found at the 3' end of mRNA.

(*iii*) Addition of **reverse transcriptase** enzyme produces cDNA strand (complementary to mRNA) by copying mRNA as the template and adding dNTPs at the 3' end of the primer. Thus, a double-stranded mRNA-cDNA complex is formed.

(*iv*) Treatment with **alkali** or the **ribonuclease** enzyme (that breaks down RNAs) helps in the removal of mRNA from this double-stranded complex to form single-stranded cDNA.

(*v*) Now, the added **DNA polymerase** utilizes cDNA as template to give double stranded cDNA molecule.

(*vi*) The hook at one end of this cDNA is removed by **S_1 nuclease.**

(*vii*) A suitable vector DNA is cut with a **restriction enzyme** producing blunt ends (not complementary sticky ends).

(*viii*) Now, the cDNA and linearised vector DNA are mixed separately with a **linker DNA** (short double-stranded DNA molecules which are actually the restriction sites of a restriction enzyme).

(*ix*) Addition of **T4 DNA ligase** (obtained from T4 bacteriophage which can ligate two double-stranded DNA molecules having blunt ends) leads to the ligation of linker DNAs at the ends of cDNA and linear vector DNAs.

(*x*) The two DNAs are treated separately with the **restriction enzyme** that recognizes and cuts the attached linker DNAs to produce complementary sticky ends.

(*xi*) Mixing of the two DNAs leads to the construction of **recombinant DNA** that is introduced into a suitable host cell. Thus, the set of all the host cells transformed with a particular cDNA, constitutes a **cDNA library** (Fig. 12.19).

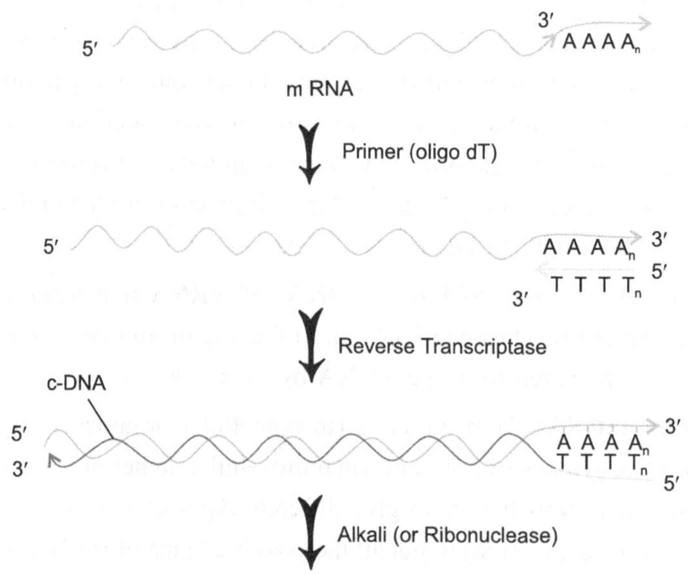

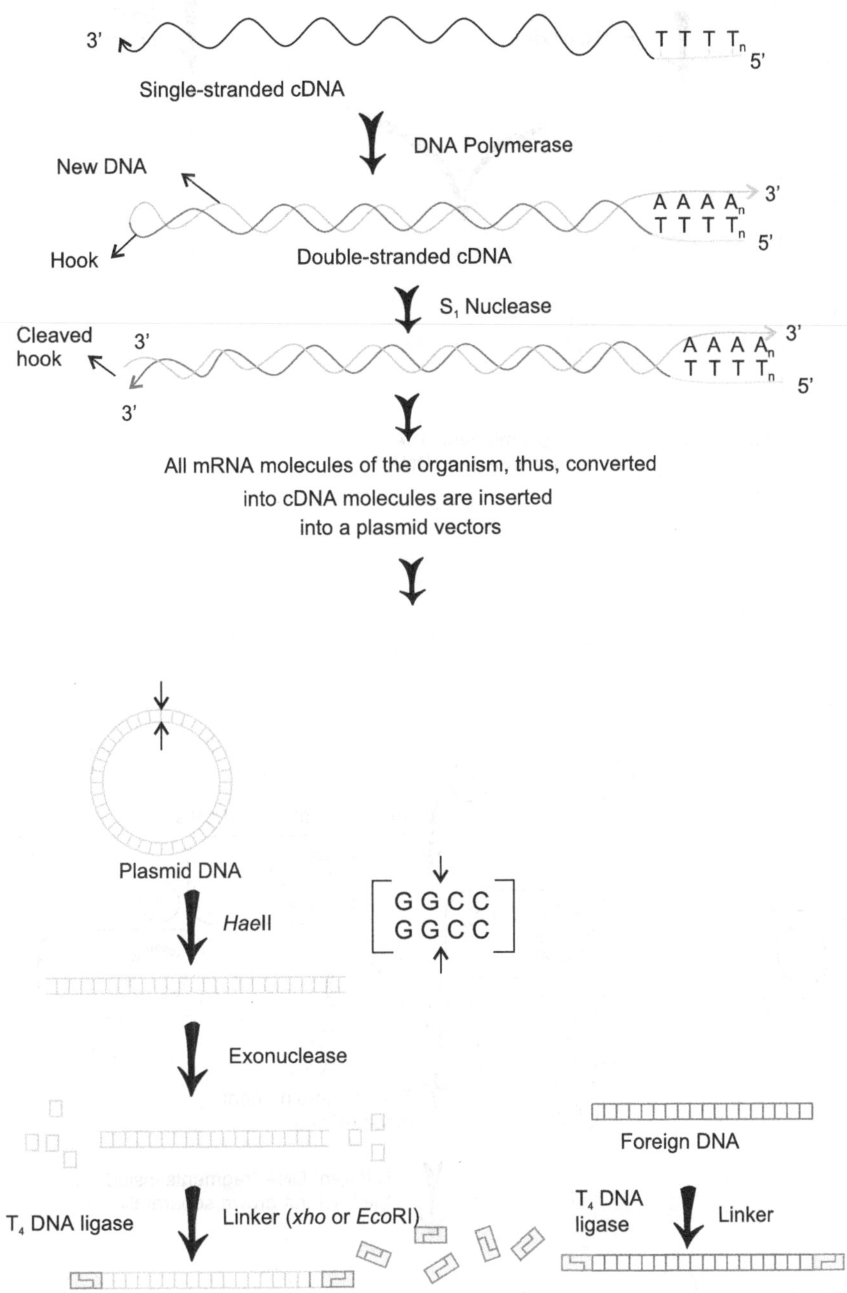
3'
T T T Tn
5'
Single-stranded cDNA
DNA Polymerase
New DNA
A A A An
3'
T T T Tn
5'
Hook
Double-stranded cDNA
S1 Nuclease
Cleaved hook
3'
A A A An
3'
T T T Tn
5'
3'
All mRNA molecules of the organism, thus, converted into cDNA molecules are inserted into a plasmid vectors
Plasmid DNA
HaeII
G G C C
G G C C
Exonuclease
Foreign DNA
T4 DNA ligase
Linker (xho or EcoRI)
T4 DNA ligase
Linker

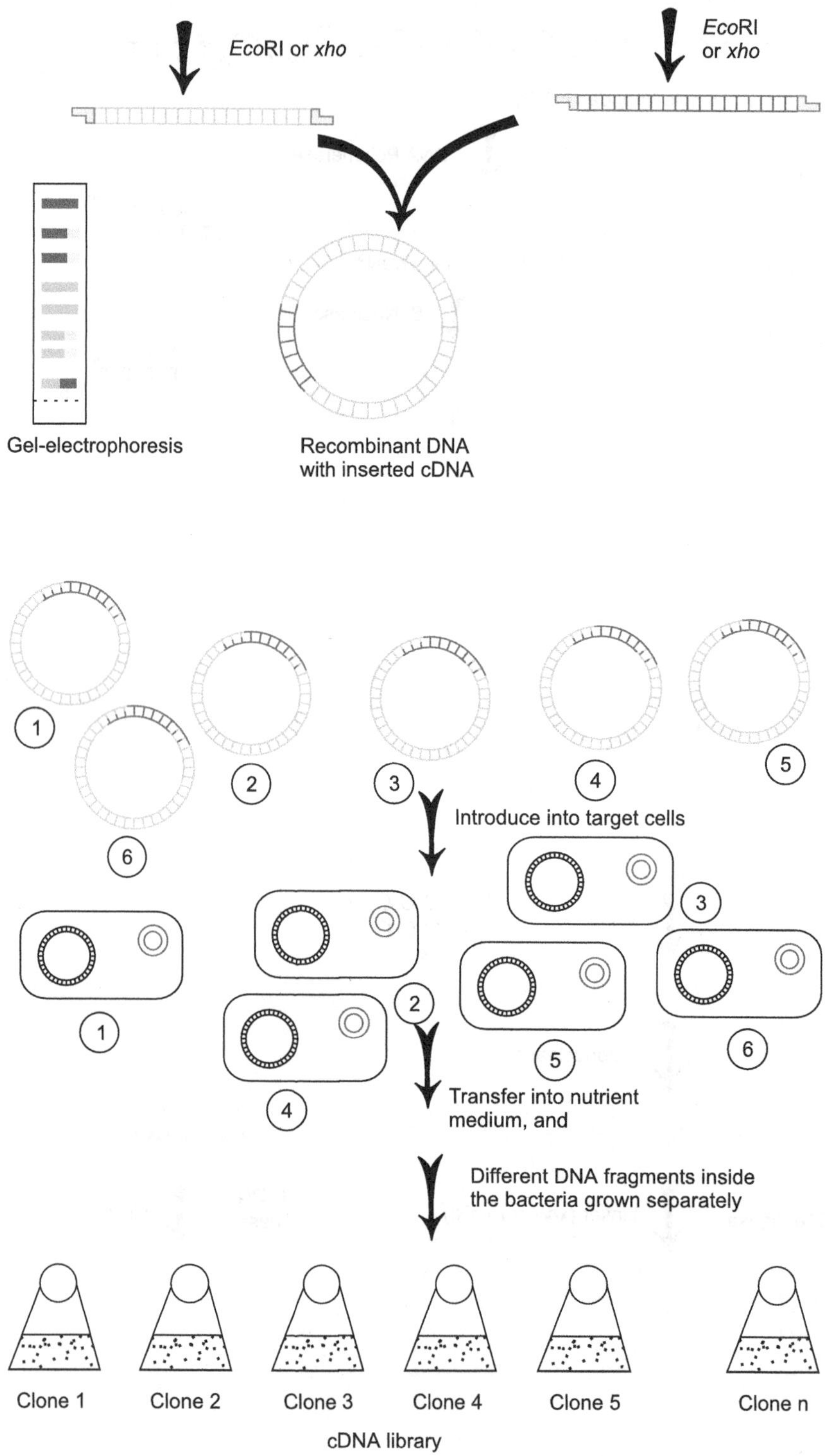

Fig. 12.19 Technique for the preparation of cDNA library.

Colony Hybridization Technique

The genomic/cDNA library can be perpetuated indefinitely and, whenever required, the desired DNA fragment, after identification by the **colony hybridization** method, can be retrieved using the same restriction enzyme that was used for making the recombinant DNA. In colony hybridization technique for screening of the desired clone from the genomic/cDNA library, following steps can be implemented:

(*i*) From the **master plate,** containing the colonies of transformed bacterial cells constituting the genomic/cDNA library, **replica plate** is made which is the impression of master plate with respect to the positions of colonies of various clones.

(*ii*) The **DNAs** of individual colonies are **denatured** (made single-stranded) *in situ* when DNA still inside the cells.

(*iii*) The denatured DNAs are then southern blotted on a **nitrocellulose membrane.**

(*iv*) The single-stranded DNA is then hybridized with a specific radiolabelled **probe** (small DNA or RNA molecules, generally single stranded) to identify the clone.

(*v*) Then it is **autoradiographed** by covering and exposing the membrane with an X-ray film in dark.

(*vi*) The identified colonies are then retrieved from the corresponding positions on the master plate (Fig. 12.20).

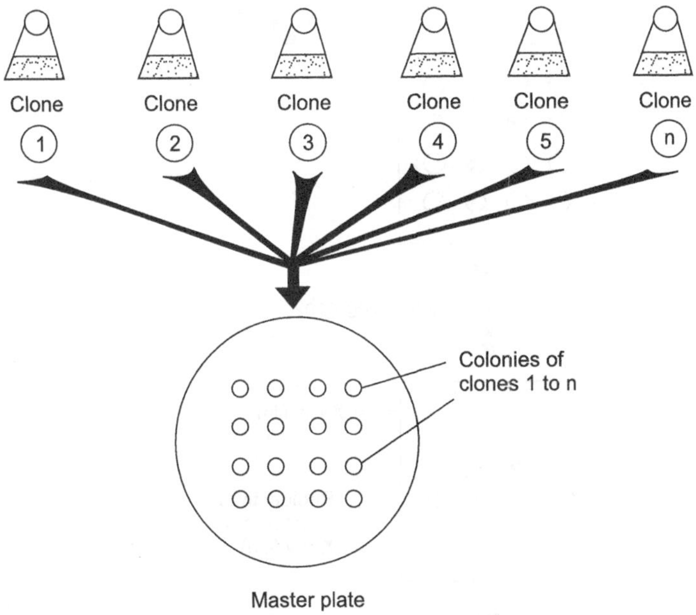

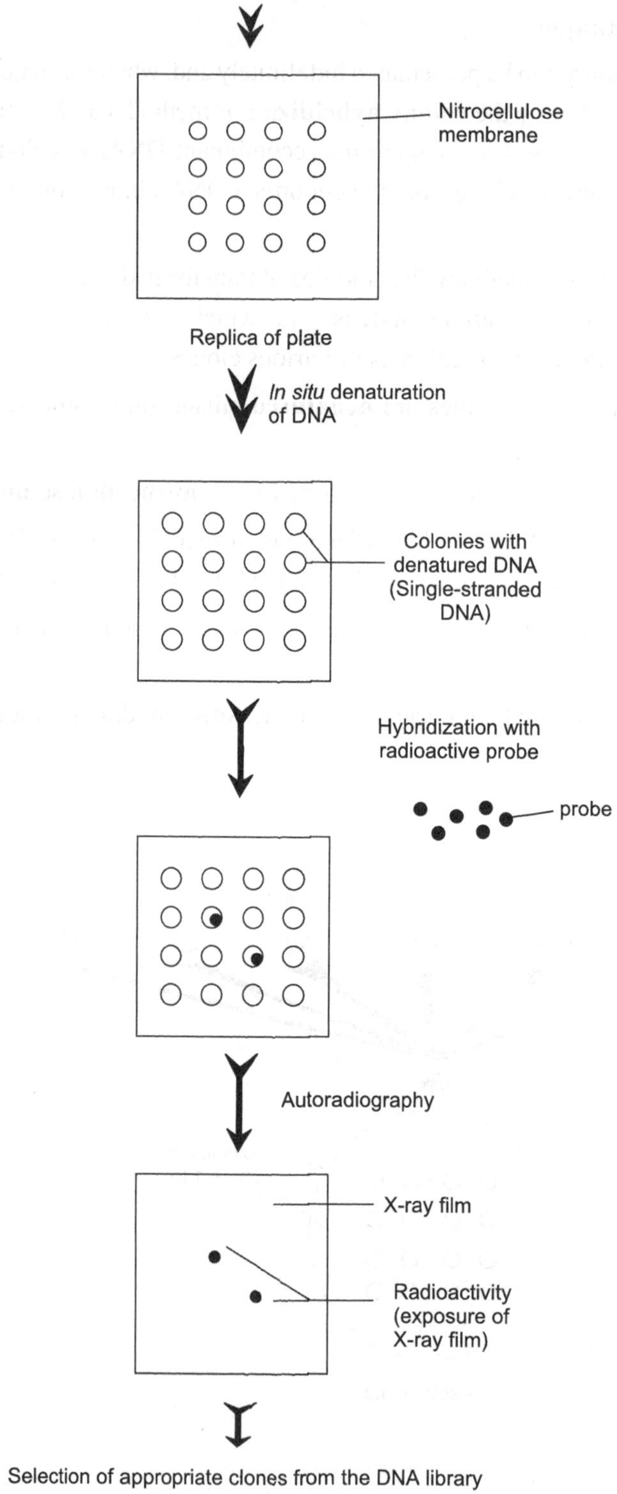

Fig. 12.20 Colony hybridization method for the selection of desired clones from genomic/cDNA library.

TRANSGENICS

Transgenic organisms contain an isolated or artificially synthesized foreign DNA or gene or extra copies of an endogenous gene (transgene) in their cell either as integrated into its genome or into a vector. Thus, they are genetically modified organisms. Foreign DNA has been successfully introduced in many microorganisms, plants and animals to utilize them for the welfare of human beings for producing desired phenotype or protein or to modify the existing biosynthetic pathway to get new product or to inhibit the expression of a gene.

1. Transgenic Microorganisms

Some important transgenic microorganisms are:

(*i*) *E. coli* containing human **insulin** gene has been used to produce insulin for the treatment of diabetes. Earlier, insulin was extracted from pigs and cattle which had some side effects due to additional contaminating compounds, whereas, the recombinant insulin, produced by recombinant *E. coli* introduced into the market in 1982, has no such problems as it is exactly similar to human insulin.

(*ii*) **Somatostatin,** a proteinaceous growth hormone, was introduced into the market in 1985. It is extremely difficult to isolate from the animals (about 0.005 g hormone is isolated from 5 lakh sheep brains). After cloning of the human growth hormone gene in *E. coli,* the yield has been increased many times (0.005 g/L bacterial fermentation medium). It is used for the treatment of hypopituitary dwarfism, which occurs due to growth hormone deficiency (with a frequency of about 1 child in 5000). The hormone improves the muscle growth and is exploited by some athletes.

(*iii*) Moreover, **interferons, blood clotting factor,** etc. are also produced by genetically engineered microbial cells or cultured mammalian cells. Interferons were discovered by two British researchers, **Issacs** and **Lindenmann** in 1957. They have been named interferons as they interfere with the virus infection by inducing degradation of viral mRNA in infected cells by the enzyme ribonuclease. Human interferons are glycoproteins (protein + sugar) and are produced by virus-infected cells to protect healthy cells. They control a large number of viral diseases, *e.g.,* common cold, cancer, etc. They have the potential application in cancer therapy as they attack cancer cells by inhibiting their growth (division) and also check the growth of viruses causing cancer virus, AIDS, hepatitis-B, etc. The protection conferred by them is non-specific (interferons induced by one virus can provide protection against other viruses also). They have not become the antibiotics of the viral infections because minute amounts of interferons are produced in the cells and, moreover, their extraction from other cellular proteins is complicated. Conventionally, they were obtained from the human blood. But currently, they are produced from cultured mouse fibroblasts or human leucocytes infected with Sendai virus and genetically engineered bacteria (*E. coli*) or yeast with inserted gene for human fibroblast interferon, which are not only more pure but much cheaper also (about 1/150 cost of the conventional method) than the conventionally obtained interferons.

(*iv*) **Recombinant vaccines** are produced through recombinant DNA technology by integrating the gene for the immunogenic proteins in *E coli,* yeast, cultured animal cells or transgenic plants. Here, DNA segment necessary for the production of either:

(*a*) **whole protein** (*e.g.,* coat protein gene for hepatitis-B has been cloned in yeast) or

(*b*) a **part of a protein** (*e.g.,* amino acid 114–160 of coat protein for foot and mouth disease virus) necessary for immunogenicity, is cloned.

But, these protein vaccines (second revolution in vaccines) are very costly and need to be stored at low temperatures. (First revolution in vaccine was the production of conventional vaccines. Third revolution in vaccine is the use of DNA in the form of vaccine that after injection into the blood expresses to produce necessary immunogenic protein.)

(*v*) *Saccharomyces cerevisiae* containing **dextrin-utilizin gene** from *Saccharomyces diasticus* by recombinant DNA technology (normally the yeast cannot utilize dextrins) increases the amount of fermentable sugars and, thus, greater amount of alcohol after the fermentation is produced. (Dextrins are small chains of glucose.)

(*vi*) **Rennet,** used in cheese industry, is traditionally obtained from the pancreas of calves but now can also be obtained from the genetically engineered microbes (*e.g., Escherichia coli*) containing the calf chymosin (rennin) gene.

2. Transgenic Plants

Man is dependent on agriculture for food and other requirements and many countries are still not self-sufficient in food production. Biotechnology is creating a revolution in this field by providing higher quality and lower cost products for continuously increasing human population. Since early times, man is trying to improve the quality and productivity of agriculturally important plants by **selection** of better varieties and employing the **traditional breeding programmes** involving sexual crosses, which are slow and difficult processes. Extensive research is also going on with forest trees, as forest resources are continually depleting because of indiscriminate deforestation. Trees have long generation times and desired traits can be achieved by the faster tissue culture and genetic engineering technologies.

The genetic engineering technology overcomes the limits of interspecific crosses and permits the introduction of foreign genes from distant species that is not possible following conventional breeding programmes. It is now possible to manipulate the genome of almost any plant species including agricultural crop plants by introducing advantageous new traits (by incorporating isolated gene from an organism or chemically synthesized gene) that either produces a new protein/other product or prevents the expression of an existing native gene. For this, many techniques can be employed.

One such important method is the use of *Agrobacterium tumefaciens* that has been used extensively to transfer the desired foreign DNA by inserting it into the Ti plasmid and then integrating into the **dicotyledonous** host plant genome. After screening of the desired recombinant cells, they are regenerated into complete plantlets, *e.g.,* Loblolly pine (used in forestry). Ti plasmids have also been used to insert 'antisense genes' to negate the functions of specific plant genes concerned with an undesirable trait. Such genes are produced by reversing the orientation of a gene in relation to its promoter (in DNA at promoter region, RNA polymerase binds to start the transcription giving rise to mRNA) so that a RNA, that is complementary to mRNA, is synthesized and after base pairing with the normal mRNA, double-stranded mRNA-antisense RNA is produced that is unable to be translated to give the protein.

This method of using *Agrobacterium tumefaciens* to manipulate the genome is unsuitable for monocotyledonous plants, like wheat, rice, maize, etc. In **monocotyledonous** plants, gene transfer is mostly achieved with the help of a gene gun by bombarding foreign DNA-coated gold or tungsten particles (micro projectiles) into the host plant cells. The foreign DNA along with microprojectiles penetrates the cell wall and the DNA is delivered into the nucleus. The recombinant cells can be regenerated into the whole plants.

The widespread use of insecticides, fungicides, herbicides (to kill the plant weeds) and pesticides has the damaging effect on the environment and human health. Many crop plants have not only been improved with respect to the pest and disease control but also to increase the yield and quality of the product by genetic engineering technology:

(*i*) **Resistance to Insect Pests:** Proteinacious toxin genes (*cry* genes) from *Bacillus thruringiensis* (Bt), that is toxic to larvae of certain insects, have been introduced into several crop plants so that they become resistant to that particular insects, *e.g.,* cotton, tomato. In India also, some insect resistant *cry* genes containing cotton varieties are grown in the fields.

(*ii*) **Resistance to Microbial Diseases:** Microbial diseases, particularly fungal and viral, cause huge losses and limit the agricultural productivity worldwide. Resistance to many viruses, by integrating the genes for viral coat proteins, have achieved in several crop plants, *e.g.*, rice.

(*iii*) **Resistance to Herbicides:** Herbicide-tolerant crop plants have been produced by genetic engineering the crop plant genomes that can reduce the application of herbicides, *e.g.*, potato, rapeseed, etc. This method of weed control is more effective, less costly and more environment friendly.

(*iv*) **Delayed Ripening Fruits:** Losses, during storage and transportation, of some crops are very high due to over-ripening and softening of fruits and vegetables. The ripening is due to some endogenous enzyme activities that can be genetically stopped or slowed down, *e.g.*, in tomato, the enzymes polygalacturonase breaks down the cell wall leading to softening of fruit during the ripening. Flavr Savr tomato has been engineered in USA by blocking the expression of polygalactarunase enzyme by introducing the antisense gene for this particular enzyme. This tomato has improved flavour and shelf life and is now marketed in USA (Fig. 12.21). Such techniques are now tried to be used in a wide variety of soft fruits. Other studies are considering the inhibition of ethylene synthesis for controlling the fruit ripening (ethylene is a plant hormone that induces fruit ripening).

(*v*) **Stress Resistance:** Cold resistant tobacco plant has been produced by introducing the gene for glycerol-1-phosphate acyl transferase from the plant *Arabidopsis*.

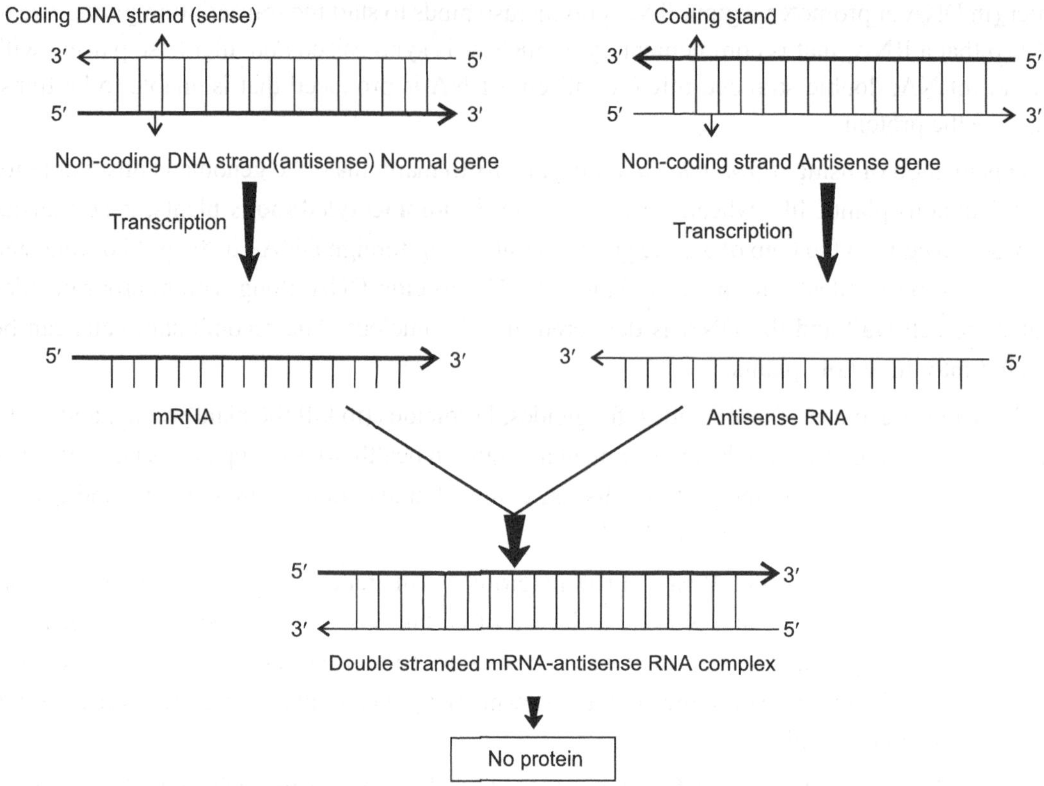

Fig. 12.21 Introduction of antisense gene for polygalacturonase enzyme for the development of Flavr Savr tomato.

Despite their beneficial effects, it has been argued that transgenic plants may be harmful to the environment as:

(*i*) There may be possibility of transfer of transgene (foreign gene) from transgenic crops to their wild relatives through pollens making them more persistent and damaging. To prevent this, transgenics should not be grown near the occurance of their wild relatives.

(*ii*) Transgenic crops may themselves become persistent weeds.

(*iii*) They may damage the environment in some unknown manner.

3. Transgenic Animals

Though transgenics have been produced in a variety of animals, most of the activities are at the experimental stage and have not commercially utilized yet. Some important transgenic animals are:

(*i*) In some animals, genes have been transferred and expressed into cultured cells to obtain important proteins, like **human growth hormone** from genetically modified mouse mammary tumour cells, **blood clotting factor VIII,** etc.

(*ii*) Genetic modification of some animals may increase the **milk** (*e.g.,* transgenic cow), **meat** (*e.g.,* by introducing growth hormone gene in fish, goat, cow, pig, etc), **wool** (*e.g.,* in sheep, etc.) production.

(*iii*) In some animals (*e.g.*, transgenic rabbits, cows, goats, etc.), important genes have been transferred to get large amount of rare and expensive pharmaceutical or biologically important proteins in their milk, urine or blood. This technology is known as **molecular farming.** Generally, the transgenes are desirable to be expressed in mammary tissue so that the protein is secreted in their milk from which it can be easily isolated.

DNA FINGERPRINTING

DNA fingerprint of an individual is the **Southern blot** of his DNA that is digested with a restriction endonuclease and hybridized with a specific radioactive probe (generally DNA). This technique, for the first time, was developed by **Alec Jeffreys (1985-86)** and colleagues in U.K. and is extremely reliable in comparison to the conventional analysis techniques of serum proteins, erythrocyte antigens (A, B, O blood group), etc. It is used with almost absolute certainty for many purposes, *e.g.*,

(*i*) Identification of **criminals** through blood stains, semen stains, hair roots, etc.,

(*ii*) Solution of dispute cases of **parentage** (earlier identified by blood groups that has many limitations)

(*iii*) Identification of disputed **immigrants** by DNA data bank that has been or is being prepared throughout the world, etc.

(*iv*) Matching of donor bone marrow cells with that of the bone cancer patient in **bone marrow transplants.**

(*v*) Detection of trait markers.

(*vi*) **Identification** of plant varieties and microorganisms for patents.

For DNA fingerprinting:

(*i*) The **DNA** samples, usually from the blood of suspected person and his or her close relatives for comparison, are obtained. DNA is also extracted from the blood stains on clothings of the murder victim, the semen from vagina of rape victims, years old bone marrow of murder victims (as in Nithari case), etc. recovered from the evidence of crime, that can be amplified by PCR (Polymerase Chain Reaction) technique, if small in quantity.

(*ii*) These DNA samples are separately digested with a suitable **restriction enzyme** that cleaves the DNAs at specific sites to produce DNA fragments of various lengths.

(*iii*) The DNA fragments thus obtained are loaded in different wells along with the standards (molecular weights of which are known and the kits are available in the market) in agarose gel slab and subjected to **gel-electrophoresis.**

(*iv*) The DNA fragments of various bands are denatured (made single-stranded) and transferred onto a nitrocellulose membrane by **Southern blotting.** Thereafter, the DNAs are fixed by baking the nitrocellulose membrane at 80°C.

(*v*) The single-stranded DNA fragments are then hybridized with the appropriate radioactive **probe** (generally single-stranded DNA) that is usually prepared from mini-satellite or micro-satellite DNA (VNTR or variable number of tandem repeats), *e.g.*, a universal probe, bkm (banded karait microsatellite DNA) probe, has been developed from the sex chromosome of banded karait that is useful in human beings, plants, etc.

(*vi*) After washing off of the free probe, the hybridized DNA fragments with the probe (due to complementary base pairing) are detected by **autoradiography,** where the nitrocellulose membrane is covered with a photographic (or X-ray) film in dark leading to its exposure at the locations of radioactive probe.

(*vii*) In DNA fingerprint, the **bands** of suspected criminal and that of the test DNA from the evidence of crime (blood stain, semen stain, etc.) thus obtained are **compared.** If the suspected person is involved in crime, the bands match perfectly. In case of disputed parentage, the DNA fingerprints of the child, the mother and the suspected father are compared where some of the band of child match with the bands of mother, whereas, rest with the father (Fig. 12.22 and 12.23).

The possibility of two persons having same DNA fingerprints is very rare.

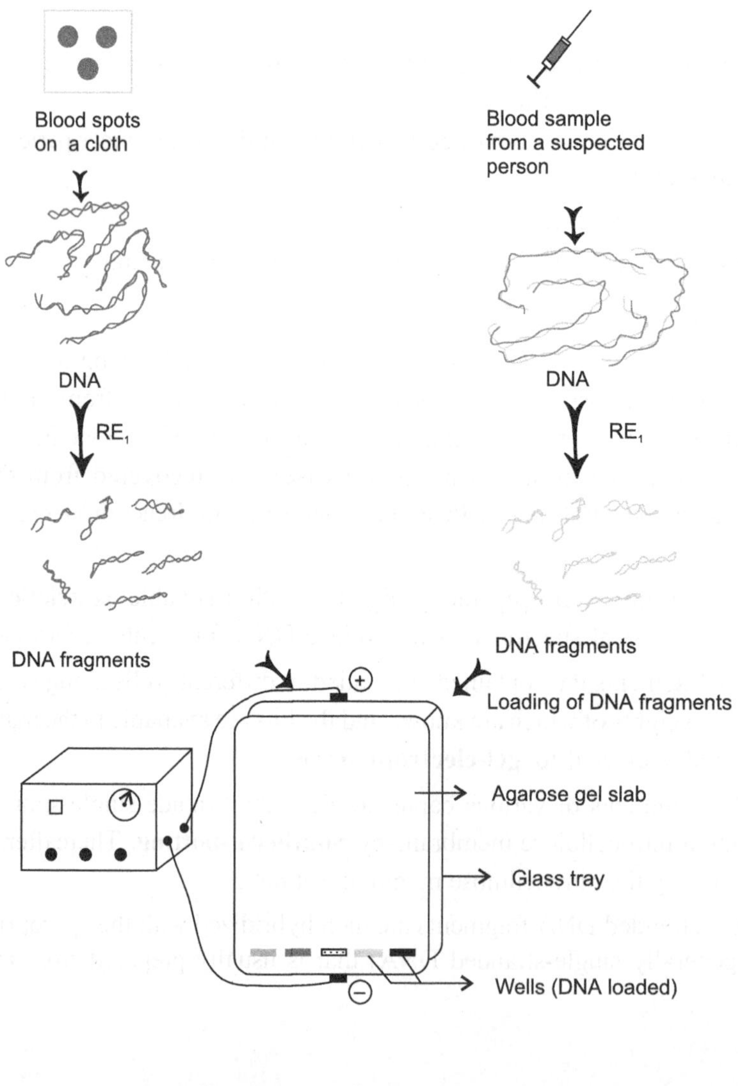

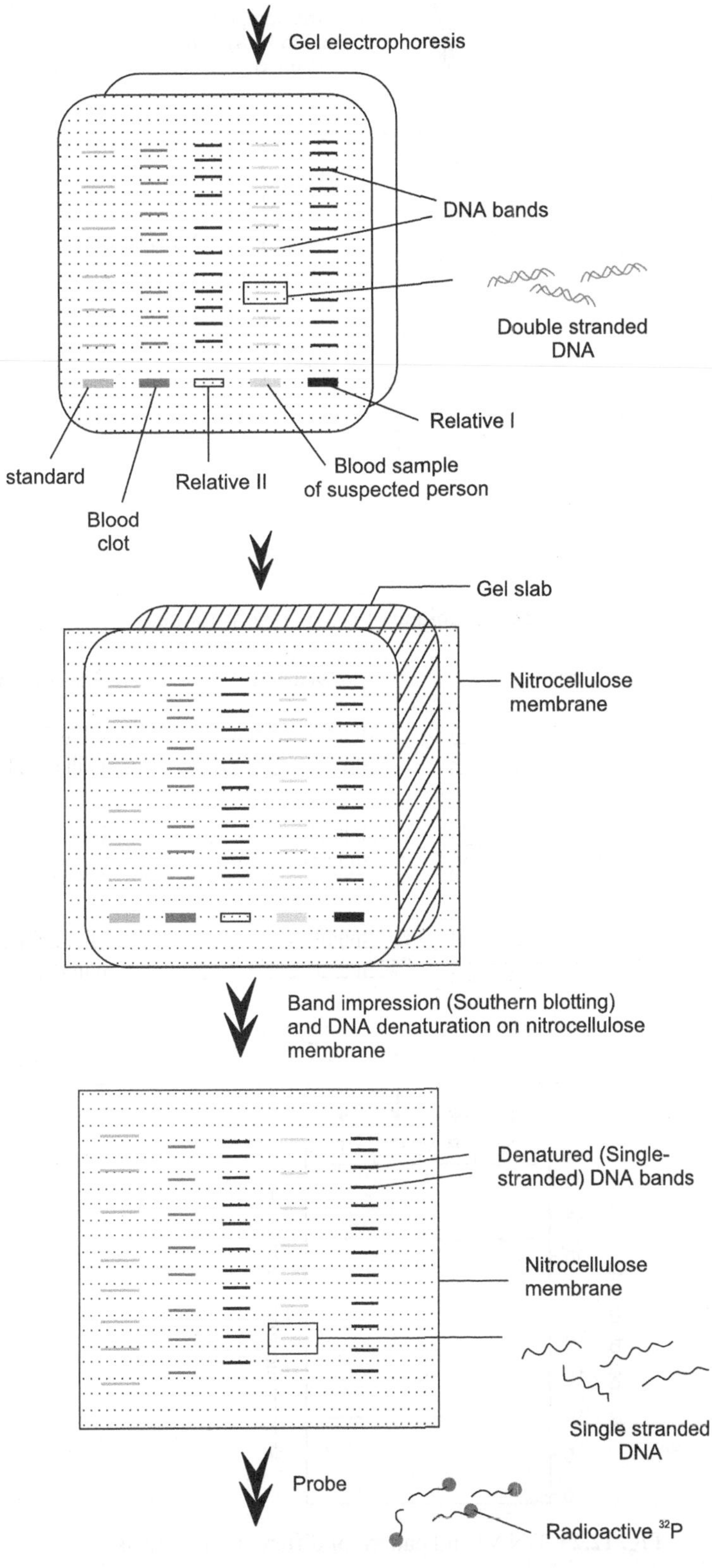

Gel electrophoresis
DNA bands
Double stranded DNA
Relative I
standard
Relative II
Blood sample of suspected person
Blood clot
Gel slab
Nitrocellulose membrane
Band impression (Southern blotting) and DNA denaturation on nitrocellulose membrane
Denatured (Single-stranded) DNA bands
Nitrocellulose membrane
Single stranded DNA
Probe
Radioactive ^{32}P

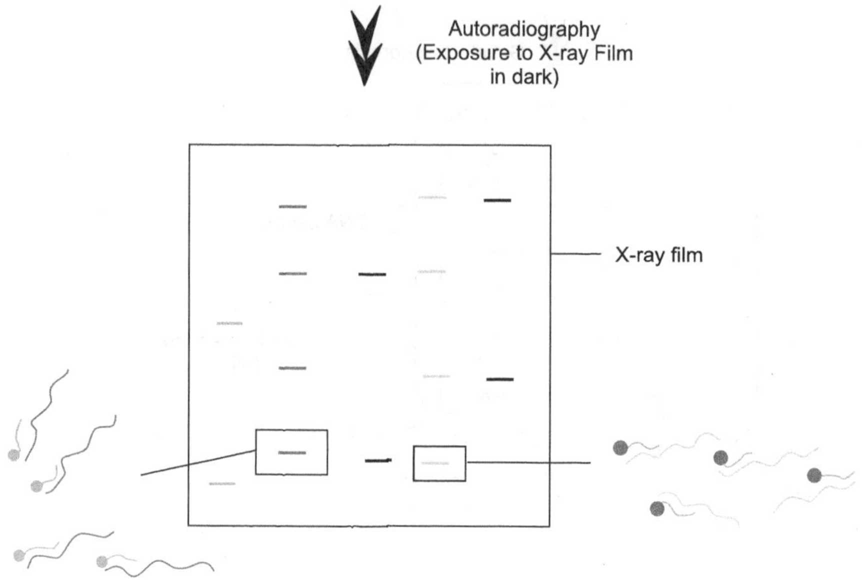

Fig. 12.22 Technique of DNA fingerprinting.

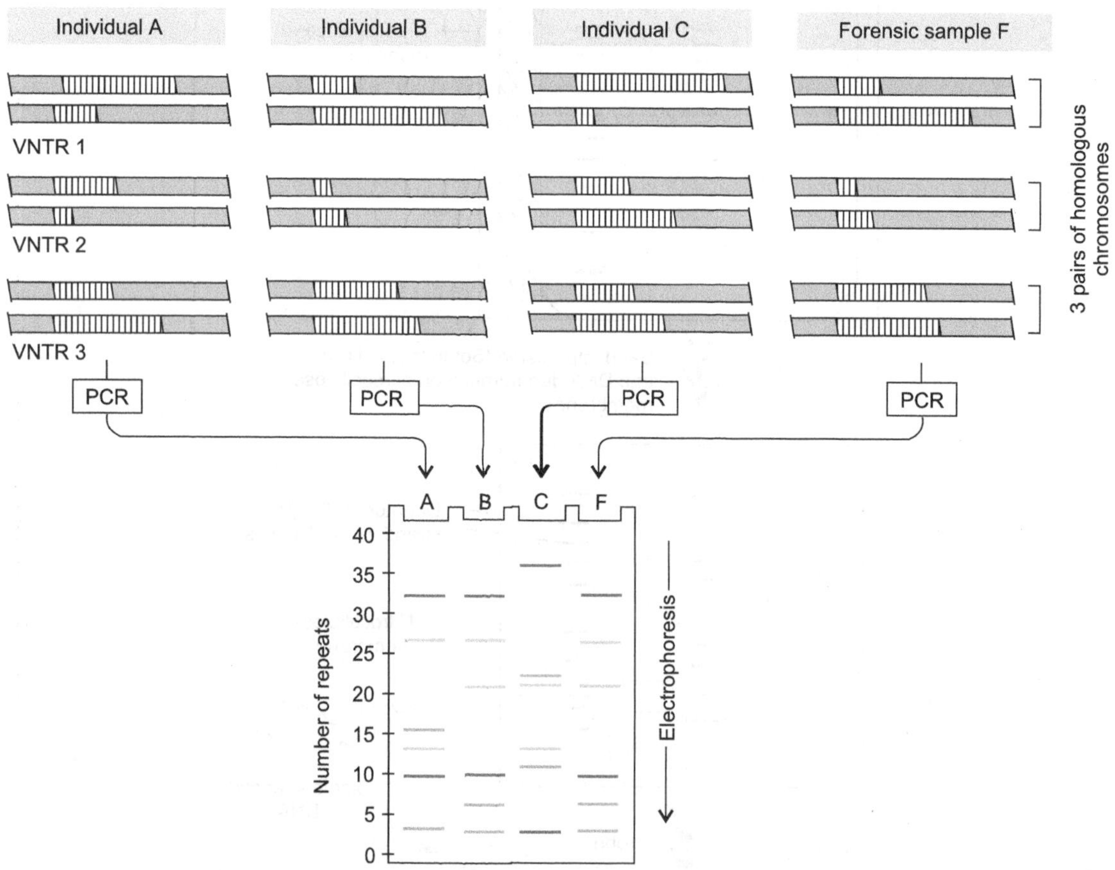

Fig. 12.23 DNA band pattern of different individuals.

GENOMICS

Genome word indicates the sum total of the genes present on the whole set of chromosomes in a cell of an organisms. **Genomics** is the study of structure and function of genome of an organism, mainly by constructing its genome map, sequencing of the genome and deciphering the function of each gene. **Genome map** is the description of structural and functional organization of the genome of an organism. The genome maps are mainly of two types:

1. **Genetic Maps (Linkage Maps):** It is prepared by measuring the recombination frequency between genetic markers (*e.g.,* **mutation sites** on the DNA) after crossing different mutants. On the basis of these recombination frequencies, genetic markers are arranged in a definite order on the DNA. The recombination frequency is directly proportional to the distance between the markers, *i.e.,* higher the frequency longer is be the distance between the two markers.

   In *Drosophila* (Fig. 12.24), certain recessive *lozenge* mutants are available that lead to the development of smaller and more elliptical eyes. The genetic map of *lozenge* **locus,** found on the X-chromosome of *Drosophila,* has been prepared on the basis of recombination frequencies obtained by crossing the 14 available mutants on this locus (Fig. 12.25). The map has 4 mutational sites, which can mutate in various ways and are separated by 0.045–0.06% recombination frequencies. The mutants belonging to the same mutational site do not recombine.

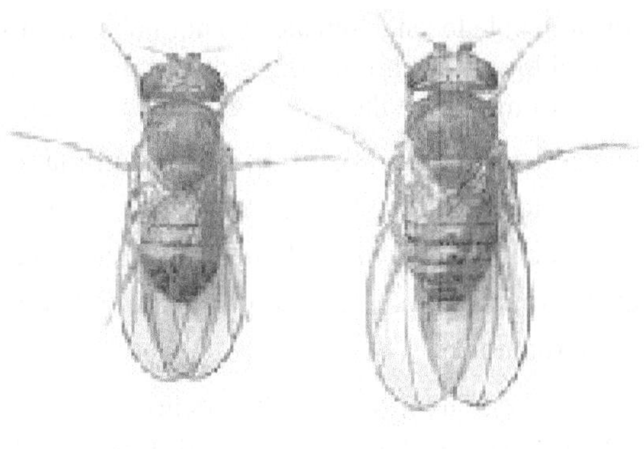

Fig. 12.24 Male (left) and female (right) *Drosophila* flies.

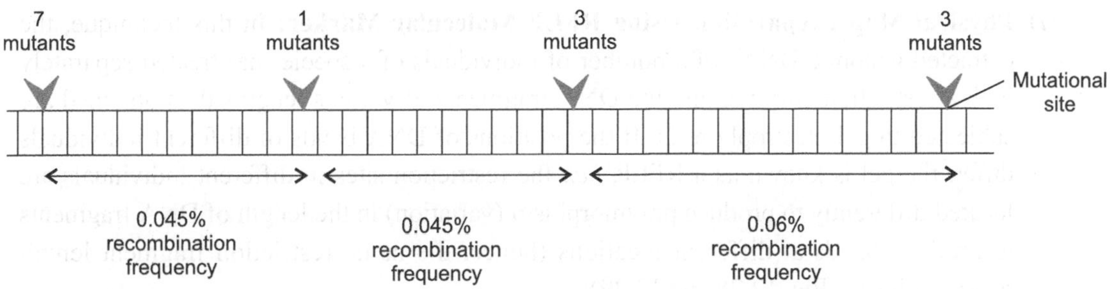

Fig. 12.25 Recombination map of *lozenge* locus, located on X chromosome of *Drosophila*.

Thus, two lozenge mutants belonging to two different (adjacent) mutational sites may recombine to produce wild (normal) phenotypes with a frequency of 0.045-0.06% (12.26 and 12.27).

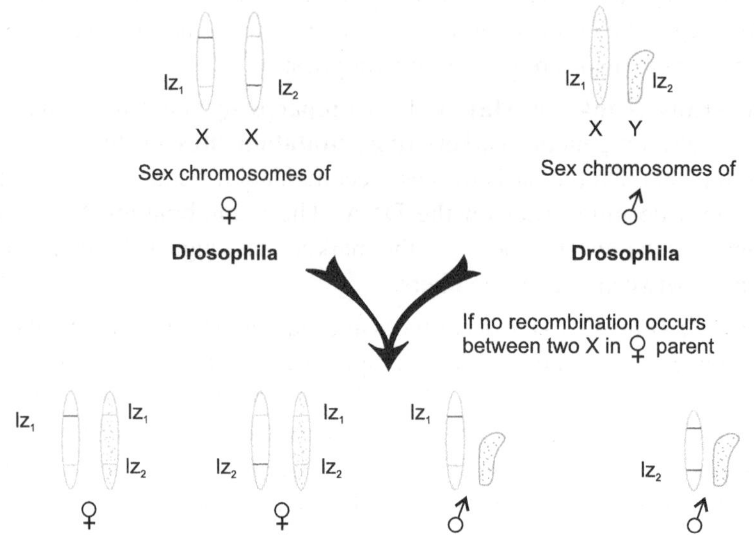

100 % mutant progeny

Fig. 12.26 A cross of double lozenge female and normal male of *Drosophila* when the two homologous X-chromosomes do not recombine during Prophase I of meiosis at the time of gamete formation.

The main problem of genetic map preparation by recombination frequency technique is the unavailability of sufficient number of genetic markers to cover the entire genome, thus, some parts of DNA cannot be mapped and gaps are created. Moreover, the distance between the two genetic markers may not exactly match the actual distance, thus, giving an idea of only relative distances.

2. **Physical Maps:** These maps are prepared by locating the more abundant **molecular markers,** *e.g.*, **RFLP** (Restriction Fragment Length Polymorphism), **RAPD** (Randomly Amplified Polymorphic DNA), that cover almost the entire genome of an organism. Moreover, they give the actual distances between the molecular markers.

(*i*) **Physical Map Preparation Using RFLP Molecular Marker:** In this technique, the extracted genomic DNAs of a number of individuals of a species are treated separately with a restriction enzyme, and the DNA fragments of various lengths thus obtained are subjected to gel-electrophoresis. If the positions of DNA bands of different individuals differ, the gel is known as a RFLP, *i.e.,* the restriction sites in different individuals are located differently to produce polymorphism (variation) in the length of DNA fragments to produce bands at different locations (hence, the name restriction fragment length polymorphism) (Fig. 12.28 and 12.29).

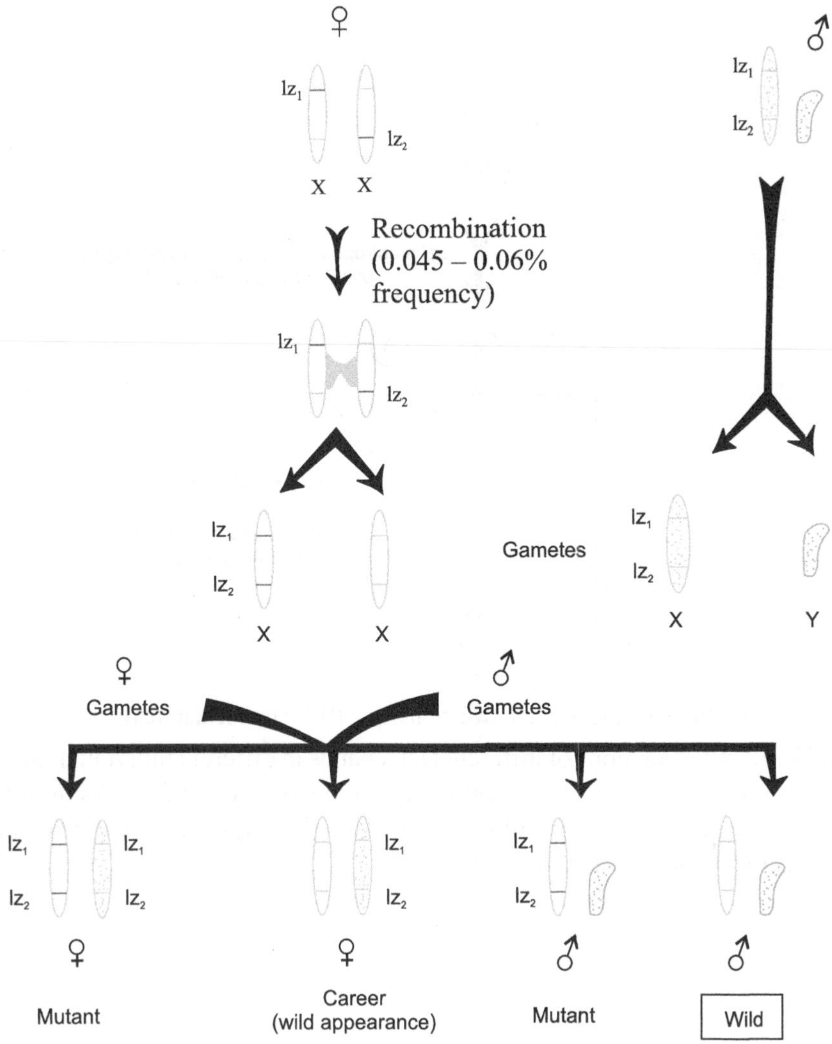

Fig. 12.27 A cross of double *lozenge* female and normal male of *Drosophila* when the two homologous X-chromosomes recombine during Prophase I of meiosis at the time of gamete formation to give 0.045-0.06 % wild progeny.

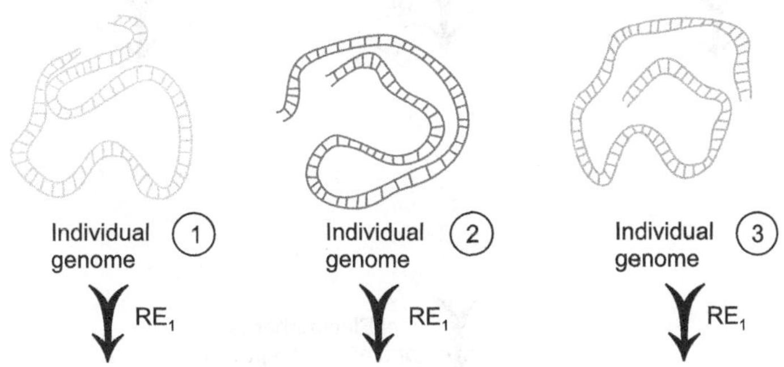

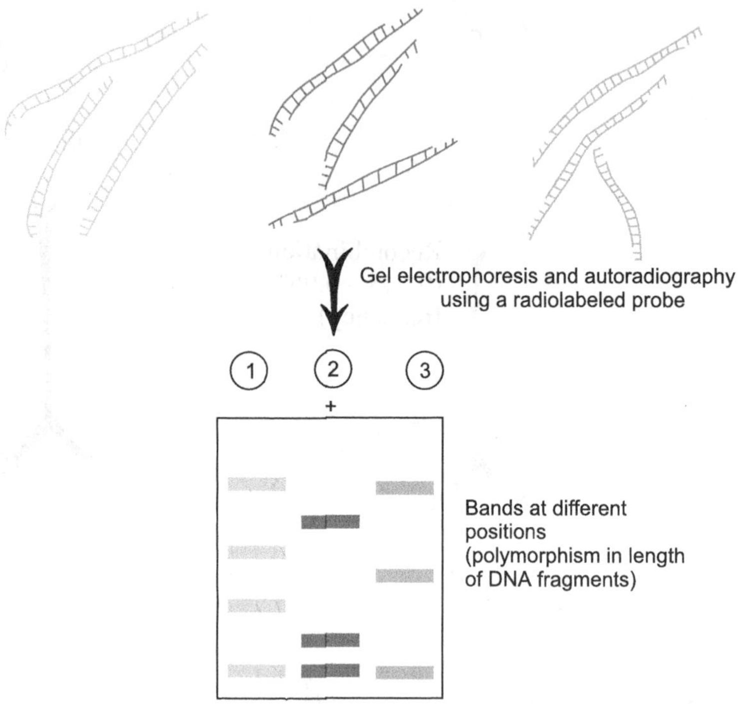

Fig. 12.28 Genomic map construction using RFLP molecular marker.

But, if the relative positions of different DNA bands in different individuals are same, then the gel is not a RFLP, *i.e.*, the restriction enzyme would cut the DNAs of different individuals at the same relative positions so that no polymorphism would be obtained in the DNA fragments of different individuals.

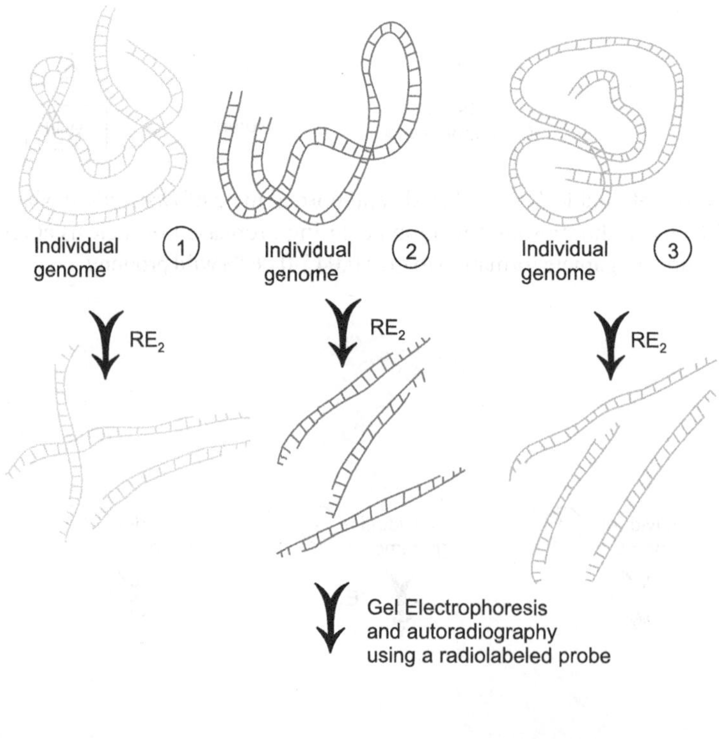

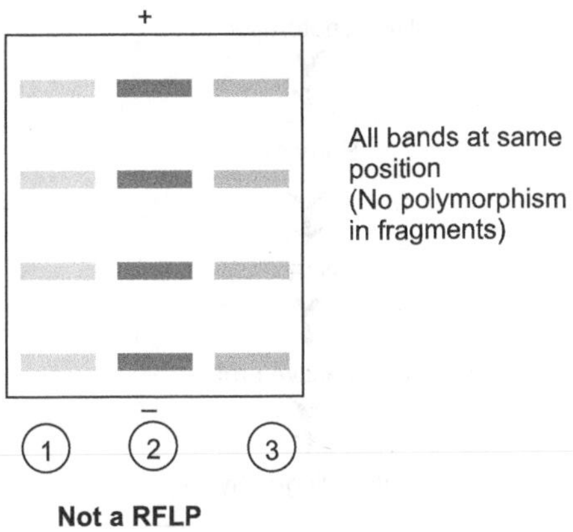

Fig. 12.29 DNA fragments of the same relative lengths in various individuals do not give RFLP.

Thus, the **restriction sites** (molecular marker) of a number of restriction enzymes (used one at a time) are mapped (located) on DNA of the species (Fig. 12.30).

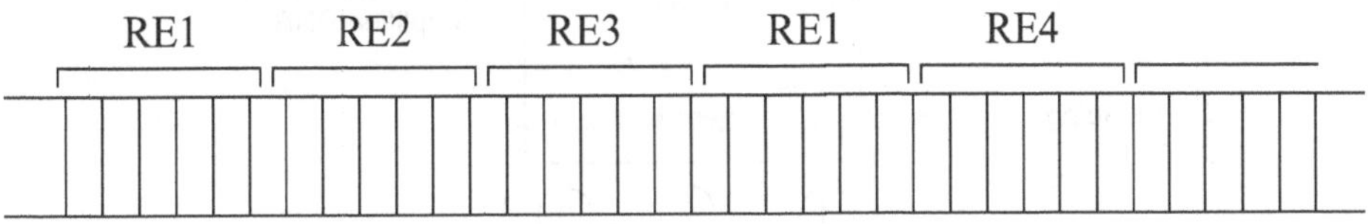

Fig. 12.30 Restriction map prepared using RFLP marker.

(*ii*) **Physical Map Preparation Using RAPD Molecular Marker:** In this method, the extracted genomic DNA of different individuals of a species are amplified separately by PCR (Polymerase Chain Reaction) technique using random **primers** (molecular markers), one at a time, the sequence of which are already known. If a particular primer finds its complementary sequence in the genome at any location, the genomic DNA would be amplified to give a single band for an individual after gel-electrophoresis. The location of the bands of amplified DNAs of different individuals may differ to give polymorphism in the amplified DNA (hence the name randomly amplified polymorphic DNA) (Fig 12.31 and 12.32).

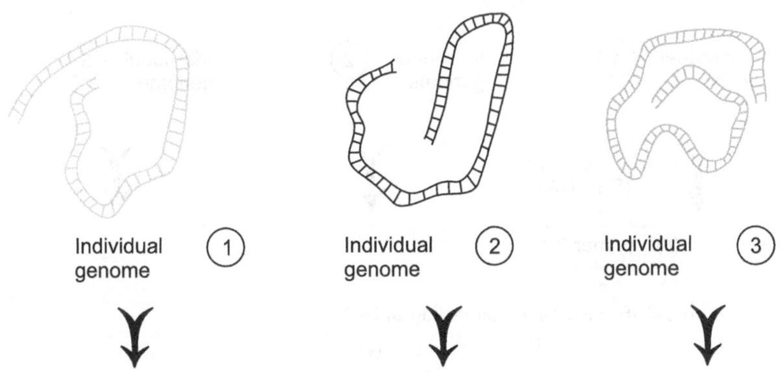

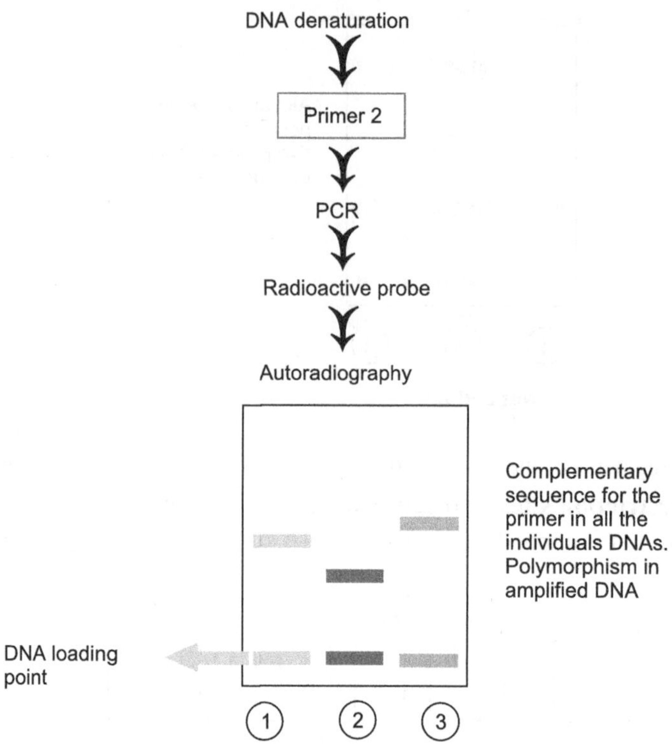

Fig. 12.31 Technique of genomic map construction using RAPD molecular marker.

But, if the primer does not find its complementary sequence anywhere in the genomic DNA, no band would be obtained as the genomic DNA would not be amplified.

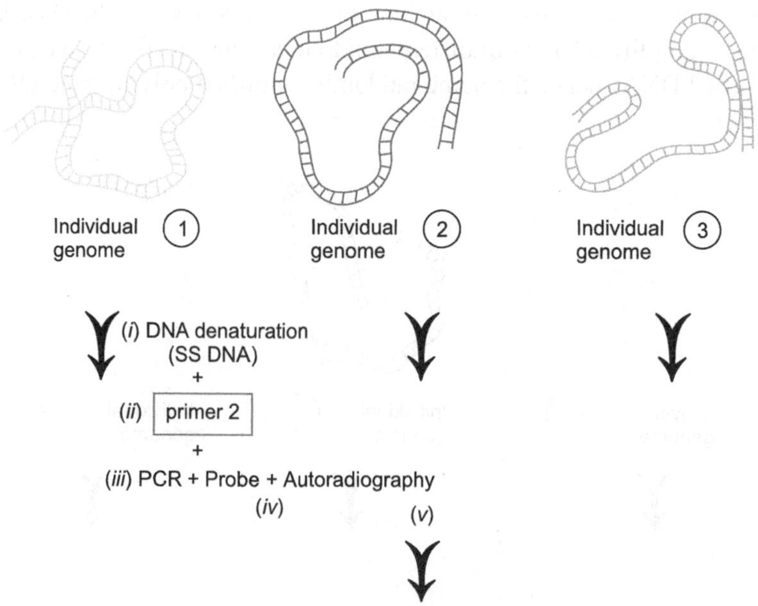

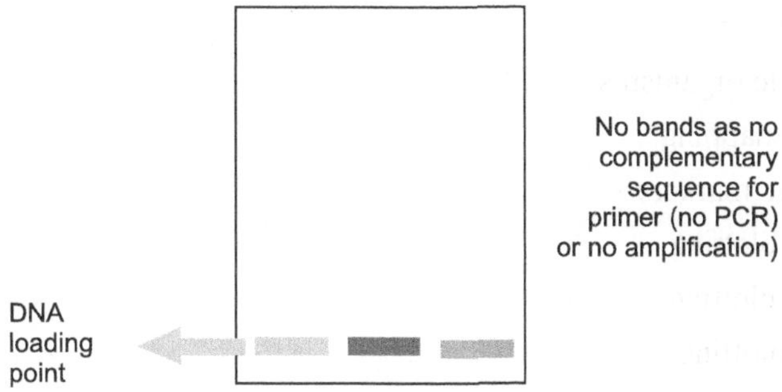

Fig. 12.32 Gel electrophoresis of PCR product of different individuals if a primer does not find any complementarity in their genomic DNA.

Thus, the sequences of a large number of random primers are located on the DNA to form the physical map of a species (Fig. 12.33).

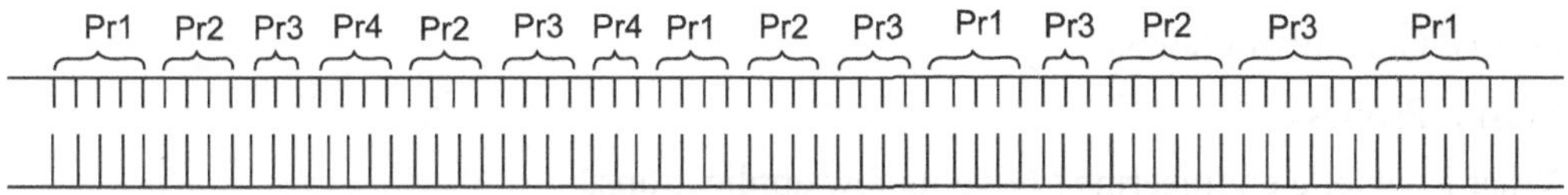

Fig. 12.33 Physical map prepared using RAPD molecular marker.

In case of human beings, genetic maps are constructed using family pedigree data (instead of crossing different human mutants to get the recombination frequency) and various molecular markers. The **human genome** project began in 1990 and it has been found that, though, the human genome is expected to contain about 100,000 genes due to its larger size, actually, it contains only about 30,000-40,000 genes located on 23 different chromosomes (haploid set) having about 3×10^9 bases. Many genes are split genes with non-coding introns, and near the centromeres the telomeres (ends of the chromosomes) the DNA contains 5-8 base pair long repetitive sequences. The mapping of human genome would help in understanding and treatment of many deadly genetic diseases.

QUESTIONS

1. Explain briefly the following:
   (*i*) Genetic engineering
   (*ii*) DNA finger printing
   (*iii*) Gene cloning
   (*iv*) Organismal cloning
   (*v*) Dolly—the first true mammalian clone
   (*vi*) Cell cloning
   (*vii*) Vectors
   (*viii*) *Agrobacterium*
   (*ix*) Human genome project

(*x*) Genomic DNA library
(*xi*) cDNA library
(*xii*) Transgenic organisms
(*xiii*) Genome mapping
(*xiv*) Molecular markers
(*xv*) Southern blotting
(*xvi*) Northern blotting
(*xvii*) Western blotting
(*xviii*) Colony hybridization
(*xix*) Dot blot and slot blot
(*xx*) Artificial synthesis of yeast alanyl tRNA gene
(*xxi*) Restriction enzymes
(*xxii*) Reverse transcriptase
(*xxiii*) Gel electrophoresis
(*xxiv*) Probes

2. What is genetic engineering? Give its few applications.

Or

What is recombinant DNA technology? Discuss its applications.

3. Describe briefly the various tools of genetic engineering or recombinant DNA technology.
4. What is cloning? Discuss the proposed applications in gene cloning.

Or

What is cloning? Give some examples of mammalian cloning.

5. What are transgenic organisms? Give some examples of transgenic animals, plants and microorganisms.
6. What are genomic and cDNA libraries? List the differences between the two libraries.
7. Discuss various methods for the identification of incorporation of foreign DNA in a cell or organism (transgenics).
8. Describe different ways for foreign DNA (or gene) transfer in plants and animals.

Chapter 13 NATURE AND SCOPE OF BIOTECHNOLOGY

Biotechnology includes all industrial processes mediated by living organisms at some step or the other. Microorganisms can be exploited for the production of industrially important biochemicals, like, various enzymes, pharmaceuticals, hormones, antibodies, which were earlier available only by sacrificing the animals. Biotechnology is also being used for a variety of crop plants and domestic animals improvement programmes throughout the world.

The word **biotechnology** is derived from two words—biology and technology, and is concerned with the exploitation of biological agents (*e.g.*, microorganisms, cultured animal or plant cells) or their components for generating useful products or for beneficial use by human beings. Although, the term biotechnology is of recent origin, man is employing microorganisms from the ancient times for making wine, vinegar, curd, leavened bread, etc., and it is known as **old biotechnology.** Production of the antibiotic, penicillin, by the fungus *Penicillium notatum* was discovered in 1930 by Alexander Fleming, and its commercial production began in 1944.

Using the sophisticated recombinant DNA technology, man is able to modify many microorganisms and some macro-organisms for creating valuable and naturally nonexistent capabilities in them, and they are employed to generate valuable products; this field is called **new biotechnology.** Biotechnology has a marked influence on human welfare, which ranges from food processing, protecting the environment, to human health, and has led to the emergence of many biotechnology companies throughout the world, like Genetech Inc. (USA), Biogen (Switzerland and USA) etc., which make use of the gene technology and, thus, 21st century would be the century of biotechnology. Some of the important areas in which biotechnology is making marked contributions are:

1. Medicines

(*i*) **Development of Vaccines:** Vaccines can be prepared from the animals by introducing isolated antigens into the blood of animals; and in conventional methods separation of these **antigens** sometimes poses the risk of contamination and may be hazardous for health. Vaccines against many diseases have been produced by cloning the genes of immunogenic proteins in suitable bacteria, *e.g.*, rabies, foot and mouth disease in cattle, typhoid, cholera, hepatitis B, malaria, etc.

(*ii*) **Valuable Drugs:** Many of the drugs (*e.g.*, insulin for treatment of diabetes and **interferons** for the treatment of some tumour viruses) were obtained only by sacrificing the animal life. But, now such drugs can be manufactured in large quantities by cloning the corresponding genes in bacterial cells. A number of other proteins, like, **blood clotting factor VIII** (for treatment of haemophilia), **human growth hormone** (for treating hypopituitary dwarfism) etc., have been synthesized by using recombinant DNA technique.

(*iii*) **Diagnosis of Diseases:** For diagnosis of different diseases, biotechnology has provided very effective and sensitive tools, like **monoclonal antibodies** and DNA **probes,** which are produced by genetically engineered microorganisms. Monoclonal antibodies through serological tests require very small time, some minutes, as against the conventional methods, which sometimes may take several days, *e.g.*, culturing of bacteria. DNA probes are even more sensitive than monoclonal antibodies and would take hours to test diseases, like, malaria, kala-azar, sleeping sickness, etc.

(*iv*) **Gene Therapy:** The techniques of gene therapy to cure genetic diseases, *e.g.*, Huntington's chorea, cystic fibrosis, are in advanced stages of development.

(*v*) **Other Notable Achievements** in the field of medical biotechnology are — birth of babies of specified sex by artificial insemination with X or Y chromosome-containing sperms prepared by sperm separation techniques, identification of parents or criminals using DNA fingerprinting which is a very accurate and reliable method and can be detected successfully from even blood or semen stains, hair roots, etc.

2. Agriculture

(*i*) **Cloning of Nitrogen Fixation (*nif*) Gene:** Nitrogen is needed as a fertilizer for the growth of crop plants. Though, plenty of nitrogen is available in atmosphere (about 78%) as nitrogen gas, but the crop plants cannot utilize it. There are some microbes that are capable of converting (fixing) this atmospheric nitrogen into ammonia, which is used by the crop plants. Cloning of the *nif* gene from a N_2-fixing bacterium, *Klebsiella pneumoniae,* is achieved in a non-N_2-fixing bacterium, *Escherichia coli,* which conferred it the N_2 fixation capacity. Efforts are also being made to change the genetic system of cereals (*e.g.*, rice) so that they may have artificially introduced *nif* genes, making them able to fix atmospheric nitrogen.

(*ii*) **Transfer of Resistance against Insects, Herbicides, Viral, Bacterial and Fungal Pathogens:** Insect-resistance gene ***bt*2** from a bacterium *Bacillus thuringiensis*, has been successfully attempted to develop transgenic plants, *e.g.,* bt cotton, which are being field tested and may be released for commercial cultivation or are planted commercially. In tobacco, TMV (a virus)-resistance has been achieved through cloning of virus coat protein genes. Several examples are now available where genes imparting resistance against fungal pathogens (*e.g. Phytophthora*) have been transferred to many crop plants (*e.g.,* potato). Glyphosate (a herbicide)-resistance gene from the bacterium *Salmonella typhimurium* was transferred to tobacco through the vector *Agrobacterium tumefaciens* (a bacterium).

3. Industry

Microorganisms have been employed since several decades for the large scale production of a variety of biochemicals—ranging from alcohol to antibiotics (*e.g.,* penicillin, streptomycin, erythromycin, etc.,) produced by various fungi, bacteria and actinomycetes as secondary metabolites, production of **single cell proteins** (SCP, total microbial biomass freed from contaminants, toxins, etc.,) from bacteria, fungi or algae (*e.g., Chlorella*, Dabur Company is making *Spirulina* tablets by the name of "Sunova") to supplement human food. **Fuel** in the form of biogas, produced mainly from cow-dung and other cheap substrates through microbial fermentation is quite popular in India.

4. Environment

Bacteria are being utilized for **detoxification** of industrial wastes, degradation of petroleum and management of oil spills (*e.g., Pseudomonas putida*). The use of pesticides, which cause pollution, can be avoided by **biocontrol** of plant diseases and insect pests by using environment friendly viruses, bacteria, fungi, etc.

Biotechnology seems to have an unlimited commercial potential to generate an unlimited range of valuable and useful products. As a consequence, Government of India has constituted the National Biotechnology Board (NBTB) to encourage research activities in biotechnology. In 1986, Department of Biotechnology was created under the Ministry of Science and Technology. In addition, Indian Council of Agricultural Research (ICAR) and Council of Scientific and Industrial Research (CSIR) are also promoting research activities in this field.

5. Enzyme Technology

Enzyme technology utilizes purified enzymes for generating useful products and services, *e.g.*, industrial processes, medicine, research (*e.g., Taq* DNA polymerase obtained from the thermophilic bacterium, *Thermus aquaticus,* is used in PCR), recombinant DNA technology, etc. Though, purified enzymes are costly, they do not produce any unwanted byproducts so that the purification of the product becomes easier. Most of the enzymes are used in **food** and **detergent** industries. Sometimes, the byproducts produced by whole cells add to the quality of the product, *e.g.*, wine and beer. Most of the enzymes are obtained from microorganisms (*e.g.* amylase from the bacterium *Bacillus* and fungus *Aspergillus*, rennet from the fungus *Mucor miehei*, etc.) as it is easier to grow them in the fermenters and to manipulate them genetically. Some enzymes are obtained from higher plants (*e.g.*, papain from papaya latex, amylase from malted barley) and animals (*e.g.*, rennet containing chymosin enzyme from the pancreas of calf).

6. Food and Beverages

Food industry is the largest industry in world and food biotechnology includes improvement in taste, consistency, colour, nutrition, safety and preservation of food. Developments in food preservation methods have made many of the seasonal foods to be available all the year round. Fruits, vegetables, cereals, meats, etc., require some degree of processing, and the relatively bulky raw agricultural products are transformed into stable, convenient and palatable foods and beverages.

Fermented foods (*e.g.*, bread, cheese, yoghurt, sauerkraut, soy sauce, tempeh, etc.,) and beverages (*e.g.,* beer, wine, brandy, whisky, tea, coffee, cocoa, etc.,) result from the action of microorganisms or enzymes on a wide range of agricultural and industrial materials with desirable biochemical changes. Fermented foods were developed before recorded history and are usually more nutritious, more digestible, with improved flavour and toxicologically and microbiologically safer. Almost 90% of revenue from biotechnology comes from the fermentation products.

QUESTIONS

1. What is biotechnology? Discuss the applications of biotechnology in the field of environment and agriculture industry.
2. **Write an essay on the 'Nature and Scope of biotechnology'.**
3. Define biotechnology and explain the meaning of old and new biotechnologies.

   Comment on the multidisciplinary nature and the commercial potential of biotechnology.
4. **Discuss the scope and importance of biotechnology in promoting human welfare.**
5. Briefly describe the initiatives taken by the Indian Government in promoting biotechnology and discuss the commercial potential of biotechnological products in India.

Chapter 14

CELL CULTURE

All biotechnological processes are performed within **bioreactors** (may be a culture vessel, an open tank or sophisticated fermenters) containing correct **medium** provided with optimum **growth conditions,** like pH, temperature, aeration, light (for photosynthetic organisms), etc., The growth of the organisms is the increase of cell material that can be measured in terms of many parameters, like mass (dry weight or fresh weight), total amount of proteins, photosynthetitic pigments, number of cells, etc., **Doubling time** is the period required for doubling the biomass (due to cell division as well as due to cell growth) which varies from organism to organism, *e.g.*, for bacteria it is about 0.25–1.00 h, for yeast 1–2 h, for plant cells 20–70 h and for animal cells 15–48 h (**generation time** is the time period required for doubling of the cell number due to cell division).

Plant cells (similar to microorganisms) are mainly grown in liquid or solidified nutrient medium for various purposes, like artificial micropropagation of certain plants, production of valuable compounds (*e.g.*, perfumes), for genetic engineering purposes, etc. (Discussed in the chapter "Biotechnology in Agriculture and Forestry").

Animal and human cell cultures are grown in liquid or solid media and these cells are used for the production of many important organic compounds (*e.g.,* vaccines, interferons, hormones, monoclonal antibodies, etc.

1. Nutrient Medium

The basic nutritional requirement for different types of cells is different and includes the components, like a carbon source (*e.g.*, glucose, lactose, starch, $NaHCO_3$, $CaCO_3$, etc.), a nitrogen source (*e.g.*, $NaNO_3$), a phosphorus source (*e.g.*, $NaHPO_4$) and other elements. Generally, the carbon and nitrogen sources are derived from cheap natural products or by-products. The liquid medium is known as broth. Solidification of the medium can be achieved by adding 1–2% agar.

Many biotechnological processes involve the growth of a single or group of inoculated or indigenous microorganisms on solid substrates, *e.g.*, agricultural and food processing industrial wastes used for biogas production, composting and mushroom cultivation, fermentation processes in the batter of *idli*, cheese, etc.

2. Bioreactor (Fermenter)

For the culturing of cells, correct environment for the growth is provided in different types of bioreactors, normally ranging between two important types:

(*i*) **Simple Bioreactors:** These are manually stirred or non-stirred, non-aseptic open containers, *e.g.*, shallow tanks made in open sunlight are used for cultivation of *Spirulina* and cyanobacterial biofertilizer sewage treatment plants, etc., (Fig. 14.1).

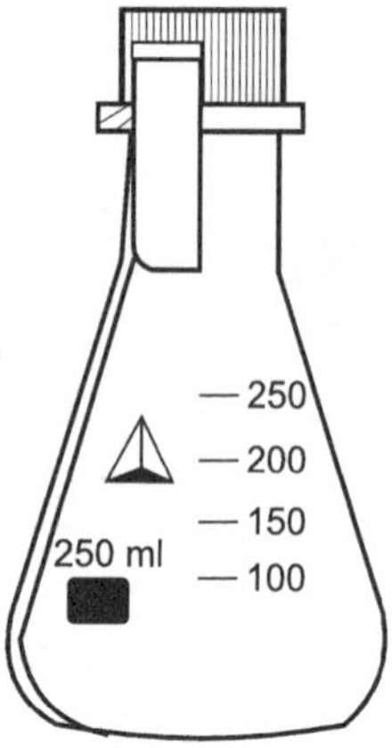

An Erlenmyer flask

Spirulina grown in artificial tanks.

Fig. 14.1 Some simple bioreactors.

(*ii*) **Complex Bioreactors:** They may be computerized, sterilized systems with controlled parameters like pH, temperature, etc., where mechanical agitation and aeration is provided by pumping, *e.g.*, used for the production of antibiotics, vitamins, etc., (Fig. 14.2). Here, the contamination by unwanted microorganisms and possibility of release of the culture organisms into the environment (particularly for genetically engineered organisms) is very small.

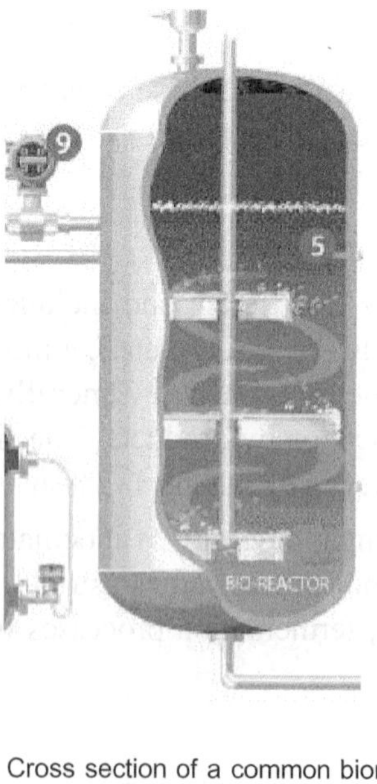

Cross section of a common bioreactor.

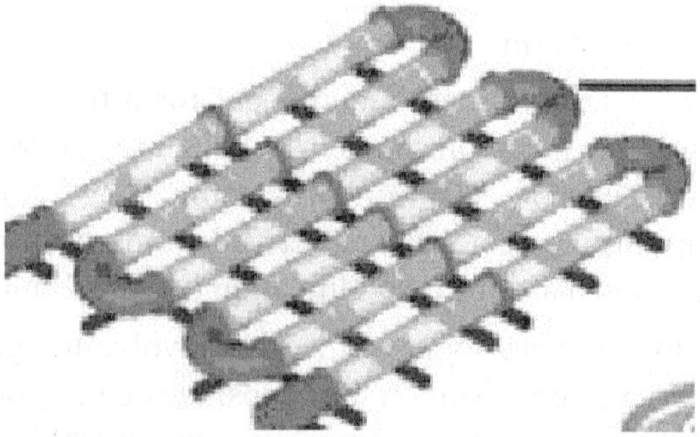

A photobiorector.

Fig. 14.2 Some complex bioreactors.

In the bioreactors, microorganisms are either **inoculated** in aqueous nutrient medium or the **indigenous microorganisms,** present in the solid medium, perform the fermentation process (*e.g.*, biogas plant, anaerobic digestion of sludge in sewage water treatment, composting, etc.). Stirring promotes growth rates by mixing uniformly the nutrients, gases and wastes, both in aerobic and anaerobic systems.

3. Culturing of Cells

For the growth of an organism, the conditions are not always ideal to permit unlimited growth. Thus, at sometime a few factors (*e.g.* a particular essential nutrient) become limiting for the growth of the cells (as the concentration of this nutrient decreases a certain level). Two main ways of growing microorganisms in the bioreactors are:

(*i*) **Batch Culture:** In this method of growth, the microorganisms are inoculated (added) into a fixed volume of artificial nutrient medium. Initially, the inoculated cells try to adjust in the new environment and do not show any growth, and this period is called **lag phase.** Thereafter, the growth occurs exponentially, proceeding with the maximum speed as the conditions are ideal for growth (optimum amount of nutrients with minimum toxic wastes), and the phase is known as **exponential phase.** As the growth proceeds, various nutrients get consumed and different metabolites (mostly toxic) are accumulated leading to continuous change of the nutrient medium, pH, etc. Gradually, the cell multiplication decreases and finally ceases due to the exhaustion of nutrients and accumulation of toxic wastes. At this time, the growth becomes stationary and the phase is known as **stationary phase.** Finally, the **death phase** starts because of decreasing metabolism and lysis of cells (Fig. 14.3).

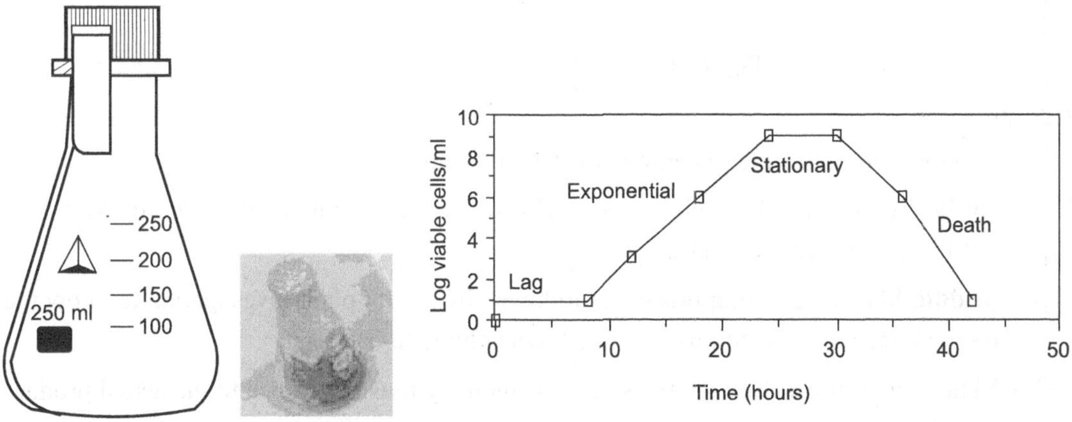

Fig. 14.3 A batch culture in Erlenmeyer flask (left) and growth curve in batch culture (right).

In industries, this system of cell cultivation is most commonly employed and the harvesting phase depends upon the type of product. When the **biomass** is concerned as the product, harvesting is done near the end of continuous exponential phase when the yield is generally maximum, whereas if a particular **metabolite** is of interest its maximum yield at a particular phase depends upon the kind of metabolite (*e.g.*, antibiotics are optimally formed during the stationary phase of growth cycle).

(*ii*) **Continuous Culture:** In this method of culturing, after the lag phase, when the exponential phase starts, fresh sterile medium is allowed to enter through an inlet with a steady flow rate and at the same time the culture (cells along with nutrient medium and wastes) is withdrawn through an outlet with the same rate, so that the volume of medium in the bioreactor remains constant (Fig. 14.4). Thus, the cells grow in the ideal growth conditions of nutrients and remain in **exponential phase** indefinitely. In industries the continuously operated systems are of limited use *e.g.*, single cell protein production, activated sludge process in oxidation ponds (secondary treatment) of sewage treatment plants, biogas plants, etc.

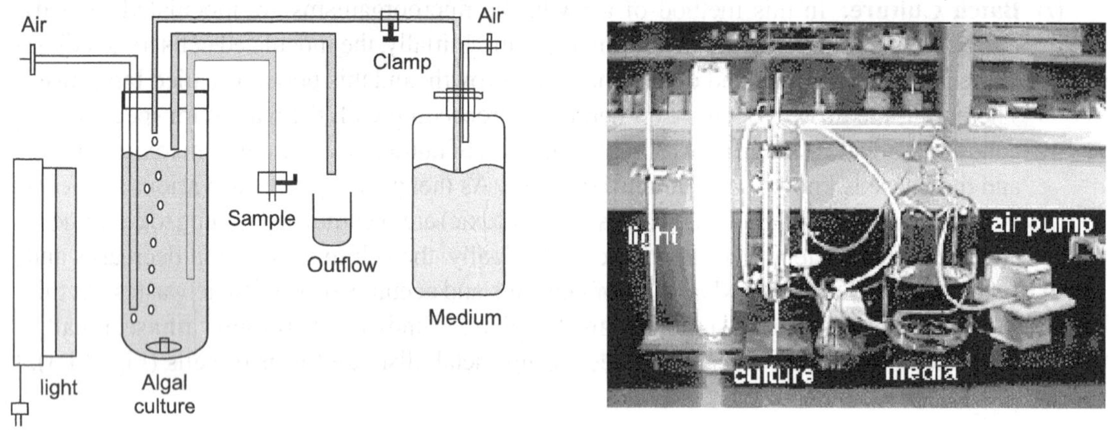

Fig. 14.4 Continuous culture devices.

4. Scale-up

Fermentation processes are generally developed in three stages:

(*i*) **Initial Stage:** Screening process is performed in small containers, like culture tubes, Petridishes, Erlenmeyer flasks, etc.

(*ii*) **Middle Stage:** Screening process is followed by a pilot plant investigation to know the optimal operating conditions in 5–200 litres volume fermenters.

(*iii*) **Final Stage:** It involves the transfer of technology thus developed to the actual production site.

5. Downstream Processing

After the growth of the cells in bioreactor, the desired end product for commercial use is extracted and purified from the biomass, and this process is known as **downstream processing,** *e.g.*, human insulin isolation from recombinant *E. coli* cells. The cost of product mainly depends upon the technique used for its production. The process involves many steps:

(*i*) **Separation of cells** from the liquid medium in bioreactor, that can be achieved by different ways depending upon the system. *e.g.*,

(*a*) Gravity sedimentation of larger heavier cells.

(*b*) Floatation of lighter cells.

(*c*) Cheap cloth filters, *e.g.*, used for *Spirulina*.

(*d*) Centrifugation of smaller cells, *e.g.*, for *Chlorella* cells, that is costly.

(*ii*) **Purification** of desired product (*e.g.*, antibiotics, enzymes, etc.) from the liquid medium can be achieved by various processes depending upon the size of the product:

(*a*) Sedimentation (10^{-1}–10^3 μm)

(*b*) Screens (10^5–10^7 μm)

(*c*) Centrifugation (10^2–10^5 μm) where the liquid sample is rotated with a high speed in tubes using the instrument, centrifuge, so that the heavier product gets settled at the bottom of the tube and the liquid remains at the upper part (Fig 14.5).

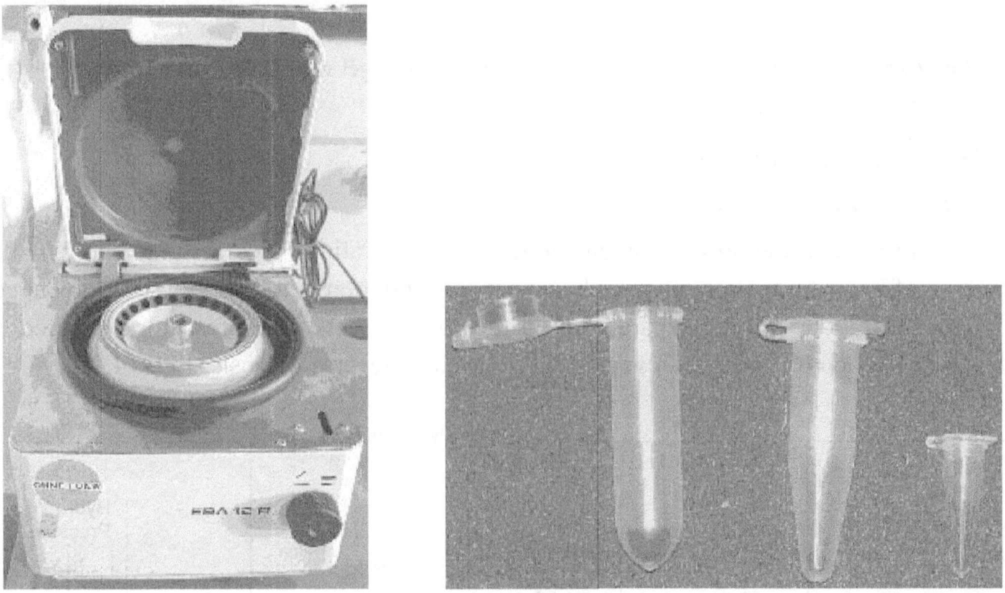

Fig. 14.5 Centrifuge machine (left) and centrifuge tubes (right).

(*d*) Microfilters (10–10^3 μm)

(*e*) Ultracentrifugation (10–10^3 μm)

(*f*) Gel electrophoresis (10–10^5 μm) where the molecules of various sizes are separated as bands by running the mixture over a solid material (*e.g.*, agarose, polyacrylamide) under an electric field (Fig 14.6).

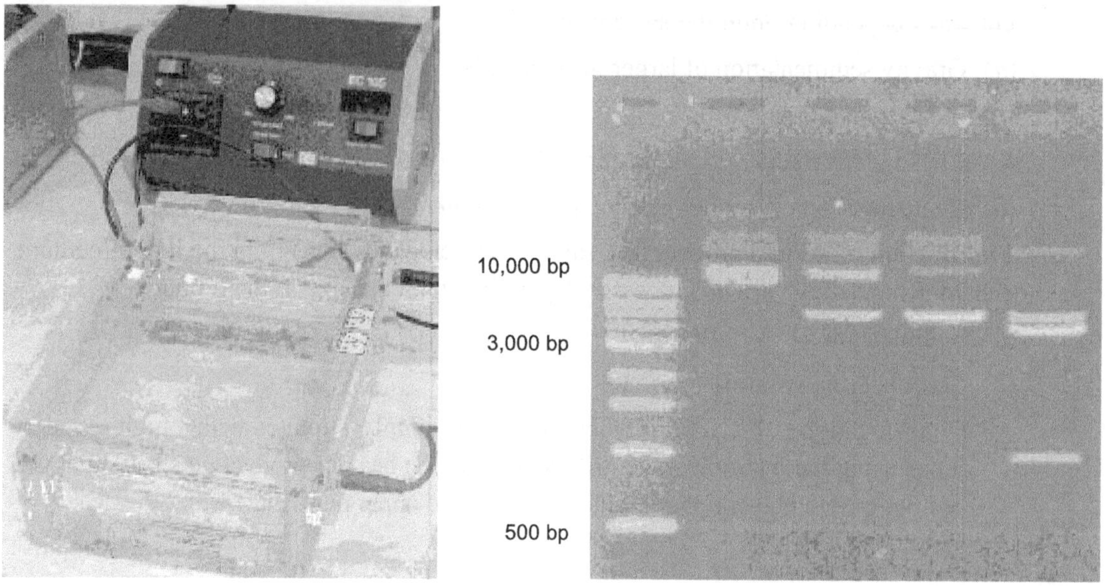

Fig. 14.6 Gel electrophoresis apparatus (left) and agarose gel with DNA bands (right).

(*g*) Distillation (1-10^3 μm)

(*h*) Ultrafiltration (1-10^3 μm)

(*i*) Reverse osmosis (1-10^2 μm)

Osmosis is the flow of solvent along the concentration / pressure gradient (*e.g.,* flow of water from a region of high solvent concentration to a region of low solvent concentration, that are separated by a membrane), but if the pressure is applied, the direction of flow gets reversed and it is called reverse osmosis (Fig. 14.7).

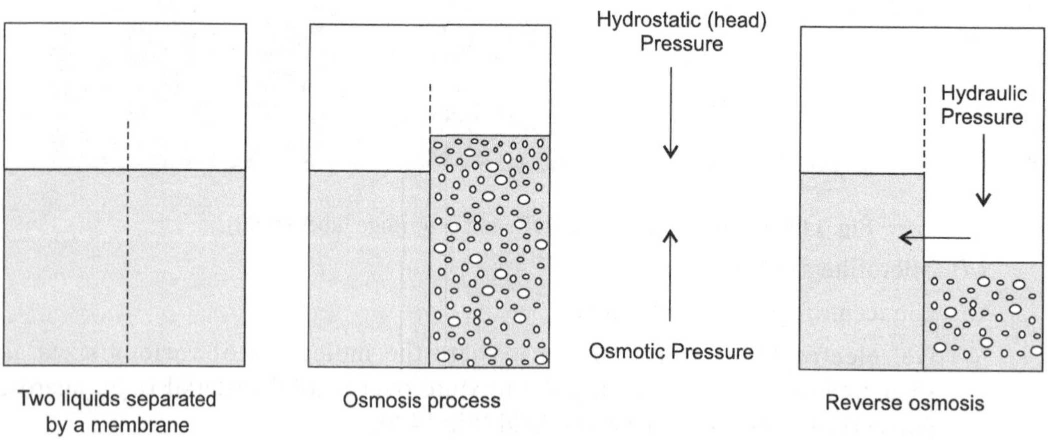

Fig. 14. 7 Mechanism of osmosis and reverse osmosis.

(*iii*) **Drying** of the final product can be achieved by many ways, *e.g.*, freeze drying, where, the product is first frozen and then allowed to evaporate at very low pressure. Lowering of pressure increases the melting point, so that increasing of temperature leads to faster evaporation of water of solid phase without getting converted into the liquid form (sublimation).When biomass is the commercial product, it can be packed after drying (either in sun or in oven) with or without any processing, *e.g.*, *Spirulina* and *Chlorella* can be used as food or feed after sun drying the biomass (Fig. 14.8).

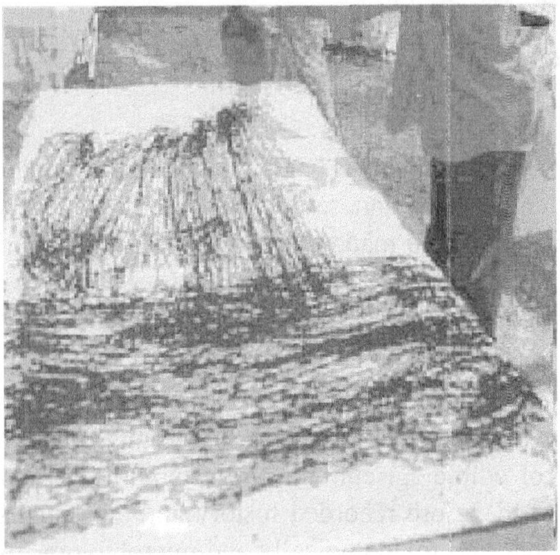

Fig. 14.8 Sun drying of harvested *Spirulina.*

Many products are sold in liquid preparations without drying.

QUESTIONS

1. What do you understand by batch culture and continuous culture techniques?
2. Write short notes on:
   (*i*) Bioreactors
   (*ii*) Downstream processing.

Chapter 15 FOOD AND BEVERAGE BIOTECHNOLOGY

Food industry is the largest industry in the world and food biotechnology includes improvement in taste, consistency, colour, nutrition, safety and preservation of the food. Developments in food preservation methods have made many of the seasonal foods to be available all the year round. Fruits, vegetables, cereals, meats, etc. require some degree of processing, and the relatively bulky raw agricultural food products are transformed into stable, convenient and palatable foods and beverages.

Fermented foods (*e.g.*, bread, cheese, yoghurt, sauerkraut, soy sauce, tempeh, etc.) and beverages (beer, wine, brandy, whisky, tea, coffee, cocoa, etc.) result from the action of microorganisms or enzymes on a wide range of edible agricultural products with desirable biochemical changes. Fermented foods were developed before recorded history and are usually more nutritious, more digestible, with improved flavour and toxicologically and microbiologically safer. Almost 90% of revenue from biotechnology comes from the fermentation products. Some of the fermentation products are:

ALCOHOLIC BEVERAGES

They are found in different forms in different regions, *e.g.*, in cooler regions of Europe, Poland and Russia production and consumption of beer from barley is more common; whereas, in warmer countries, like Spain, Greece, Italy and France, wines derived from grapes are more common.

The starting material normally contains either:

(*a*) **sugary** materials, like apple juice (to get **cider**), grape, pear, palmyra (to get **toddy** in India and South East Asia) juices, molasses (to get distilled **rum,** and **gin** is obtained from distillation of rum in the presence of juniper berries), honey, sweet potato, etc., or

(*b*) **starchy** materials like barley, rice, rye, sorghum, beet root, potato, etc., and the starchy materials need to be hydrolyzed to simple sugars by some saccharifying agent, like, grain malt, *Aspergillus oryzae*, *Rhizopus*, *Mucor*, *Bacillus*, etc., (the process is known as saccharification) before the fermentation process.

When the surgary substrates are incubated with suitable microorganism (generally yeast) after fermentation an acidic, nutrient-poor liquid containing a few per cent to about 16% (or more) ethanol is formed which does not allow growth of most of the contaminating microorganisms. These beverages require a period of storage (aging) that leads to improved flavour and taste.

Distillation process increases the concentration of ethanol to about 40–50%, *e.g.*, whisky, brandy, vodka, gin, rum.

Most regularly used fermenting organism is the yeast species, *Saccharomyces cerevisiae, S. carisbergensis* or *Zygosaccharomyces* sp., which can utilize simple sugars, like glucose and fructose to metabolize them to ethanol. The selected species is tolerant to high concentration of ethanol:

$$C_6H_{12}O_6 \longrightarrow 2C_2H_5OH + CO_2$$

(*i*) **Wine:** Wine means fermented grape juice and historically, it is a Middle Eastern and European drink. France, Italy and Germany are its major producers. It is obtained from fruit juices, mainly from the wine grape, *Vitis vinifera*, which contains 15–25% sugar. Some of the important types of wines are:

(*a*) ***Red Wine*** is formed from the fermentation of crushed whole black grapes (with skin).

(*b*) ***White Wine*** is produced from the black grapes without skin or green grapes.

(*c*) ***Rose Wine*** is produced from black grapes with some skin.

(*d*) ***Dry Wine*** is obtained from the complete fermentation of the sugar.

(*e*) ***Sweet Wine*** retains some sugar.

(*f*) ***Sparkling Wine (champagne)*** is obtained from secondary fermentation of wine containing 1% fermentable sugars and yeast in sealed bottles. The CO_2 produced, due to fermentation of added sugars by yeast, gets dissolved under pressure in the bottle. After the fermentation, the water in the neck of bottle is frozen to form an ice-plug. The cork plug with attached yeast is removed and replaced by a new cork plug. Thus, the CO_2 is not allowed to escape from the bottle.

(*g*) ***Fortified Wines (distilled)*** in which additional alcohol is added after the fermentation, thus, raising the alcohol content to about 20%, *e.g.*, brandy.

For the formation of wine, first grapes are crushed and the juice, thus, obtained, called must (containing glucose and fructose), is sterilized by SO_2 to control the natural fermentation by the contaminating yeasts. Then, it is inoculated with the desired strain of yeast, *Saccharomyces cerevisiae* vem ellipsoideus, and is subjected to fermentation (of the simple sugars, glucose and fructose), in tanks or bioreactors for a few days to 14 days (no saccharification is required). After fermentation, it is stored for months or years (aging) for chemical changes to improve the flavour, aroma and for clarification by settling of impurities. The alcoholic contents of wines ranges from 7–15%.

Moderate wine consumption is beneficial for coronary heart diseases. **Champagne** and **brandy** are obtained from distillation of wine.

(*ii*) **Beer:** It is produced from starchy substrates, *e.g.*, barley corn, potato, etc. Additional carbohydrate sources, called adjuncts, may also be added. There are five major steps in the manufacture of beer:

(*a*) ***Malting***: Dried barley seeds are soaked in water and spread on the floor for germination. This results in the formation of amylases (starch-degrading enzymes) and proteases (protein-degrading enzymes) and other enzymes by the embryo. The germinated seeds are subjected to slow heating upto a temperature of 80°C for the embryo [which otherwise would utilize the reserve starch (a polymer of glucose) and proteins in the seeds for its own growth] whereas, the enzymes are not spoiled and remain active even at high temperature. The seeds are dried and powdered. This product is called **malt,** which contains maltose, glucose, dextrins (short chains of glucose), hydrolyzed proteins, unhydrolyzed starch and proteins, etc.

(*b*) ***Mashing*:** The dried powdered malt is added with warm water (55–70°C) to initiate further hydrolysis of unhydrolyzed starch (process known as saccharification) and proteins by amylases and proteases, respectively, in the malt. The content is called **mash,** which mostly contains the fermentable sugars like glucose and galactose, dextrins, protein breakdown products, growth factors, minerals, etc.

(*c*) ***Formation of Beer Wort*:** The content is filtered and the filtrate is given some more time for the saccharification process. Sometimes extra amylase (from the fungus *Aspergillus oryzae* or some bacteria) is also added to the filterate to enhance the hydrolysis of starch. The product is known as **beer wort** and mostly contains fermentable sugars like, glucose, galactose, etc.

(*d*) ***Fermentation*:** The beer wort is boiled with hops (papery scales of female flowers of *Humulus lupulus*) to stop saccharification, for development of flavour, aroma and colour and to impart mild antibacterial activity. After cooling, the content is added with commercially available culture of the yeast, *Saccharomyces carisbergensis* (modern yeast, settles at the bottom after fermentation) or *Saccharomyces cerevisiae* (traditional yeast, floats at the surface after fermentation). The yeast may contain dextrin-utilizing gene from *Saccharomyces diasticus* by recombinant DNA technology (normally the yeast cannot utilize dextrins) to increase the amount of fermentable sugars. Thus, greater amount of alcohol after the fermentation for 5–10 days is produced. Finally, the growth of yeast is limited by anaerobic condition, ethanol (yeast action ceases after 18% alcohol) and drop of pH. The product is known as **fermented wort,** which predominantly contains ethanol.

(*e*) ***Conditioning*:** The fermented wort is chilled to 0°C for a period of two weeks to many months that leads to the clarification and removal of harsh flavour and undesirable characters. Now, the contents are stored at normal temperature for 8–10 months to further improve the quality and eliminate the off flavours. Sometimes papain (a protease, obtained from papaya latex) is added to hydrolyze the proteins and to remove the haziness. The yeast is finally removed (that either settles at the bottom or floats at the surface, depending upon the species) and the beer is either pasteurized at 60°C (20 min) to kill any leftover yeast cells and other microbes, or it is filtered to remove them. The final product is known as **beer,** which contains 3–6% alcohol.

Vodca and **whisky** are obtained from the distillation of beer.

VINEGAR

Vinegar means 'sour wine' and its production involves two successive fermentation processes:

(*i*) Conversion of sugary (*e.g.*, grape juice) or starchy material (*e.g.*, barley) to ethanol by *Saccharomyces cerevisiae:*

$$C_6H_{12}O_6 \longrightarrow 2C_2H_5OH + CO_2$$

(*ii*) After removal of the yeast, conversion of ethanol to acetic acid is achieved by the bacterium *Acetobacter aceti*:

$$C_2H_5OH + O_2 \longrightarrow CH_3COOH + H_2O$$

The species of *Acetobacter* for vinegar production should be tolerant to ethanol. The vinegar contains 4% acetic acid and small amount of alcohol.

Some species of *Acetobacter* should be avoided that continue the oxidation of acetic acid to CO_2 :

$$CH_3COOH + O_2 \longrightarrow 2CO_2 + 2H_2O.$$

DAIRY PRODUCTS

(*i*) **Yoghurt:** Prepared by inoculation (addition) of concentrated milk either with a sample of yoghurt (traditional) or commercially available culture of *Lactobacillus bulgaricus* and/or *Streptococcus thermophilus* (modern practice). Fermentation of lactose to lactic acid at 42-45°C for 4-5 hours drops the pH to 4 that leads to the precipitation of protein casein and thus, production of the yoghurt (*dahi*).

(*ii*) **Cheese:** Many kinds of cheese are produced which involves three steps:

(*a*) Curdling of the milk by inoculating it with starter culture of *Streptococcus lactis* or *S. thermophilus* and/or *Lactobacillus lactis* or *L. bulgaricus*. The culture ferments the lactose sugar of milk to lactic acid in about 6 hours, which leads to the precipitation of casein to form curd. Curd may also be prepared by adding rennin (rennet), which contains the enzyme chymosin (a protease), to milk. Rennet is traditionally obtained from the pancreas of calves but, now, can also be obtained from some naturally occurring microbes, like the fungus *Mucor miehei* or the genetically engineered microbes (*e.g.*, *Escherichia coli*) containing the calf chymosin (rennin) gene. Combined actions of proteolysis (by rennet) and acidification (by culture lactic acid) leads to the coagulation of casein.

(*b*) Separation of the protein casein from curd by heating it at 37°C (the liquid separated is known as whey). This heating process leads to the inactivation of rennin also. The protein is then molded into characteristic shapes and applied with salt by floating it in salt solution. This checks the contamination by undesirable microbes.

(*c*) Ripening of the cheese is allowed for 2–12 months at 2–10°C by microorganisms that are already present in cheese or by adding the desirable commercial microbes (*e.g.*, *Penicillium* sp., *Lactobacillus* sp., *Streptococcus* sp., depending upon the kind of cheese to be prepared). This gives flavour and aroma and improves the physical properties of the cheese due to degradation of lactose (into lactic acid), proteins and fats (into butyric acid). Sometimes, proteases from *Penicillium* or *Bacillus amyloliquefaciens* and lipases from *Aspergillus* or *Mucor* are also added.

The byproduct whey can be used to produce ethanol by the yeast *Kluyveromyces* sp.

(*iii*) **Butter:** It is obtained from the churning of pasteurized sweet or sour milk, which separates the fat globules from the liquid. The cream thus obtained, after the pasteurization at 71°C for 30 min., to kill the microbes, is either soured naturally or by adding the starter culture of *Streptococcus lactis* or *Leuconostoc*. This increases the butter yield and also improves the aroma and flavour. Salt is added to inhibit the contaminating microbial growth.

(*iv*) **Low Lactose Milk:** The milk and whey contain about 5% lactose. In many parts of world, people are not tolerant to lactose as they lack the enzyme lactase which breaks down the lactose sugar into glucose and galactose. Such persons suffer from diarrhoea or stomach ache after taking the milk. Lactase obtained from the yeast *Kluveromyces fragilis* or the fungus *Aspergillus niger* is used to treat the milk (conversion of lactose to glucose and galactose) so that they can be safely consumed by such persons.

LEAVENED BREAD

The Baker's yeast, *Saccharomyces cerevisiae,* (strain is selected to produce abundant CO_2 quickly) culture is mixed with bread dough to oxidize some of its sugar into CO_2 under aerobic condition at 25–26°C which is responsible for the leavening or rising of the dough. Sometimes, amylases (that break downs starch into simple sugars, which is used by yeast) is added to increase the CO_2 production. Baked foods have extended shelf life.

$$C_6H_{12}O_6 \longrightarrow CO_2 + H_2O$$

OTHER FERMENTED FOODS

(*i*) **Idli:** It is a favourite dish of southern India which is prepared by soaking the rice and black gram in water. After grinding, the batter is added with a little salt and, then, is allowed to ferment for 10–12 hours. The fermenting microorganisms are predominantly *Leuconostoc* sp., *Pediococcus*, and *Streptococcus* sp. The batter increases in volume and gets sour due to formation of lactic acid and acetic acid along with CO_2.

(*ii*) **Pickles:** Pickles are also the result of fermentation of vegetables and fruits to produce lactic acid by the salt tolerant *Lactobacillus* sp. in 6–8 weeks. The added salt inhibits the growth of unwanted microorganisms.

(*iii*) **Fermented Soybean:** In Asia the cooked soybean (*Glycine max*) is fermented with *Rhizopus oligosporus* to yield **tempeh.** Soybean can also be fermented along with wheat flour by *Aspergillus oryzae* to produce **soy sauce.** Fermentation breaks down the starch, proteins, pectins, etc. to simpler products.

NON-ALCOHOLIC FERMENTED BEVERAGES

(*i*) **Tea:** The tea leaves are crushed to release various enzymes, the activity of which improves the colour and aroma of tea. After drying, the product is packed.

(*ii*) **Coffee and Cocoa:** The pulp of coffee and cocoa fruits is removed by natural fermentation by bacteria and fungi. This leads to the development of flavour and aroma in the seeds. Finally, the product is dried and powdered.

ORGANIC ACIDS

(*i*) **Citric Acid:** It is prepared by the fermentation (partial oxidation) of molasses, corn, beets, etc., by the fungus *Aspergillus niger.* Ciric acid is used as a preservative, to enhance flavour and to prevent oxidation of various components in soft drinks, confectionary, jams, preserved fruits and in medicinal preparations.

(*ii*) **Lactic Acid:** It is produced by the fermentation of whey, molasses, starch (corn and potato) by *Lactobacillus bulgaricus*. It is used as an acidulant in confectionary, fruit juices, canned meats and vegetables.

SWEETENERS

Sweeteners are used in soft drinks, confectionary, jams, ice creams, etc. and some important sweeteners are:

(*i*) **Aspartame:** It is a low calorie sweetener used in soft drinks. It is a dipeptide (two amino acids) made up of the amino acids phenylalanine (obtained by fermentation) and aspartic acid. It is 150 times sweeter than sucrose (the common sugar) (Fig. 15.1).

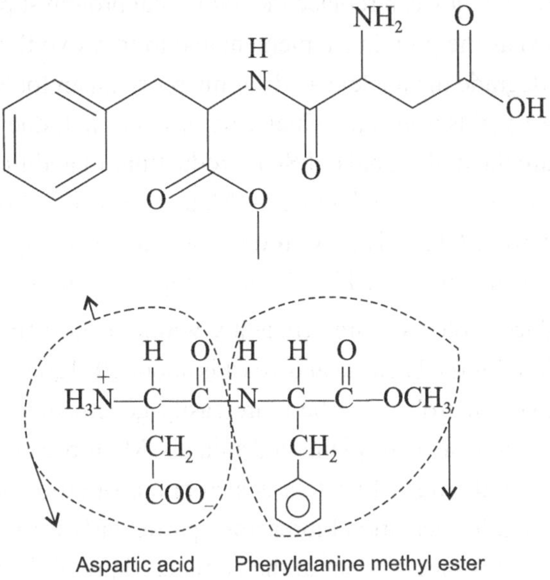

Fig. 15.1 Chemical structure of aspartame.

(*ii*) **Thumatin:** It is a protein extracted from the berries of the plant, *Thaumatococcus danielli*. It is the sweetest compound known and is 3000 times sweeter than sucrose. (Saccharin, that is obtained chemically, is 300 times sweeter than sucrose)

Diagnosis of Food Contamination

Major food contaminating microbes are identified by DNA or RNA probe technology or by ELISA (Enzyme Linked Immunosorbent Assay) technique using commercially available monoclonal antibodies or probes for the contaminating microbes.

Single Cell Protein

Explosive population growth is a major problem of the world, particularly for the developing countries. Conventional agricultural practices are unable to supply sufficient food, particularly the proteins, despite increasing productivity. Through new agricultural practices high protein cereals have been developed. The use of processed microbial biomass, which are usually single-celled or filamentous in structure as the source of protein, is called single cell protein (SCP). The biomass is called single cell protein as it is rich in protein (more than 50% of the dry weight). The interest in SCP was generated to compensate for the protein deficiency, particularly in the developing countries.

People have recognized the nutritional value of mushrooms, yeast, many bacteria and algae from the time immemorial, *e.g.*, people from the Lake Chad in Africa and the Lake Texcoco in Mexico have been harvesting the naturally growing blue-green alga, *Spirulina*, and using it as food after drying in the sun. During the last three decades, interest is growing in using microbes for food production, particularly for feeding animals, poultry and in aquaculture for the farming of shrimps, prawns, fish,etc., which in turn would improve the human nutrition.

SCP are easier to store and they can replace the traditional protein supplements, like fishmeal and soya meal. They are used as the protein supplement and to improve flavour of food for human consumption. Processing to degrade the nucleic acids of microorganism for SCP is required because metabolism of DNA and RNA yields uric acid, which causes stones in kidney. The quality as well as the quantity of the proteins are the major goals of SCP production. In addition to proteins, microbes also contain carbohydrates, fats, vitamins and minerals. Many companies throughout the world have been involved in the production of SCP, and many products are commercially available, *e.g.*, 'Sunova' capsule containing *Spirulina* is manufactured by Dabur Company in India.

Microorganisms produce proteins more efficiently within a short time period than the farm animals, *e.g.*, the doubling time for the bacteria and yeast is about 20–120 min, whereas, for a young cattle is 1–2 months. Moreover, microbes can be more easily genetically modified for a desirable amino acid composition in comparison to the plants and animals. Microorganisms have relatively high protein content and the nutritional value of protein is also good. Microorganisms can be grown in large amounts in relatively small fermentation bioreactors, particularly through continuous cultures, all the year round, where growth is independent of the climate. Many of the microorganisms can be cultivated on a wide range of low value raw materials or waste materials, particularly low value water, which would also help to reduce the pollution.

A unique aspect of SCP technology is the problem of safety, nutritional value and acceptability of the product. The raw material used for the production of SCP is the main safety hazard, *e.g.*, presence of carcinogenic hydrocarbons, heavy metals and other contaminants as well as the toxin production by certain fungi. Sanitation and quality control procedures must be maintained to avoid the contamination by pathogenic or toxicogenic microorganisms. Toxicological testing of the final product must be thoroughly performed. In addition, the odour, taste and texture of the product are also equally important.

First industrial production of SCP, *Candida utilis*, occurred during World War I by Germany and it was used in soups and sausages. A variety of substrates, like inorganic carbon (*e.g.* CO_2), industrial effluents (*e.g.*, confectionary effluents, whey, molasses) and low cost organic materials (*e.g.*, cellulosic wastes, like straw, starch hydrolysate) are used for SCP production.

The microorganism used for SCP production must be non-pathogenic, should have good nutritional value, should be easily and cheaply produced on large scale, toxin-free, fast growing and easy to separate from the medium. Some of the important microorganisms used for SCP production are:

(*i*) **Algae:** Most commonly used algae are *Chlorella* (a unicellular green alga), *Scenedesmus* (a colonial green alga) and *Spirulina* (a filamentous blue-green alga) which are photosynthetic and are generally grown in open tanks or ponds. They utilize CO_2, sunlight and a few inorganic nutrients for their growth. Algal SCP has about 60% crude protein with good amino acid composition. They are suitable for protein-rich feed supplement for domestic animals. *Chlorella* and *Scenedesmus* (Fig. 15.3) have long been used as food in Japan, while, *Spirulina* in Africa and Mexico. *Chlorella* is commercially produced in Japan to be used in yoghurts, ice-cream and breads, while *Spirulina maxima* in Mexico is used as animal feed. *Spirulina* is harvested by filtration method. The major disadvantages of algae as SCP are the risk of contamination and costly recovery methods, especially for unicellular algae, green algal cellulosic cell walls that are not digested by human beings etc. The CFTRI (Central Food Technology and Research Institute) in Mysore produces *Spirulina* tablets of 1 gm weight to be used as food and feed (Fig. 15.2).

Spirulina filaments — *Spirulina* tablets

Fig. 15.2 *Spirulina* grown in artificial tanks (top left) and sun-drying of harvested *Spirulina* (top right), *Spirulina* filaments as seen under the microscope (bottom left) and *Spirulina* tablets (bottom right).

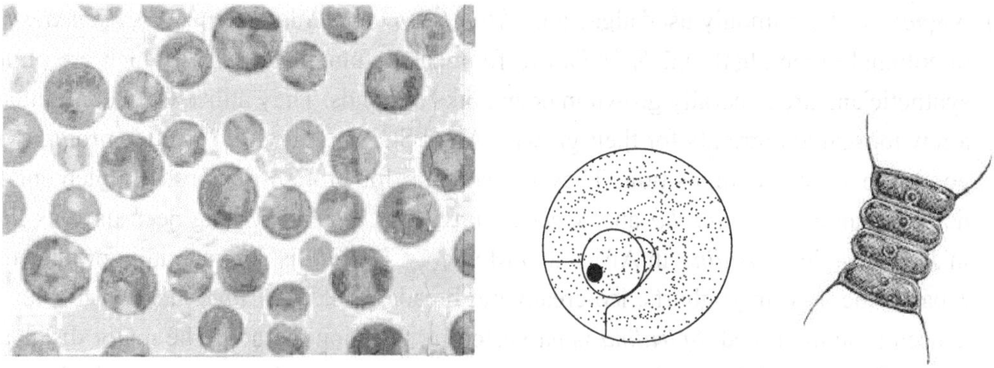

Chlorella sp. A single Chlorella cell Scenedesmus sp.

Fig. 15.3 Animal feed, *Chlorella* and *Scenedesmus*.

(*ii*) **Fungi:** Some fungi used as SCP are unicellular (yeasts), whereas, others are filamentous in structure.

(*a*) ***Yeasts*:** Members of *Saccharomyces cerevisiae* (Baker's yeast), *Candida utilis,*

(Torula yeast) and *Kluyveromyces fragilis* are widely used as SCP. They have 55-60% protein with good amino acid balance and also rich in vitamin-B. They are used both for human food and animal feed supplementation. During their production, the risk of bacterial contamination is low and they are harvested by centrifugation.

Saccharomyces cerevisiae is grown on molasses and is used commercially for the fermentation of dough in bakeries, and, thus, eaten indirectly as the component of food. *Candida utilis* is also used commercially in U.K., USA, Russia, Europe in soups and sausages and grown on various substrates, like confectionary effluents, ethanol and sulphite liquor.

Kluyveromyces fragilis is basically grown on whey and is used commercially in France. The yeast secretes lactase enzyme, which helps in the digestion of the milk sugar lactose into glucose and galactose.

Yeasts have advantages of larger size (easier to harvest), lower nucleic acid content, high lysine and ability to grow at low pH.

(*b*) ***Filamentous Fungi*:** The commonly used filamentous fungi as SCP are:

Fusarium graminearum, *Chaetomium*,etc. which are grown on starch hydrolysate and cellulosic wastes, respectively (Fig. 15.4). They are usually grown as submerged cultures and have protein content of 50–55%. Harvesting of these fungi is rather easy, generally by filtration method. The main problem associated with these fungi are their slower growth, and contamination by yeast.

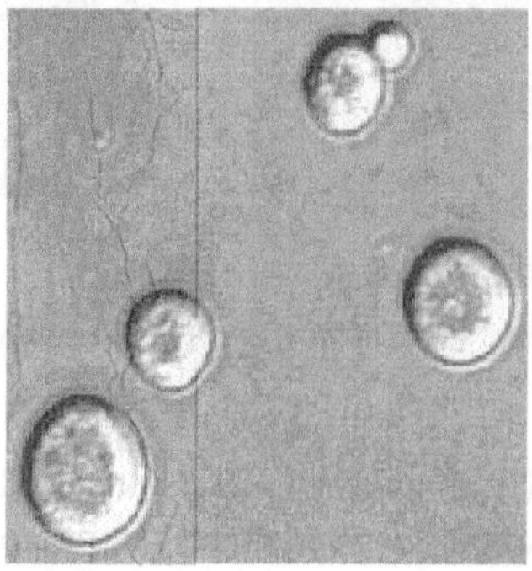

Fig. 15.4 Cells of *Saccharomyces cerevisiae*.

Mushrooms are the fruiting bodies of certain large fungi belonging to the group Basidiomycetes and are rich in proteins, vitamins and other nutrients.

Moreover, they are devoid of starch and are suitable for diabetic individuals. *Agaricus bisporus* is the common white button mushroom and accounts for over 70% of total mushroom production. It is grown on moistened paddy or wheat straw compost in wooden trays. After white cottony mycelial (filaments) growth, the compost bed is covered with a 1-2 cm thick layer of soil and sand. The mushrooms (fruiting bodies) are harvested at the button stage (Fig. 15.5).

Fig. 15.5 *Agaricus bisporus.*

Lentinula edodes is the second most cultivated mushroom variety in the world and over 90% of its production occurs in Japan.

Some of the other species of edible mushroom are *Pleurotus* and *Volvariella*.

(*iii*) **Bacteria:** A large number of bacterial species have been evaluated for SCP production using wide variety of substrates. They have very high growth rates and are used at commercial level, *e.g. Methylophilus methylotrophus* grows on methanol. They have over 80% protein. The risk of contamination by pathogenic bacteria is high during cultivation, moreover, their recovery is also problematic. They are recovered by flocculation and floatation combined with centrifugation. Bacteria have high nucleic acid content relative to yeast and fungi.

Sun-drying of SCP is cheap but it reduces its quality. Heat treatments are used during the final stages of harvesting to inactivate heat-sensitive organisms and to reduce RNA content. The biomass may be further processed or even the protein may be isolated and purified from the biomass. SCP can be stored and shipped over long distances.

SCP processes are mostly capital and energy intensive, most processes must be conducted under sterile conditions in expensive equipments. The future of SCP would depend upon the reduction of production costs and improvement of quality. However, the main limitations for SCP products for human use are sociological and their major role will be in animal feed supplements.

QUESTIONS

1. Describe the role of various microorganisms and enzymes in food and alcoholic beverages industries.

   Or

   Discuss the use of various microorganisms in dairy industry.

   Or

   Discuss the role of various microorganisms in the production of alcoholic beverages.

   Or

   Briefly outline the steps in the production of following, discussing the role of various micro-organisms and enzymes:

   (*i*) Leavened bread
   (*ii*) **Yoghurt**
   (*iii*) **Cheese**
   (*iv*) **Beer**
   (*v*) **Wine**
   (*vi*) Distilled alcoholic beverages
   (*vii*) **Vinegar**
2. **Define single cell protein. Briefly describe the production of various SCPs.**
3. Write short notes on:
   (*i*) Rennet in cheese production
   (*ii*) Cloudiness of fruit juices and wines
   (*iii*) Use of lactase in dairy industry
   (*iv*) Mushrooms
   (*v*) **Fermented foods and beverages**
   (*vi*) **Single cell protein**
4. Discuss the role of the following enzymes in food industry:
   (*i*) Proteases
   (*ii*) Amylases
   (*iii*) Lactase

Chapter 16 ENZYME TECHNOLOGY

The term enzyme was given by F.W. Kuhne (1878). They are generally made up of **proteins** and catalyze biochemical reactions. They are specific and have three-dimensional structure with highly specific one or more **active sites** for the recognition of substrates. The active site (lock) of the enzyme binds with its substrate (key) as "lock and key model", given by Fisher (1890) (Fig. 16.1).

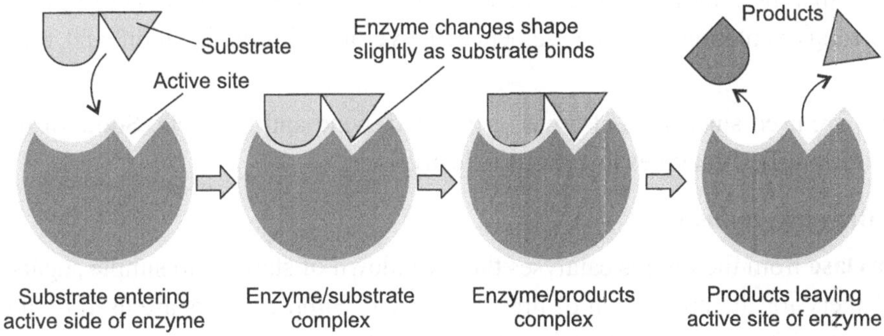

Fig. 16.1 Mechanism of enzyme action.

Some enzymes also have non-protein parts, called **prosthetic groups,** that can be inorganic (**cofactor,** *e.g.* nitrogenase having Mo, hydrogenase having Ni, catalase having Fe^{2+}, etc.) or organic (**coenzymes,** *e.g.* coenzyme A) molecules. Some RNA molecules, called **ribozymes,** also have catalytic function reported by Cech (1986), which catalyze either their own cleavage or cleavage of other RNA and DNA molecules. Some antibodies, called **abzymes,** not found naturally, have also been reported to have catalytic activities, *e.g.* monoclonal antibodies.

Enzymes, required in very low amount, increase the rate of reaction by10^3–10^6 times by lowering the activation energy. They do not take part in the reaction and operate under mild conditions of temperature, pH, pressure, etc. **Carbonic anhydrase,** catalyzing the reaction:

$$CO_2 \quad + \quad H_2O \quad — \quad H_2CO_3$$

is the fastest enzyme enhancing 10^7 times reaction rate in comparison to the uncatalyzed reaction.

The first isolated and crystallized enzyme was **urease** (J. Sumner, 1926), and **insulin** was the first enzyme of which the amino acid sequence was given by Sanger (1953).

Enzyme technology utilizes purified enzymes for generating useful products and services, *e.g.*, industrial processes, medicine, research (*e.g. Taq* DNA polymerase obtained from thermophilic bacterium, *Thermus aquaticus,* used in PCR), recombinant DNA technology, etc. Though, purified enzymes are costly but they do not produce any unwanted byproducts, thus, the purification of the product becomes easier. Most of the enzymes are used in **food** and **detergent** industries. Sometimes, the byproducts produced by whole cells add to the quality of the product, *e.g.* wine and beer. Most of the enzymes are obtained from microorganisms (*e.g.* amylase from the bacterium *Bacillus*, fungus *Aspergillus*, rennet from the fungus *Mucor miehei*) as it is easier to grow them in fermenters and manipulate them genetically. Some enzymes are obtained from higher plants (*e.g.* papain from papaya latex, amylase from malted barley) and animals (*e.g.* rennet containing chymosin enzyme from the pancreas of calf) also.

Enzymes are used since ancient times, *e.g.* for bread, cheese, wine and beer making, etc. In 1950s purified enzymes are either used in **freely suspended** form or in **immobilized** form, where the enzymes are fixed to some suitable inert carrier material, *e.g.* agarose, calcium alginate, polyacrylamide, carbon, clay, membrane, etc. When enzymes are used in immobilized form, the separation of enzyme from the product becomes easier and the immobilized enzyme can be used repeatedly. But sometimes, immobilization of the enzymes results in the loss of activity due to immobilization in a wrong orientation (*e.g.* involvement of active site in immobilization) and, thus, increase in cost.

Enzymes are used since ancient times, *e.g.* bread, wine and beer, etc. Some of the modern applications of the purified enzymes in various industries are:

1. Food and Brewing Industry

(*i*) **Amylase** from the fungus catalyses the breakdown of starch into simple sugars in dough in bakery for the production of bread, buns, etc. Simple sugars are used by the yeast to generate CO_2 that gets entrapped to raise the dough.

Amylase is also used in brewing for the conversion of starchy material to simple sugars that are utilized by yeast for beer (ethanol) production under anaerobic conditions.

Amylases are also used to convert starch into glucose and various syrups.

(*ii*) **Proteases** also help in bakery and brewing for the conversion of complex proteins to simpler short peptides and amino acids.

(*iii*) **Chymosin,** which is a protease (in rennet), is used in dairy industry for cheese production to coagulate protein casein in milk.

(*iv*) **Lipases** are used to enhance ripening of cheese.

(*v*) **Cellulase** is used in the production of fruit juices for the removal of cellulose impurities.

(*vi*) **Glucose isomerase** is used to convert glucose into fructose syrup that is used in soft drinks.

(*vii*) **Invertase** is used in chocolates where the centre of chocolates is filled with sucrose and the enzyme invertase. After some time, sucrose gets converted into invert sugar (glucose + fructose), which being more soluble and sweeter than sucrose, makes the centre of the chocolate soft and sweet.

2. Detergent Industry

(*i*) **Amylase** (from *Bacillus*, *Aspergillus niger*) is used to remove resistant stains of starch residues with dirt on the cloths.

(*ii*) **Proteases** (from *Bacillus*, *Clostridium*) are used to remove proteinaceous spots of milk, blood, etc. in association with dirt. They cause allergic reactions, therefore, the protease granules are encapsulated in soluble wax coating.

(*iii*) **Lipases** from *Bacillus* to remove oily spots on clothes.

(*iv*) **Cellulases** from fungi are used for digesting away small fibres raised from the cotton fabric (made up of cellulose) with associated soil particles without damaging the major fibres.

3. Medical and Pharmaceutical Field

(*i*) **Amylase, lipase, proteases, papain** are used to cure digestive disorders due to starch, oil and proteins, respectively.

(*ii*) **Trypsin** is used to dissolve blood clots.

(*iii*) Many enzymes are used in disease diagnosis tools, *e.g.* ELISA, diabetes diagnosis kit that utilizes the enzymes **glucose oxidase** and **peroxidase** to detect the presence and quantity of glucose in blood and urine, various biosensors, etc.

(*iv*) **Lysozyme** obtained from hen's egg albumin (egg white) digests away bacterial cell wall and is used as antibiotic.

(*v*) **Lactase** (from *Kluyveromyces lactis*, *Aspergillus niger*) treated milk is safely used by those people who are sensitive to lactose as their digestive system does not produce lactase enzyme and, thus, they cannot digest lactose of milk.

4. Leather Industry

Protease from pancreas is used to remove hair from hides.

5. Textile and Wool Industry

Papain is sometimes used to remove overlapping and emerging scales of wool (made of protein), thus, to give silky appearance and to avoid shrinkage of the wool. But this treatment is very costly.

6. Energy Field

Cellulase from the fungi *Trichoderma viridi* and *Aspergillus niger* is used to treat cellulosic raw plant materials for ethanol (used to blend petrol, called gasohol) production by the yeast.

Efforts are also being made by **genetic engineering technology** to obtain improved enzymes or to create new enzyme activities by making suitable changes in amino acid sequences of the enzymes.

QUESTIONS

1. **What are enzymes? Describe in brief the industrial applications of various enzymes.**
2. Discuss the various applications of immobilized enzymes giving suitable examples.
3. Write short notes on:
   (*i*) Biosensors
   (*ii*) Substrate saturation
   (*iii*) Enzymes in medicine
   (*iv*) Advantages and disadvantages of enzyme immobilization
   (*v*) Advantages and disadvantages of purified enzymes
   (*vi*) Mode (mechanism) of action of enzymes
   (*vii*) Abzymes and ribozymes
4. **Describe the role of various enzymes in food industry.**

Chapter 17 ENVIRONMENTAL BIOTECHNOLOGY

Continuously increasing population is leading to the **intensive agriculture** (to get greater amount of food and other products on the same limited land by the use of improved varieties of crop plants, which require higher amount of chemical fertilizers, insecticides, herbicides, etc.), **urbanization, industrialization, deforestation,** and **excessive utilization of various resources** (fossil fuels, underground water, etc,). This leads to the degradation of our environment by generating various kinds of liquid, solid, gaseous and hazardous wastes, particularly in the **developed countries,** where, only about 15% population consumes about 85% resources to generate about 85% of the wastes (most of which are non-biodegradable) in comparison to the developing countries. On the other hand, in **developing countries** higher population, limited technological knowledge and finance facilities for the treatment of wastes are the major causes of environmental degradation. Earlier, when the population was small the environment was sustainable and the biodegradable wastes generated were easily decomposed by the decomposers. But now, the amount of waste generated is so great that the limited natural capabilities of the environment are unable to decompose it properly, particularly the **toxic non-biodegradable** ones, leading to the accumulation of such wastes in food chains (**biomagnification** of insecticides, like, dichloro diphenyl trichloro ethane or DDT, herbicides, etc.) as well as in the surroundings. Large-scale production and application of **synthetic chemicals** in most industrialized countries is a problem of serious concern causing serious health hazard. Organic chemicals that cannot be easily degraded by microorganisms or totally resistant to biodegradation are known as recalcitrants, *e.g.*, lignin; whereas, synthetic compounds (foreign substances) not formed by natural biosynthetic processes are called **xenobiotic** compounds (that can be recalcitrants) and often have toxic effects, *e.g.*, insecticides DDT, BHC (benzene hexa-chloride), etc.

The scientific and public awareness for managing a good environment is increasing since 1960s and the application of biological system, particularly the microorganisms (either single or group of microorganisms), to clean the environment (waste treatment) as well as to minimize the release of various wastes in the environment is known as **environmental biotechnology.** All environmental biotechnological processes make use of metabolic activities of microorganisms. Many successful biological processes have now been developed for water, gas, soil and solid waste treatments and some of them are:

WASTEWATER AND SEWAGE TREATMENT

Sewage is the used water from domestic, industrial or agricultural communities, surface run off, etc. and contains 0.1–0.5% **inorganic** and **organic** (proteins, amino acids, fats, fatty acids, carbohydrates, etc.) matters in suspended and soluble forms. Domestic wastes include detergents, antiseptics, pesticides, human excretements, various types of **microorganisms** (aerobic and anaerobic) like, bacteria, fungi, viruses, protozoa, algae, etc. (many are pathogenic) to which a variety of industrial organic and inorganic wastes are added that depends upon the type of industries. The microorganisms found in wastewater help in decomposition of organic matter in the sewage. During the course of sewage decomposition there is shift in the types of microorganisms and initially **aerobic** microorganisms predominate and later **anaerobics** (specially methanogenic bacteria, which convert H_2 and CO_2 to CH_4 that can be entrapped and used as an energy source). Fermentation products and reduction of NO_3^- and SO_4^{2-} give rise to noxious odours, tastes and colours. Replenishment of dissolved O_2 from atmosphere and by photosynthetic O_2 evolution can be slower than the utilization rate of O_2 by heterotrophic microorganisms in the presence of abundant organic substrates, thus, reducing the DO of natural waters. Small amounts of sewage discharged through drainage systems (sewers) into a water body (river, lake or ocean) gets diluted due to inherent self purification capacity without significant deterioration of water quality, but large amounts of liquid wastes produced due to increased human populations do not get diluted sufficiently, leading to anaerobic conditions (high biochemical oxygen demand or BOD) because of consumption of O_2 by the aerobic microorganisms and unacceptable deterioration of water quality. This leads to the death of aerobic plants and animals and foul odour in water. Decomposition of these dead organisms causes an additional O_2 demand. At much slower rates the sewage may also percolate to groundwater table.

The **BOD** determination (by incubating about 1 litre of polluted water for 5 days at 20°C in dark and assaying the amount of O_2 used for oxidation of biodegradable organic material by an O_2-electrode) provides the best information concerning the strength of sewage and the efficiency of sewage treatment processes. Other methods of pollution level estimation are **COD** (chemical oxygen demand, that gives an idea of chemically oxidisable organic matter. The water sample is incubated with a known amount of $K_2Cr_2O_7$ for about 2 hours and the unused $K_2Cr_2O_7$ is estimated by titrating with ferrous ammonium sulphate), **biological indicators** (relative abundance of resistant or absence of certain sensitive organisms in the polluted water), **PO_4^{3-}** level, **NH_4^+**, quantity of **suspended solids** by weighing the amount of filtered or settled solids/litre water sample, etc.

Municipal waste water:

(*i*) Total solids = 700 ppm
  → Dissolved = 500 ppm
  → Suspended = 200 ppm

(*ii*) BOD = 300 ppm

(*iii*) COD = 400 ppm

(*iv*) N = 40 ppm
  → Organic = 15 ppm
  → NH_3 = 25 ppm

(*v*) P = 10 ppm
  → Organic = 3 ppm
  → Inorganic = 7 ppm

(*vi*) Grease = 100 ppm

Efficient waste (sewage or industrial) collection through drainage system (sewers) and specific treatment processes are employed since early times to reduce the biologically degradable organic compounds before discharging the wastewaters into natural water bodies. It is the largest biotechnological industry and some of the important technologies (the choice of method depends largely upon the area), depend upon the metabolic activities of mixed microbial populations, thus, leading to complete elimination or reduced incidences of various infectious waterborne diseases (typhoid, cholera, dysentery, polio, hepatitis, etc.) due to elimination of pathogens (by competition and predation by other aquatic populations, adsorption and settling), involve physical, chemical or biological processes. Organic nutrients are utilized and mineralized by heterotrophic microorganisms. NH_4^+ is nitrified and along with other inorganic nutrients utilized by algae and aquatic plants. The usual operation in wastewater treatment is the reduction of BOD associated with suspended and dissolved organic matter. Only a small percentage of suspended and dissolved organic matter is actually mineralized during wastewater treatment and most of the organic material is removed by settling that may create disposal problem. Occasionally, it may be followed by the removal of inorganic nutrients and recalcitrant organics prior to discharging the effluent into natural water bodies. After a prolonged septic period, O_2 diffusion eventually re-aerates the water and allows the mineralization of accumulated fermentation products that give rise to algal bloom. If there is no further disturbance, secondary succession may eventually restore the aquatic water body to its former state. The primary and secondary treatment processes aim to reduce BOD of sewage prior to its discharge and this complex and highly successful system with aerobic coupled with anaerobic decomposition processes involves three steps (Fig. 17.8):

(*i*) **Primary Treatment:** It is the physical treatment in which a portion of larger suspended and floating solids, grit, and grease from sewage is removed by various mechanical devices, like passing the sewage through **screening** devices (the screened solid is collected and disposed by incineration or burial), traps and skimming devices (for the removal of grease). After screening, the sewage is passed through grit chambers to allow heavier solids (*e.g.* gravel, grit) to settle down. After this, the sewage is passed through the **sedimentation** tanks at a slower rate to settle out the solid matters as much as possible. The settled solids are drawn off from the bottom. For typical domestic sewage, primary treatment removes 30–40% of BOD and secondary treatment is necessary for acceptable BOD reduction. If the liquid waste mainly contains solids and little dissolved organic matter, primary treatment may remove 70–80% of BOD that can be disposed off safely. (If only primary treatment of sewage is to be performed, the liquid after chlorination is discharged into the natural water body).

(*ii*) **Secondary Treatment:** In secondary treatment, a small portion of the dissolved organic matter is mineralized, whereas, the larger portion is converted from a dissolved state to removable solids. The combination of primary and secondary treatments reduces the BOD by 80–90% to produce effluents with less than 20 ppm BOD. This is the biological treatment and several methods are employed. These are:

(*a*) **AEROBIC DECOMPOSITION**

The microorganisms are either **attached as surface films** or they are **homogeneously suspended in sewage water.** Organic materials are oxidized to inorganic materials such as CO_2, SO_4^{2-}, NO_3^-, PO_4^{3-} and water by applying any of the technologies:

SUSPENDED CELL SEWAGE TREATMENT SYSTEMS

(*i*) ***Oxidation Ponds*** **(*Lagoons*)**: These are shallow ponds, about 5 feet deep, where aeration of wastewater for oxidation is accomplished by diffusion through wind action and photosynthetic activity of algae. Oxygenation is usually incomplete with consequent odour problems. Bacteria digest and oxidize the organic material in sewage, whereas algae utilize CO_2 and other substances resulting from bacterial action and through photosynthesis they produce O_2 needed for aerobic bacterial action. Because of this interaction oxidation ponds reduce 75–90% BOD. This is low cost but inefficient system requiring long retention times. This process is recommended for small communities in rural areas for warmer climates where suitable and sufficient land is available (Fig. 17.1).

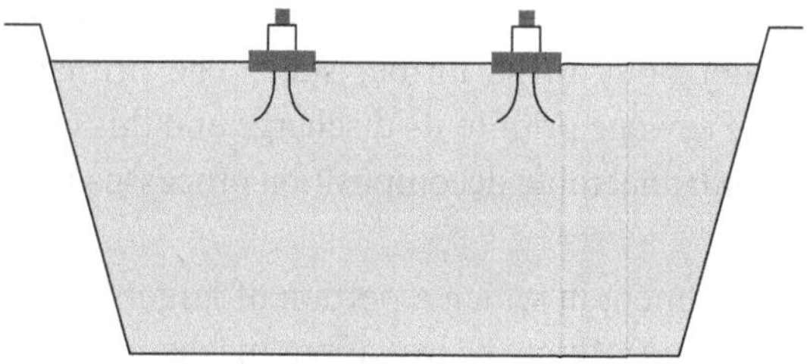

Fig. 17.1 Oxidation pond (top) and sedimentation tank (bottom).

(*ii*) ***Activated Sludge Process***: It is an efficient and popular biological treatment process of sewage treatment, where, the sewage is aerated vigorously for 4–8 hours to promote the growth of aerobic microorganisms, where, aeration is provided by mechanical stirring or air injection. The finely suspended and colloidal matter along with many microorganisms forms aggregates (flocs). After repetition of this process several times, the excess sedimented floc thus obtained is known as **activated sludge** and it is made up of micro-organisms along with organic matter. The liquid is passed into a settling tank to settle out the sludge. A portion of this sludge is returned to the previous chamber to maintain a sufficient population of microorganisms for digestion of new batch of sewage (hence the name of the process) and rest is pumped to an anaerobic digestion chamber [In this anaerobic chamber, organic compounds get decomposed anaerobically to produce soluble substances and various gases like, about 70% CH_4, 25% CO_2, H_2 and N_2. This gas mixture can be used as **fuel** and the sludge obtained after digestion can be used as **manure** after drying it in sun. The liquid from the settling tank does not contain any organic matter (higher BOD removal) but contains NO_3^-, SO_4^{2-}, PO_4^{3-}, etc. and a few pathogens and can be discharged into natural water body. In this technology, energy is needed for aeration of the wastewater.] or incinerated or composted or disposed in landfills.

Advanced activated sludge treatment systems with series of aerobic and anaerobic tanks also remove inorganic nitrogen (in aerobic tanks by nitrification $NH_4^+ \rightarrow NO_2^- \rightarrow NO_3^-$ for 1–3 hours and in anaerobic tanks by denitrification process $NO_3^- \rightarrow N_2$ for 0.5–1 hour) and PO_4^{3-} (uptake of soluble PO_4^{3-} by bacteria during aerobic period). Toxic heavy metals can also be taken up by the microbial populations and removed.

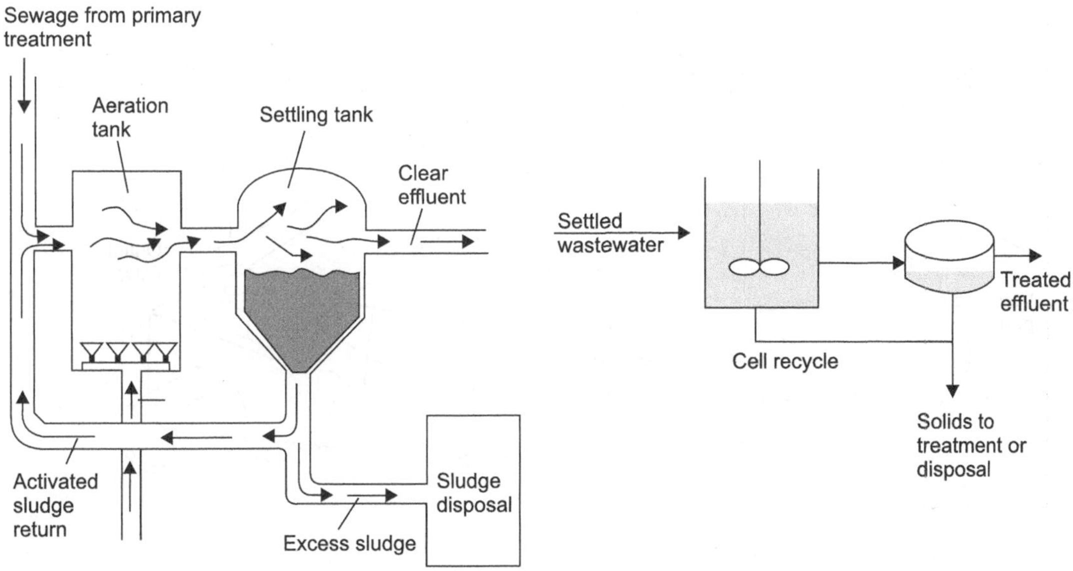

Fig. 17.2 Activated sludge system.

FIXED FILM SEWAGE TREATMENT SYSTEMS

(*iii*) ***Trickling Filters***: It is simple and relatively inexpensive device consisting of 8–12 feet deep bed of stones at the surface of which the liquid sewage is sprayed either by a series of sprinklers or by a rotating sprinkler. The spraying is done intermittently to permit the proper aeration in between the stones to maintain the aerobic condition. The stones become coated with a film of microorganisms that oxidize the organic matter in wastewater as it seeps slowly through the bed. Thus, at the bottom of the filter, relatively clear water is collected that is drained off. Sewage may be passed through two or more trickling filters or recirculated many times through the same filter. The drawback of this system is that nutrient overload may lead to excess microbial slime, reducing the aeration and percolation rates and it becomes necessary to renew the filter bed (Fig. 17.3).

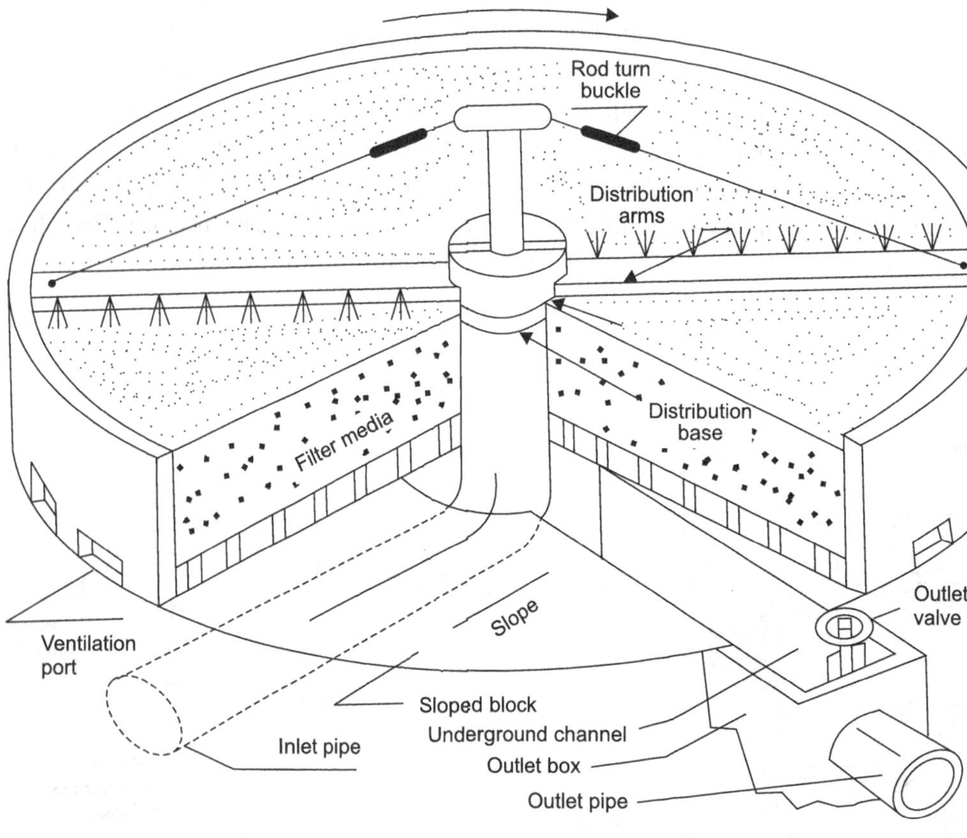

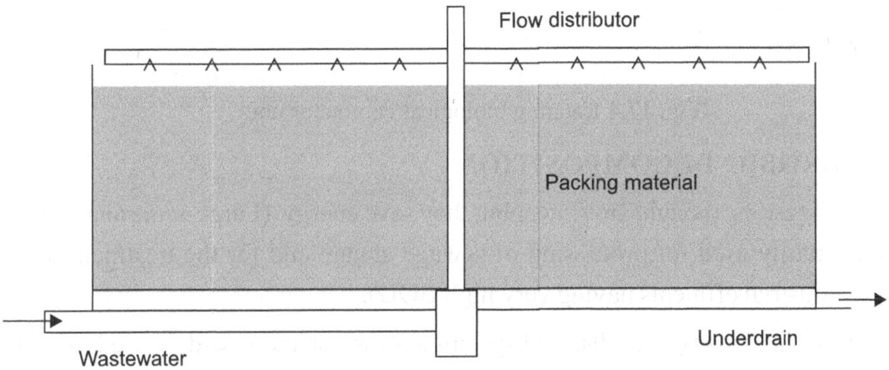

Fig. 17.3 Trickling filter device.

(*iv*) ***Rotating Biological Contractors (Discs)*:** This system consists of a set of 2–3 m diameter rotating discs (1–2 rpm) made up of wood, metal or plastic that are oriented vertically to the direction of flow of wastewater and are partially submerged in the water. The microbial film adhering to the disc decomposes the organic matter in the wastewater aerobically where the rotation of discs provides aeration. The microbial film grows to a thickness that prevents the nutrients to reach the inner microbial cells leading to their death and detachment that can be removed by settling. It has also high BOD removal efficiency, requiring less space and energy for rotating the discs but requires higher initial investment (Fig. 17.4).

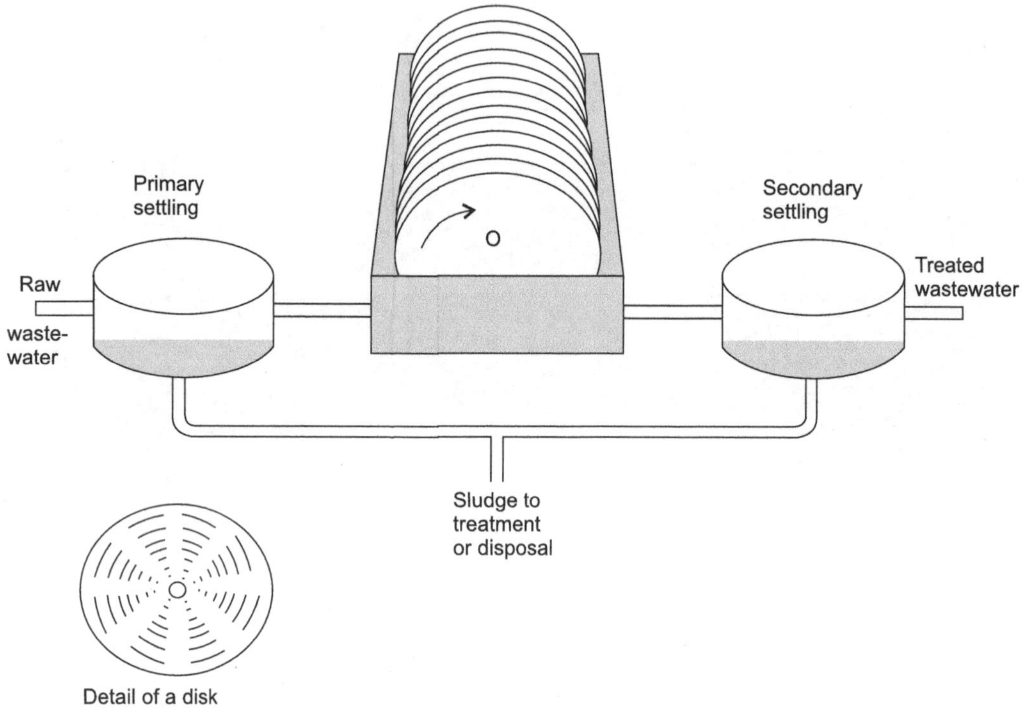

Fig. 17.4 Rotating biological contractor unit.

(*b*) **ANAEROBIC DECOMPOSITION**

These digestors, though, are slower but they save energy. (Large-scale anaerobic digestors are primarily used for processing of sewage sludge and for the treatment of some very high industrial effluents having very high BOD).

(*i*) ***Septic Tank*:** This type of disposal system is simplest and is widely used in rural areas by individual families where public sewers are not available. It is made up of metal or concrete chamber below the ground level to which all the domestic wastes flow. Here, the wastewater is retained long enough to permit limited anaerobic digestion of various organic materials by anaerobic bacteria to produce simpler chemicals and gases, like H_2S, CH_4, CO_2 and H_2 and the residual solids settle to the bottom of the tank. This water has still high BOD and is odourous. The clarified effluent from the tank is disposed off under the soil surface (leaching field) through perforated pipes where dissolved organic matter undergoes oxidative biodegradation. The process does not eliminate pathogenic microorganisms. The undecomposed solid, called sludge, is pumped out periodically to prevent clogging of tank and the drain field (leaching ground). After drying, the sludge can be used as manure (Fig. 17.5).

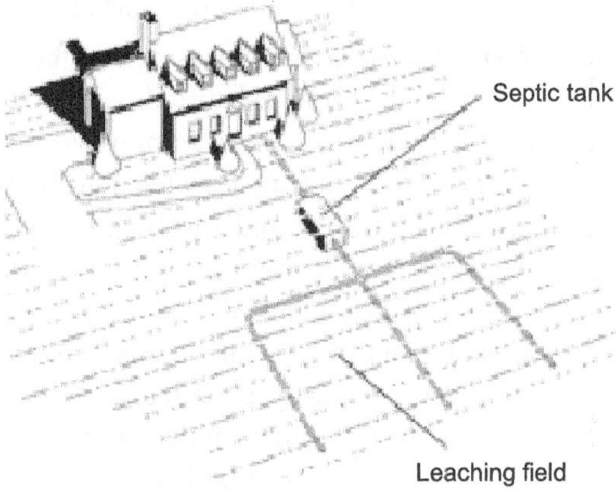

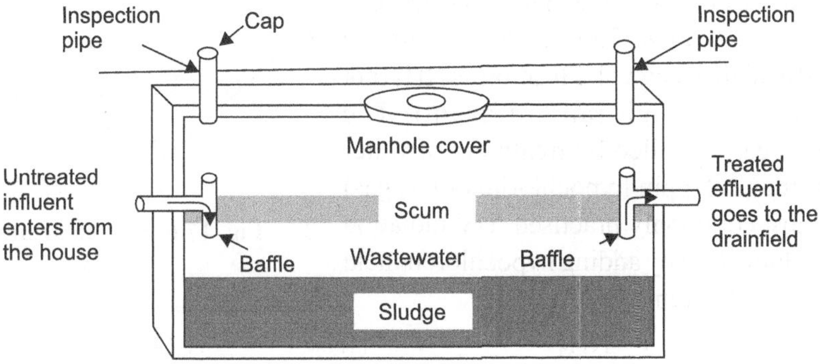

Septic tank

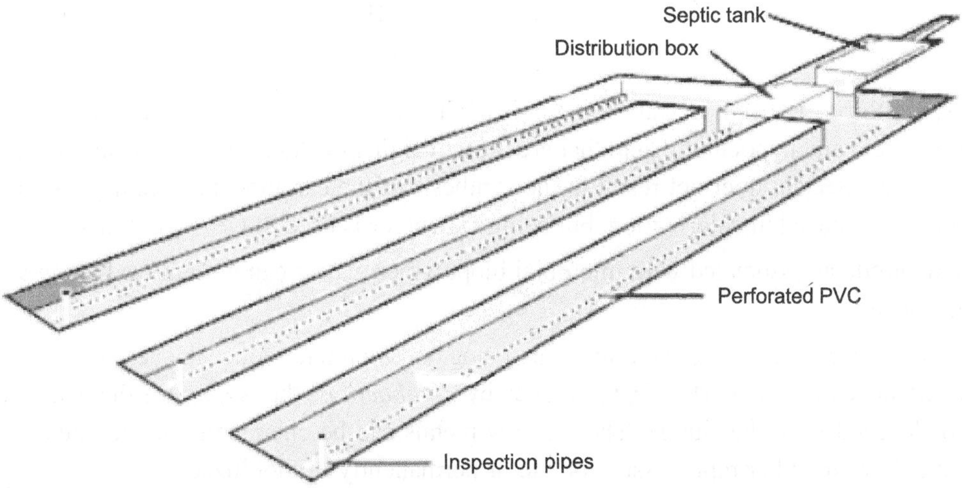

Absorption field

Fig. 17.5 Septic tank device.

(*ii*) ***Imhoff Tank***: It is an improved design of septic tank with a separate settling compartment from where the stabilized solids are removed (from the bottom of the tank) (Fig. 17.6).

(*iii*) ***Tertiary Treatment***: This is mainly the chemical treatment to remove the turbidity due to presence of nutrients and, being expensive, generally applied when the wastewater is to be used for drinking purpose and rarely for the disposal purpose. The inorganic salts in water after secondary treatment may stimulate growth of aquatic plant life, *e.g.* algae. Some of these salts can be removed by chemical treatments, *e.g.* addition of alum that induces the precipitation of phosphate salts, which can be settled out. Activated carbon filters are used to remove non-biodegradable organic pollutants from secondary treated industrial effluents. NH_4^+ can be removed at high pH by volatilization as NH_3 (but it can return through rainfall again). The pathogens can be killed by treatment of water by chlorine (2–5 ppm hypochlorite or Cl_2 gas) that is most commonly practised, UV radiation or O_3. Chlorination by adding hypochlorous acid (HOCl) can also remove NH_3

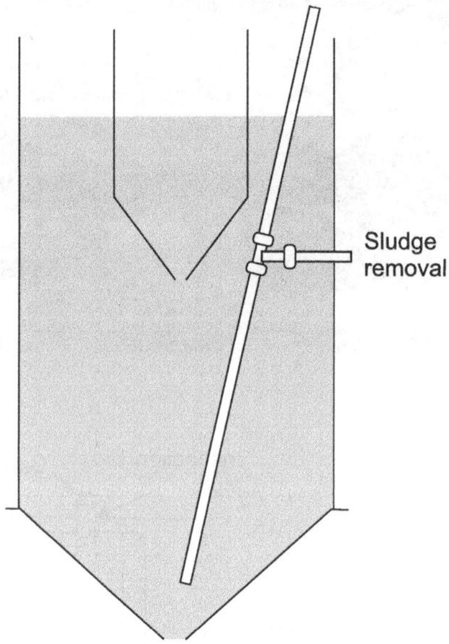

Fig. 17.6 Section of an Imhoff tank.

$$HOCl + NH_3 \longrightarrow NH_2Cl + H_2O$$

$$NH_2Cl \xrightarrow{HOCl} NHCl_2 \xrightarrow{H_2O} N_2$$

Hypochlorite is used as $Ca(OCl)_2$ or NaOCl. The Cl_2 gas reacts with H_2O to produce hypochlorous acid and hypochloric acid that are the actual disinfectants as they are strong oxidizing agents. O_3 as disinfectant doer not result in the synthesis of the carcinogens, like trichloromethane compounds, as produced in chlorination, but being expensive is used in USA and Europe.

Toxic products associated with microbial biopolymer matrix can be removed along with the microbial biomass.

Processes similar to microbial mining are being considered to reduce the heavy metals. Many organisms produce acids, *e.g.,* (H_2SO_4) produced by *Thiobacillus,* that solubilize the heavy metals, that can be leached from the sludge. These heavy metals can be chemically precipitated from the leachate and they are either reprocessed for use or permanently immobilized.

(*iv*) ***Sludge Treatment***: The settled or screened out solid waste from the primary treatment (sludge) and the activated sludge (solid waste containing large amount of microbial biomass is digested in a tank mainly anaerobically to reduce its odour, volume and pathogens.

Under anaerobic condition, organisms, like *Desulfovibrio desulfuricans* reduce SO_4^{2-} to S^{2-} (H_2S), *Thiobacillus denitrificans* NO_3^- to N_2 (denitrification), where SO_4^{2-} and NO_3^- act as acceptor of electrons instead of O_2 in anaerobic respiration. The methanogenic bacteria produce CH_4. Thus, soluble substances and various gases, like about 70% CH_4, 25% CO_2, H_2 and N_2 are produced. This gas mixture can be used as **fuel** and the sludge obtained after digestion can be used as **manure** after drying in sun. The sludge can also be landfilled or aerobically composted. Because of noxious odour, these digestors should be constructed at a sufficient distance from the residential areas (Fig. 17.7).

Fig. 17.7 Anaerobic digestor for sewage sludge.

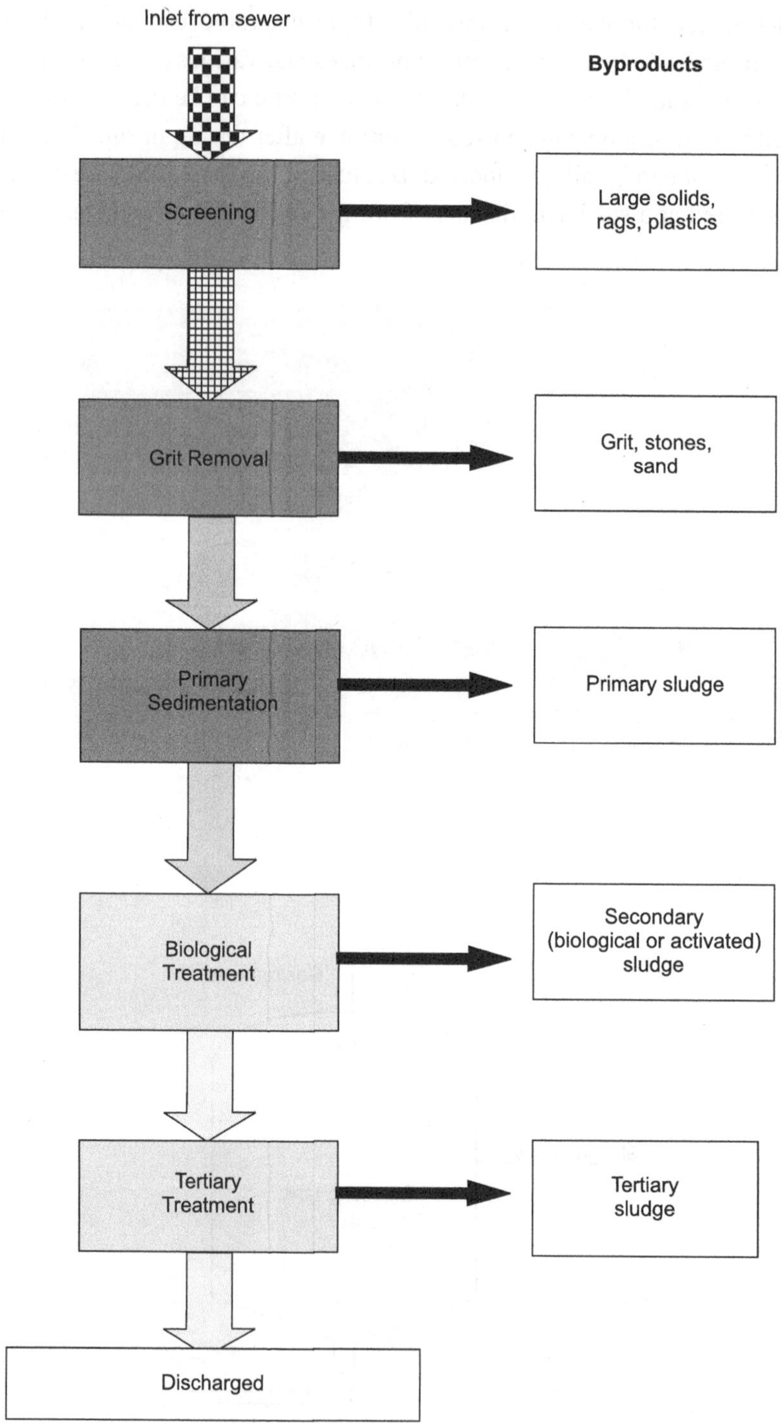
Inlet from sewer
Byproducts
Screening
Large solids, rags, plastics
Grit Removal
Grit, stones, sand
Primary Sedimentation
Primary sludge
Biological Treatment
Secondary (biological or activated) sludge
Tertiary Treatment
Tertiary sludge
Discharged

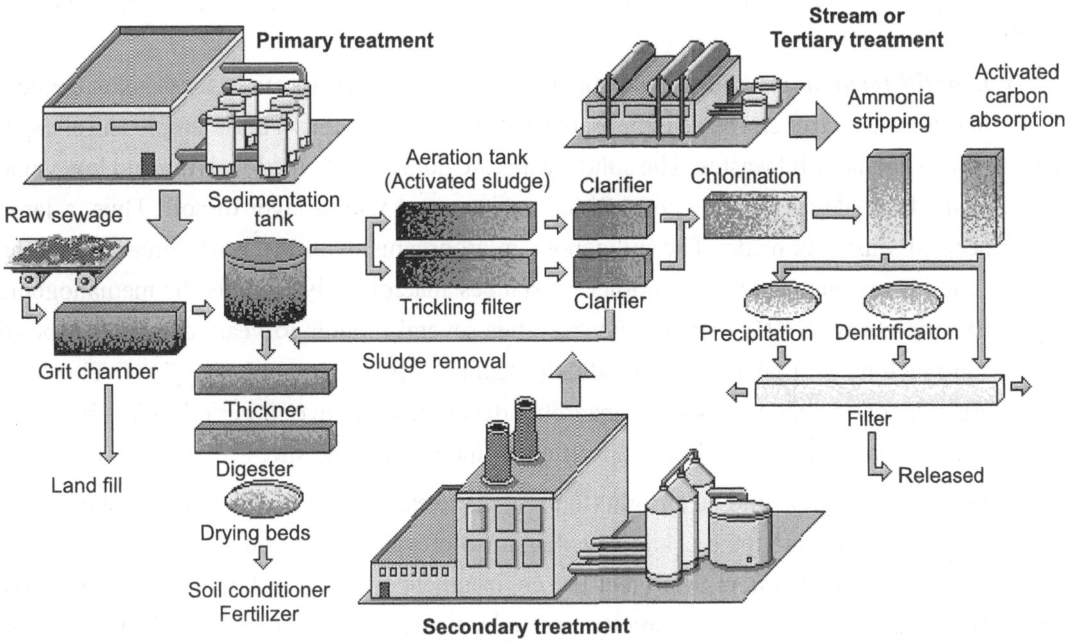

Fig. 17.8 Sewage treatment system.

2. Biological Filters to Clean the Air

Biofilters are used for the treatment of gases and they are made up of some solid support (that is suitable for the growth of microorganisms, *e.g.* peat) or a liquid-gas system, where, the liquid and gas phases are separated by a membrane containing appropriate microorganisms. The polluted gas is passed through the biofilter where various microorganisms convert the hazardous gaseous pollutants into non-hazardous compounds.

3. Solid Waste Management

The various kinds of solid wastes, like municipal waste and sludge, agricultural, mining, industrial wastes, may be biodegradable (paper, food wastes, sewage wastes, etc.) or non-biodegradable (glass, plastic, many insecticides, etc) and their disposal is problematic. The **reduction** in the use of various resources (*e.g.* minimum packaging, avoiding disposable items, plastic bags, etc.), **reusing** some of the products (*e.g.* fly ash or the fine particulate matter produced mainly from thermal power plants can be used for the construction of roads and bricks; glass and plastic bottles, etc.) and **recycling** of many items (*e.g.* paper, plastic, glass, metal items, etc.) may not only help in the reduction of pollution but also in reduced consumption of energy. **Incineration,** the burning of combustible municipal waste in a furnace in the presence of O_2 at very high temperature (815°C for 1 hour) producing CO_2 and H_2O, though reduces the volume of the waste to 90% but generates CO_2

and other gases and fly ash in the air. Some of the safer technologies used for the management of solid wastes are:

(*i*) ***Landfill Technology*:** Landfills are gigantic anaerobic bioreactors used for the management of solid wastes that can be decomposable or non-decomposable. This is simple and cheaper way to handle solid wastes. The solid wastes are dumped into the low lying and low value lands. Everyday's waste is compressed and covered with a layer of soil. Thus, a large heap of wastes is made. The indigenous microorganisms in the waste break down the complex organic matter into simpler substances anaerobically. Finally the methanogenic bacteria generate CH_4 that commences after several months to years of waste disposal and reaching to a peak value, gradually declines after several (5–10) years. Thus, the organic content undergoes slow anaerobic decomposition over a period of 30–50 years. This CH_4 can be used as fuel by inserting a pipeline into the site.

The landfill sites should be airtight (to prevent the escape of CH_4 that is a greenhouse gas and other gases into the air) and water tight (to prevent the leaching of toxic substances into underground water) to protect the environment and this technology, if improperly managed, can be unsightly, smelly and unhygienic. Toxic wastes (like heavy metals, insecticides, etc.) or toxic products generated by anaerobic decomposition can create problems by run off to nearby water bodies and lands, seepage into the underground aquifers and by affecting the microbial decomposition in the landfill site. Therefore, the sites should be properly constructed with impermeable lining at the bottom to avoid leahates to damage surrounding land and water. The underground aquifer contaminated with leachates remains polluted for many years because of slow movement, minimal self-purification process, low microorganisms, O_2 and nutrients of underground water-table. Remedial actions are inadequate and expensive, *e.g.*, pumping out the aquifer followed by its treatment or *in situ* treatment by oxygenation or nutrient supplementation. Therefore, prevention of groundwater contamination is the most effective remedy. An impermeable clay cap prevents water infiltration and leaching of pollutants. The recovery of CH_4 needs investment for installation of collection pipes and pumping systems and for sealing of landfill to prevent the escape of methane. Finally, the leachate collected at the bottom can be pumped out for treatment (Fig. 17.9).

The limited numbers of suitable disposal sites in urban areas are rapidly becoming filled. Premature construction on landfill sites may result in structural damage to the buildings because of subsidence and explosion hazard due to seepage of CH_4 into the basements. CH_4 seepage may also damage planting on disposal site.

Land filling

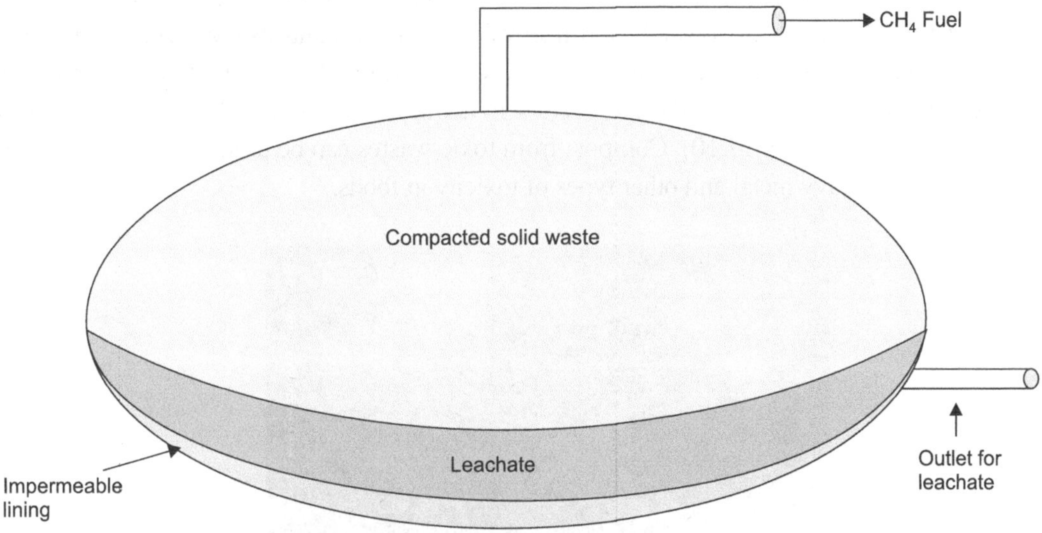

Fig. 17.9 Landfill technology.

(*ii*) ***Composting***: It is an attractive alternative to landfills for the decomposition of solid, domestic and agricultural wastes and has environmental advantages (landfill operations may have lower costs but the groundwater contamination due to land filling, composting process is favoured). Only organic wastes are sorted for decomposition by separate collection of garbage. The waste can be ground and mixed with sewage sludge (that may contain heavy metals, like Cd, Cr, etc.) or wood chips and composted. Traditionally at small scale, most of the solid wastes are composted. In urban areas and due to large-scale agriculture the other treatments of wastes are expensive. It involves the aerobic microbial decomposition by bacteria and fungi of solid organic wastes from domestic, agriculture and food industry, which gets converted into sanitary humus-like material. The waste is converted into huge piles or is dumped into a pit lined with straw or wood chips and is covered with dried leaves or a layer of soil to control odour. The waste is watered once or twice a week to keep it moist (50–60% moisture is optimum as excessive moisture interferes with aeration) and the aeration is provided by regular mechanical turning (after every 14 days) of the waste to facilitate proper oxidation of organic matter into CO_2, H_2O and organic byproducts. Self-heating raises the temperature to 55–60°C in 2–3 days and then, thermophilic microorganisms are involved in the aerobic decomposition. In well-aerated or oxygenated piles the temperature may rise to 70-80°C. Thereafter, the temperature declines gradually and mesophilic microorganisms now decompose the organic matter. These alternate thermophilic and mesophilic phases are repeated several times. The high temperatures thus attained help in killing of most of the pathogens present in fecal matter and sewage sludge. Aeration or turning may be adjusted to prevent the excessive heating. Periodic water spraying can also help to reduce the temperature. Inside the heap, O_2 concentration is about 5 times lesser than the atmospheric concentration. C : N ratio should be greater than 4 : 1 (lower N restricts microbial decomposition, whereas, excessive N leads to volatilization of NH_3 to lower the fertilizer value of compost and also cause odour problem). Composting is completed in about 3 weeks. Uniform and more stable compost can be obtained more rapidly in the bioreactors in about 2–4 days, which requires curing for about a month (Fig. 17.10). Compost from toxic wastes can be used in parks and gardens to prevent heavy metal and other types of toxicity in foods.

Fig. 17.10 A heap of compost

Problem of bad odour arises due to the presence of various S and N containing organic compounds in the waste. The escape of various gases produced during decomposition can be inhibited by allowing the decomposition in closed chambers and using biofilters or gas scrubbers. The **compost** thus, obtained is reduced in bulk and used as manure for soil improvement and cultivation of mushroom *(Agaricus bisporus)*. Presence of high amount of cellulose and lignin (*e.g.* in wood, straw,etc.), which are relatively resistant to decomposition, slow down the compost formation.

Solid waste can be inoculated with commercially available cultures of earthworm, which help in the breakage of large pieces of organic waste into smaller particles to be easily decomposed by the indigenous bacteria and fungi in about 45 days. The compost thus obtained is rich in nutrients and is called **vermicompost** (Fig. 17.11).

Fig. 17.11 Vermicompost production.

4. Bioremediation

Bioremediation (or biotreatment or bioreclamation or biorestoration) is a new technology and involves application of biological system (mostly microorganisms) for the treatment of various hazardous wastes, particularly the non-biodegradable ones (xenobiotic). These microorganisms either degrade and detoxify the pollutant or accumulate them in their body. Two types of bioremedial technologies, which employ generally a group of microorganisms, can be applied:

(*i*) After prolonged exposure to a particular pollutant, some of the indigenous microbes develop ability to utilize or degrade that pollutant due to spontaneous mutation or acquiring some plasmid DNA from other microorganisms by conjugation. At the affected site, if some of the **nutrients** are limiting for the growth of such microorganisms, the addition of such nutrients (like urea (N), single super phosphate (P), etc.) may help in the **promotion of growth** of these microbes. This technology was successfully applied by using fertilizers to clean the oil spillage in the ocean due to accident of the oil tanker Exxon Valdez at the Alaskan beaches in 1989 where the oil was decomposed to less harmful products that became the part of the food chain. Similarly, the soils contaminated with highly toxic industrial recalcitrant chemical, polychlorinated biphenyls, can be dechlorinated by the microbial activity.

(*ii*) In another technology, if the population of the indigenous microbes involved in cleaning the pollutant is limiting, the **microbial** samples are isolated from the polluted site and **cultivated artificially in bioreactors.** The biomass thus, obtained is used for the re-inoculation of the polluted site to build up the desired population size. This technology has been found to be successful at some polluted sites.

(*iii*) Some microbial **cultures** are **commercially** available for this purpose, *e.g.*, the white rot fungus, *Phanaerochyte chrysosporium,* helps in the degradation of lignocellulosic materials, *Pseudomonas putida* can cure oil spills, *Pseudomonas fluorescens* biodegrades HCN. In this approach the introduced microbes must be able to survive and compete with the indigenous microbes in the new environment.

(*iv*) Alternatively, the microorganisms may be **genetically engineered** to develop the ability to degrade the pollutant molecules and at present, the research is in progress in this field and would need years to be widely used for the environmental applications.

Though, the technology is **slow** as compared to physical clean up methods and is applicable only in certain environmental conditions only for **certain chemicals** and sometimes, the added chemicals may **contaminate the environment,** it is simple and cheaper and leads to the conversion of toxic substances to non-toxic ones.

5. Microbial Mining

By using certain microbes, *e.g. Thiobacillus ferrooxidans*, many commercially important metals like Cu, Au (gold), U,Fe, etc. can be extracted economically from the poor grade ores containing very small quantity of metal sulphides (ores) (the conventional metallurgical processes would be uneconomical with such ores). Though, this method of metal extraction is slow, it does not affect the environment (no evolution of NO_2 and SO_2 as in convetional metallurgical operations). The heap of the crushed ore (*ex situ* extraction), *e.g.* CuS, or the mining site by drilling the hole (*in situ* extraction) is inoculated with the desired bacterium, *e.g. Thiobacillus ferrooxidans* (a chemoautotroph that derives its energy by the oxidation of sulphides to sulphates) along with some nutrients are pumped into the hole. This leads to the oxidation of S^{2-} in CuS ore to SO_4^{2-} (H_2SO_4) and water soluble $CuSO_4$ following a series of reactions is formed while percolating through the orepile by the bacterium:

(*i*) $CuS + 2O_2 \longrightarrow \mathbf{CuSO_4}$

(*ii*) (*a*) $CuS + Fe_2(SO_4)_3 \longrightarrow \mathbf{CuSO_4} + 2FeSO_4 + S\downarrow$

(*b*) $S_8 + 12O_2 + H_2O \longrightarrow H_2SO_4$ + ATP (*T. ferrooxidans* requires low pH for growth)

(*c*) $2FeSO_4 + H_2SO_4 \longrightarrow Fe_2(SO_4)_3 + H_2$ + ATP

The $CuSO_4$ thus, collected from the bottom of the ore heap (or pumped out from mine site) is drained to recovery plant where it is allowed to flow through the surface of Fe metal leading to the production of pure Cu metal and $FeSO_4$ due to displacement reaction:

$$CuSO_4 + Fe \longrightarrow FeSO_4 + Cu\downarrow$$

Alternatively, many microbes (*e.g.* bacteria *Bacillus* and *Streptococcus*, algae *Chlorella* and *Anabaena*, etc.) can accumulate inside their cells the desired rare metals (*e.g.* Mn, Au, U, etc) by growing them in very dilute metal solutions (*e.g.* many industrial effluents) from such microbial biomass, the metal can be extracted economically.

6. Desulphurisation of Coal

Generally the coal contains 0.2–8.0 % S but, many poor quality coals contain higher amount of S, sometimes exceeding 11%. Burning of such poor quality S-containing coal leads to the release of SO_X into the atmosphere. Such coal can be cured by inoculating the mine sites through drilling the hole with the desired microbe (*e.g. Thiobacillus ferrooxidans*) and water. After microbial growth the S^{2-} (mainly FeS) is be oxidised to SO_4^{2-} that is leached out.

7. Biocontrol Agents

Discussed in the chapter "Biotechnology in Agriculture and Forestry".

8. Biofertilizers

Discussed in the chapter "Biotechnology in Agriculture and Forestry".

QUESTIONS

1. Write short notes on the following:
   (*i*) **Sewage treatment**
   (*ii*) Biofertilizers
   (*iii*) Biocontrol agents
   (*iv*) Recovery of metals by biomining
   (*v*) Anaerobic treatment of wastewater and sewage sludge
   (*vi*) Xenobiotic compounds
   (*vii*) **Landfill technology**
   (*viii*) Activated sludge treatment
   (*ix*) **Treatment of solid wastes**
2. **Discuss the applications of biotechnology in the field of environment giving suitable examples.**
3. Define waste and pollutants. **Briefly describe the various sources of wastes and discuss the hazards from them.**

Chapter 18 *Biological Fuel Generation*

The world depends for its energy requirement upon three major fossil fuels—coal, natural gas and oil, which would become exhausted completely in near future (**coal** may last for 200 years, **natural gas** and **oil** for 40 years). Realizing energy crisis during 1970s when the oil prices increased massively and due to environmental awareness, the scientists are trying to find the alternative non-conventional cleaner sources of energy, like wind, solar, tidal and wave, geothermal and bio-energy. Photosynthetic organisms convert dilute solar energy into more concentrated chemical energy (*e.g.,* carbohydrates) with 3–4% efficiency, making the biomass (2×10^{11} tonnes C fixation/year). Photosynthetically derived **biomass** (which does not have much calorific value), like forest, agricultural and animal residues and domestic organic wastes can be converted by fermentation into clean renewable fuels (like alcohol and CH_4) possessing quite higher energy value. With time, continuous depletion of fossil fuels is making them increasingly expensive and in such situation **biofuels** may become more economic. Biofuels can be used for cooking, generation of electricity or in automobile engines. The CO_2 released during their combustion is entrapped in the photosynthesis (which also produces O_2) for biomass production and, thus, do not damage the environment as the fossil fuels do. Except for biogas, other biofuels are not economical.

The biomass for energy production can be obtained either by cultivating the energy crops (which would require considerable agricultural land and would compete for the food production) or by utilizing agricultural and other organic wastes (*e.g.*, bagasse, molasses, straw, that would help in the waste management also). As there would be competition between biomass energy production and food production on limited agricultural land, particularly in developing countries, obtaining energy from the waste materials would be more promising. Alternatively, the energy crops can be better grown on waste lands unsuitable for food crops.

Forest biomass can also be used for fuel production but slow metabolism of lignocellulosic materials poses problem. 'Steam explosion technology' can be employed for partial breakdown of such complexes in which wood is first exposed to high pressure steam and subsequently pressure is released to cause explosive shredding of wood. Thus, wood gets converted into fermentable substrate.

Moreover, algae, cyanobacteria, aquatic plants can also be used as biomass for fuel production as they are not very commonly used as food resources (though microalgae are difficult to harvest). In addition, productivity in natural habitats is generally limited by low nitrogen and phosphorus, and cyanobacteria are potentially attractive sources as many of them fix N_2 gas. In open seas the nutrients are made available by natural or artificial upwelling forces to enhance biomass production.

The biomass can either be used directly (burning of dried fuel wood, cow dung cake, etc.) or can be converted to more usable fuels achieved by biological (*e.g.*, fermentation) or chemical (chemical hydrolysis, destructive distillation, gasifiction) means or by a combination of both. Some of the important bioenergy are:

1. Bioethanol

Among biofuels, ethanol is the most promising alternative automobile fuel because of its good fuel properties (flash point 45°C and octane number is 89) and availability of huge amount of raw material that is needed for its production. It can be obtained by fermentation of either **sugary** (*e.g.*, molasses, bagasse, sugarcane, sugarbeet) or **starchy** (*e.g.*, corn, wheat, barley, sorghum, potatoes, cassava roots, etc.; but by saccharification they are required to be first converted to simple fermentable sugars) materials by the yeast ***Saccharomyces cerevisiae*** (Fig. 18.1) at 25°C and pH 4.5–4.7 under anaerobic condition or by the anaerobic bacterium ***Zymomonas mobilis*** (that gives higher yield of ethanol than the yeast). Co-culturing of amylolytic *Aspergillus niger* with *S. cerevisiae* converts starch to ethanol.

$$\underset{\text{(starch)}}{(C_6H_{10}O_5)_n} \xrightarrow[\text{(amylase)}]{\text{Saccharification}} \underset{\substack{\text{(disaccharide)}\\ \text{(malt)}}}{C_{12}H_{22}O_{11}} \xrightarrow{\text{maltase}} \underset{\text{(monosaccharide)}}{C_6H_{12}O_6}$$

$$\underset{\text{(sucrose)}}{C_{12}H_{22}O_{11}} \xrightarrow{\text{Invertase}} 2C_6H_{12}O_6 \xrightarrow{\text{yeast}} 2C_2H_5OH + 2CO_2$$

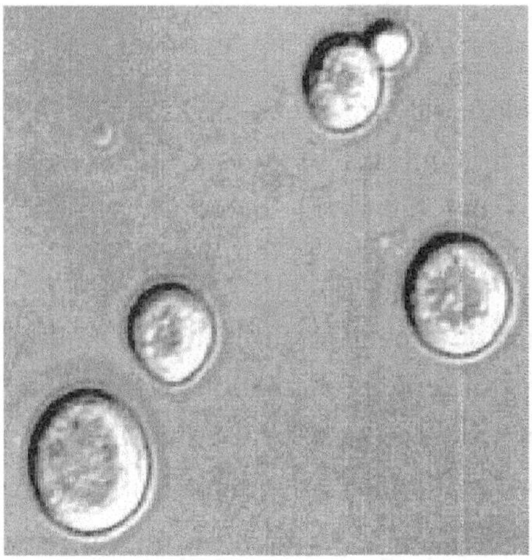

Fig. 18.1 Cells of *Saccharomyces cerevisiae.*

About 225 litres ethanol/ton of molasses can be obtained. To make the availability of necessary fermentable sugars, most raw materials require some degree of pretreatment. With sugarcane this treatment is minimal (milling), whereas cassava roots (25–38% starch) require saccharification (conversion of starch into simple sugars) by hydrolysis either by acids or by enzymes (amylases).

Cellulosic raw materials (*e.g.,* wood and straw) require more extensive pretreatment, which needs increased energy input *e.g.*, treatment with cellulase (obtained from the fungus *Trichoderma reesei*) to convert cellulose into glucose.

The distilled ethanol (green petrol) burns completely and can be used as a partial or complete substitute for motor fuel, **fossil petrol** (gasoline, flash point 13°C). It is energy efficient, produces 57% lesser (toxic) CO, 64% lesser hydrocarbons and 13% less NO during combustion in vehicle engines than petrol. Historically, ethanol and methanol were used as motor fuels in Europe prior to World War II either in pure form or as mixed with petrol. **Gasohol** is **10–20% ethanol** in gasoline that can be used in existing engines without any modification. Ethanol of very low moisture content is only suitable for blending with petrol, as the alcohol and gasoline get separated in the presence of water. About 95% ethanol is obtained by the **distillation** process (ethanol has boiling point 87°C as against 100°C for water helps in the distillation process) which is further distilled to about 100% ethanol by adding small amount of benzene. The proper distillation of ethanol makes the bioethanol costly. Since 1980s, it is extensively used in **Brazil** (in about 30% vehicles) and is obtained from sugarcane and cassava where the climate suits for the growth of these crops. Here, the ethanol is obtained mainly by batch fermentation of the substrate. In **USA** it is obtained from corn. Though at present, the ethanol's price is about twice to that of petrol, the oil prices would continue to increase, whereas, with technical advancement the production price of ethanol would decrease; thus a time will come when the prices of the two would become almost same and the bioethanol would substitute for some of the fossil petrol. Field trials and testing are continuing in many countries.

Biotechnology is involved in the production of more efficient microorganisms by **genetic engineering** technology for improved ethanol production, with extra necessary enzymes, resistance to high alcohol concentration (as ethanol produced limits the growth of yeast), etc. This would lead to the substantial reduction of cost of ethanol production.

2. Methanol

Methanol, which produces lesser hydrocarbon, NO_x and CO than the fossil fuels in automobile engines, can be obtained from fermentation of organic wastes and can be used as fuel in vehicles either in pure form or as 85% methanol blended in petrol or as15–20% in diesel **(diesohol).**

3. Biodiesel

Considerable interest is being developed for utilization of biodiesel in Europe, where, plenty of agricultural land can be utilized for the energy plantation. Many plants yield oils **(lipids)** and many others **hydrocarbons,** which after processing can be used to substitute for the fossil diesel. Vegetable oil based fuels are most important to replace the diesel.

(*i*) ***Edible Vegetable Oils***: Diesel engine without any modification can run on many unblended edible oils like, peanut oil (used in Paris in 1990, flash point 237°C), sunflower, rapeseed (10% in diesel), corn, linseed, soybean (having flash point 219°C, flash point of fossil diesel is 46°C), coconut oils (scientists in Australia and Europe have conducted tests). It was found that fuel consumption in automobiles was slightly higher for these oils in comparison to the fossil diesel. Since these edible oils are used in large scale for food preparation, more attention should be given for the use of non-edible oils for the purpose of energy use.

(*ii*) ***Non-edible Vegetable Oils*:** Some of the important non-edible vegetable oils that can be used as fuel for diesel-vehicles are *Jatropha*, *Jojoba*, *Azadirachta indica* (neem), *Croton*, *Garcinia*, *Salvadora*, etc. In Thailand *Jatropha* seed oil has been successfully tested to run the diesel engine. This plant can grow on marginal nutrient poor soils. Trees are planted at a distance of about 2 m, which start producing nuts after 4–5 months and can survive for about 50 years. The seeds contain 30–60% oil to yield about 2,000 kg oil/ hectare land (Fig. 18.2).

Fig. 18.2 *Jatropha* plants

For the extraction of oils, the seeds (edible or non-edible) are subjected to a number of preliminary treatments to produce the meal from which the oil is extracted by solvent extraction using hexane or trichloroethylene. The crude oil thus obtained is then refined to be used as biodiesel.

(*iii*) ***Hydrocarbons*:** Many plants belonging to the families Euphorbiaceae (*e.g.*, *Euphorbia*) and Asclepidaceae (*e.g.*, *Calotropis procera, Asclepias Copaifera Multijuga*). *Copaifera multijuga*, a tropical legume tree with N_2-fixing root nodules and preferably grows in arid lands that are unsuitable for the growth of crop plants. It is most promising source of biodiesel. The latex produced contains mainly water and emulsified 30% hydrocarbons (mainly C_5H_8) that can be processed into biodiesel. After collection, the latex is dried and processed into diesel-like substance. *Euphorbia lathyris* is the most important species for this purpose.

Many algae (*e.g.*, green alga *Botryococcus braunii* can accumulate 75% hydrocarbons and grows in abundance in Sambhar Salt Lake in Rajasthan) (Fig. 18.3), fungi and bacteria (*e.g.*, *Micrococcus leuteus*) also accumulate high concentration of hydrocarbons. Through genetic manipulation, about 10-fold increase in hydrocarbon in *M. leuteus* has been achieved.

Fig. 18.3 Bloom (top) and colonies of *Botryococcus braunii* as seen through microscope (bottom).

The exploitation of hydrocarbon as biodiesel is at the experimental stage.

4. Biogas

Methane is a valuable and economical energy source that is obtained by the fermentation of organic wastes from agriculture (including cattle dung), domestic, food processing industries sewage sludge in biogas reactors or landfill by the anaerobic digestion where CH_4 is captured. It is obtained by a group of indigenous anaerobic microorganisms optimally at about 30°C and helps not only in generating energy and manure production but also in the management of solid wastes (has multiple benefits). At the first stage a group of microorganisms lead to the **solubilization** of complex organic molecules, like cellulose, fats, proteins, etc. to produce simple soluble products. These soluble products are then converted to **organic acids** (mainly acetic acid and formic acid) and H_2 by another group of microorganisms. In the final stage, these acids are converted (or decomposed) by **methanogenic bacteria** (*e.g.*, *Methanobacterium*, *Methanobacillus*, *Methanosarcina*, etc.) to CH_4 and CO_2 using H_2 that is generated microbially.

$$\underset{\substack{\text{(Complex carbohydrates)}\\\text{(Polysaccharides)}}}{(C_6H_{10}O_5)_n} \longrightarrow \underset{\substack{\text{(Simple carbohydrates)}\\\text{(Disaccharides)}}}{C_{12}H_{22}O_{11}} \longrightarrow \underset{\substack{\text{(Simple carbohydrates)}\\\text{Monosaccharides}}}{C_6H_{12}O_6}$$

$$C_6H_{12}O_6 \xrightarrow[\text{condition}]{\text{Anaerobic}} \text{Organic acids} + CO_2 + H_2$$

$$4H_2 + CO_2 \xrightarrow{\text{Methanogens}} \mathbf{CH_4} + 2H_2O$$

$$\text{Organic acids} + CO_2 + H_2 \xrightarrow{\text{Methanogens}} \mathbf{CH_4} + H_2$$

$$[C_6H_{12}O_6 \longrightarrow \text{Organic acids} \longrightarrow 3CH_4 + 3CO_2]$$

The production of CH_4 depends upon the nature of waste, *e.g.*, the presence of lignin (in most agricultural and urban wastes), which is not easily biodegradable and affects the CH_4 production. When CH_4 is obtained from cattle dung, it is called **biogas** (gobar gas) and is used for cooking purpose in rural areas of India, China (largest user having over 7 million biogas units) and Pakistan. It generates lesser smoke than cow dung cake burnt in conventional chulhas. For its production, animal dung is mixed with water and allowed to ferment in underground airtight bioreactors of family-size or community-size that are made up of bricks or steel. Gas formed gets entrapped and raises the inverted metal drum covering the surface of waste from where it is withdrawn by pipelines. Biogas is a mixture of 50–80% CH_4, 15–45% CO_2, 5% H_2O and traces of gases like N_2, H_2S, etc. (Fig. 18.4). About 30–50% of urban waste can be converted to CH_4. Purification of CH_4 from biogas has more calorific value, but is expensive and unsafe.

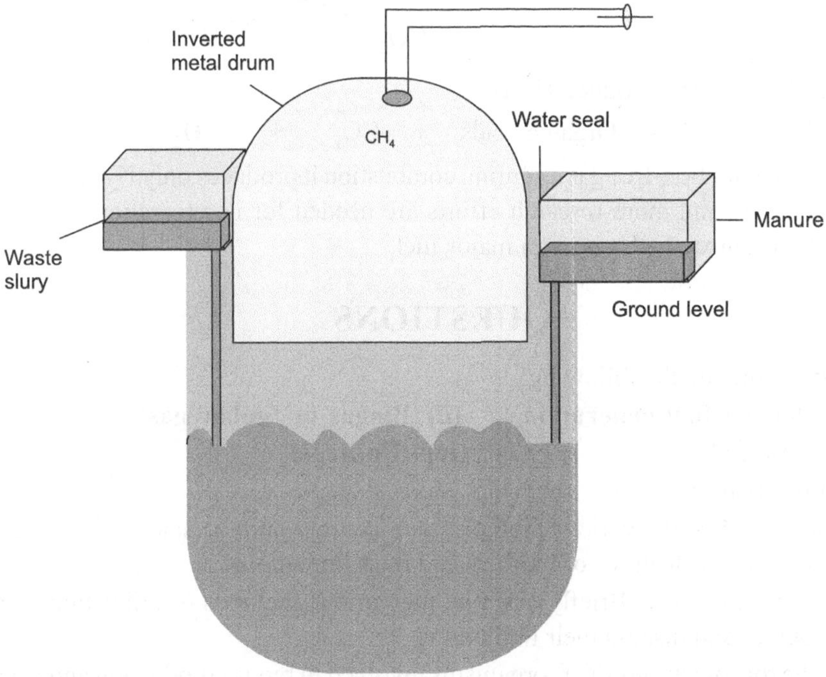

Fig. 18.4 Biogas plant.

The solid or liquid residue after CH_4 production is rich in nutrients (N, P and K) and can be used as **manure** for crop plants. Methane gas, the main component of biogas, can be used for the generation of mechanical, electrical (used as fuel to generate steam for driving electricity turbines as in California, USA, UK, Timarpur in Delhi) and heat energy (cooking). Under favourable conditions, 10 kg of dry organic matter can produce $3m^3$ of biogas which can provide 3 h of cooking or lighting.

Methane gas also exists in the atmosphere as greenhouse gas (next to CO_2) and of this total methane, 20% is derived from natural wetlands, 20% from rice fields and 15% from enteric fermentations in animals by microbial action, which can be captured and used as fuel.

5. Biohydrogen

Phototrophic microorganisms are capable of significant amount of H_2 production. Hydrogen is highly energetic and many photosynthetic (*e.g.*, *Rhodospirillium*, *Rhodopseudomonas*, etc.) and non-photosynthetic bacteria (*e.g.*, anaerobic *Clostridium* reduces H^+ ions by fermentation), green algae (*e.g.*, *Scenedesmus*, *Chlorella*, *Chlamydomonas*), cyanobacteria (*e.g.*, *Anabaena variabilis* that can produce H_2 and CO_2 from fructose sugar in the presence of light and anaerobic condition created by Ar gas), etc. possess hydrogenase enzyme (which is sensitive to O_2) and produce H_2 efficiently. The photosynthetic organisms produce H_2 from H_2O in the presence of sunlight under certain conditions. *In vitro*, it can be obtained by isolated chloroplasts in the presence of water, light and the hydrogenase enzyme:

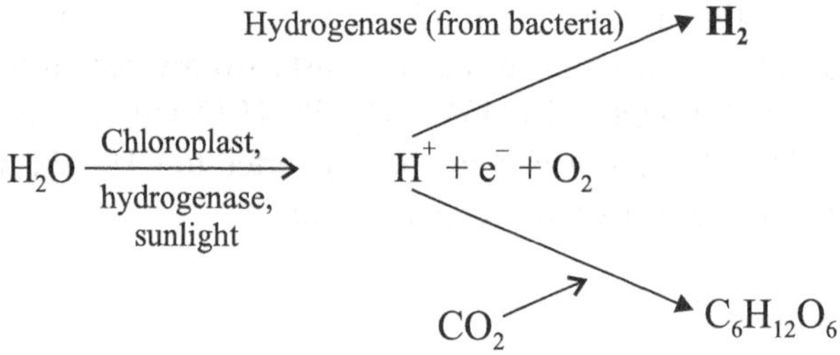

The anaerobic bacteria produce H_2 as:

$$C_6H_{12}O_6 \longrightarrow \text{Organic acids} + CO_2 + \mathbf{H_2}$$

Though, it is a pollution-free gas as during combustion it produces only H_2O, but its production at present is very costly and more research efforts are needed for its utilization as an as attractive and economical fuel. Currently, H_2 is not a major fuel.

QUESTIONS

1. Write short notes on the following:
   (*i*) **Biological fuel generation** (*ii*) **Biogas or 'gobar gas'**
   (*iii*) **Bioethanol** (*iv*) **Biodiesel**
   (*v*) **Biohydrogen**
2. Define biofuel. List the various biofuels, their sources and the status of their exploitation. Discuss the various features of biofuels and their limitations.
3. What are energy crops? Briefly describe the various methods of utilization of biomass as energy sources and discuss their limitations.
4. Explain the role of various microorganisms involved in biogas production and list the various advantages and disadvantages of biogas as a fuel.

Chapter 19

BIOTECHNOLOGY IN AGRICULTURE AND FORESTRY

Human is totally dependent on agriculture and many countries are still not self-sufficient in food production. Biotechnology is creating a revolution in this field to provide higher quality and lower cost products to meet the requirement of continuously increasing human population. Since early times, man is trying to improve the quality and productivity of agriculturally important plants by **selection** of better varieties and employing **traditional breeding programmes** involving sexual crosses, which are slow and difficult processes. Extensive research is also going on with forest trees, as forest resources are continually depleting fast because of indiscriminate deforestation. Trees have long generation times and desired traits can be achieved by faster tissue culture and genetic engineering technologies.

TISSUE CULTURE TECHNOLOGY

This technique employs the isolation of a few cells from the concerned plant that are grown and maintained in artificial conditions using nutrient media containing certain plant hormones. The cells either grow in the **suspension** form (where shaking facilitates aeration, mixing of medium and breakage of cell aggregates) or undifferentiated mass of cells called **callus,** in flasks or tubes. The cultured cells can be used in various ways in the field of agriculture and forestry:

(*i*) The valuable **metabolites** (*e.g.,* quinine used for the treatment of malaria, jasmine for fragrance) are obtained commercially by growing the cells in large bioreactors.

(*ii*) Alternatively, these cultured plant cells can be induced to develop shoots and then roots to produce complete **plantlets** by selecting correct hormone (auxins and cytokinins) concentrations (ability of a plant cell to regenerate into whole plant is known as totipotency). These plantlets after removal from the culture vessels, finally, can be grown in the fields. Thus, a large number of clones of slow-growing rare plants can be obtained rapidly either by avoiding the formation of embryo (*e.g.,* ornamental plants, fruit trees like banana). (This technique is known as **micropropagation**) (Fig. 19.1)

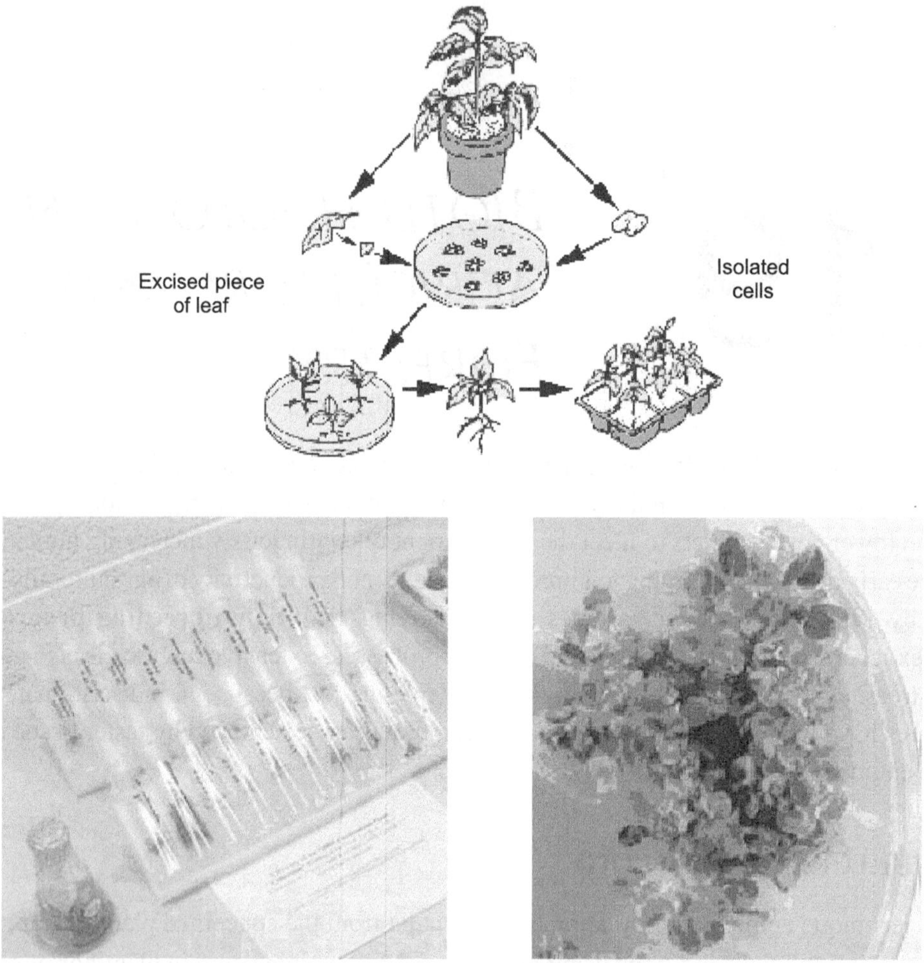

Fig. 19.1 Micropopagation of plants.

(*iii*) The cultured cells after the formation of **somatic embryos** can germinate to produce the whole plant, *e.g.*, oil palm, date palm etc. These somatic embryos can be encapsulated within gel like substances (calcium alginate) containing nutrients, fungicides and bactericides to check the microbial contamination and covered with a waxy coating to prevent the drying of these artificial seeds. Such seeds are known as **artificial seeds.**

(*iv*) This technique is also used for the **germplasm conservation** in the form of cultured cells or tissues (instead of seeds) and production of **transgenic plants** (genetically modified) after genetic manipulation of the cultured cells.

(*v*) Sometimes, in interspecific crosses the hybrid embryos in the seeds are unable to grow completely because of degeneration of endosperm (part of the seed containing stored food) or some seeds lack stored food (*e.g.*, orchids) or seeds are not able to germinate due to the presence of some inhibitors in the seeds. But, by **embryo culture** technique the excised embryo can be *in vitro* allowed to complete the development and finally grown into plantlets.

(*vi*) Using **anther** (contains male gametes) **culture** technique, that was first used in India in producing haploid *Datura* plants, large number of haploid plants can be obtained from the pollen grains. These haploids (having one set of chromosomes) are sterile and are used to produce superior quality homozygous (having two copies of the same set of chromosomes) plants by treatment with colchicines (inhibits the spindle formation during cell division so that all the chromosomes remain in the same cell and not distributed into the two daughter cells) that are used in the plant breeding programme, *e.g.*, in cereals. Haploids can also be used in the isolation of mutants (recessive mutants would be expressed easily as they have just one set of genes) and chromosome number in these haploid mutants can be doubled by treatment with colchicine. Haploids can also be produced by **culturing** the unfertilized **ovaries** or ovules.

(*vii*) Many types of plants can also be regenerated from protoplasts that are produced by digesting away the cell wall by the enzymes pectinase and cellulase and then maintaining them in a suitable osmotic medium, *e.g.*, potato, tobacco, tomato etc.

(*viii*) Somatic hybrids can be produced by fusion of **protoplasts** of two species in the presence of polyethylene glycol (PEG) or by exposure to a very brief high voltage electric current that cannot be produced normally by sexual hybridization process. After cell wall formation, the hybrid cells divide to produce plantlets, *e.g.*, somatic hybrid (that produced fertile flowers) between a non-flowering variety and a flowering variety of potato, somatic hybrid produced between rice and carrot etc. These somatic hybrids may be used for gene transfer, production of useful allopolyploids (*e.g.,* the allopolyploid AABB produced from the diploid hybrid AB), etc.

(*ix*) ***Somaclonal Variation***: In *in vitro* culturing of cells, the genetic variability increases with duration of culture to give rise to different kinds of cells. This property is known as somaclonal variation and may arise due to changes in chromosome number or structure, spontaneous mutation, alteration of gene expression, gene amplification, etc. From these different kinds of cells, cells with desirable properties can be screened out, *e.g.*, disease resistance (potato plants resistant to Early Blight disease caused by the fungus *Alternaria solani* and Late blight disease caused by the fungus *Phytophthora infestans*), herbicide resistance, improved nutritional quality, adaptation to stress (saline soil tolerance in *Citrus*, low temperature tolerance in tobacco, toxic metals), increased biosynthesis of plant products of medicinal or industrial importance etc.

GENETIC ENGINEERING (TRANSGENIC PLANTS)

The genetic engineering overcomes the limits of interspecific crosses and permits the introduction of foreign genes from distant species that is not possible following conventional breeding programmes. It is now possible to manipulate the genome of almost any plant species including agricultural crop plants by introducing advantageous new traits (by incorporating isolated gene from an organism or chemically synthesized gene) that produces a new protein/other product or by preventing the expression of an existing native gene following various techniques.

Agrobacterium tumefaciens has been used extensively to transfer the desired foreign DNA by inserting it into the Ti plasmid and then integrating into the **dicotyledonous** host plant genome. After screening of the desired recombinant cells, they are regenerated into the complete plantlets, *e.g.*, Loblolly pine used in forestry. Ti plasmids have also been used to insert 'antisense genes' to negate the functions of specific plant genes concerned with an undesirable trait. Such genes are produced by reversing the orientation of a gene in relation to its promoter (in DNA at promoter region RNA polymerase binds to start the transcription giving rise to mRNA) so that a RNA, that is complementary to mRNA, is synthesized and after base pairing with the normal mRNA, double-stranded RNA is produced that is unable to be translated to give respective proteins (Fig. 19.2).

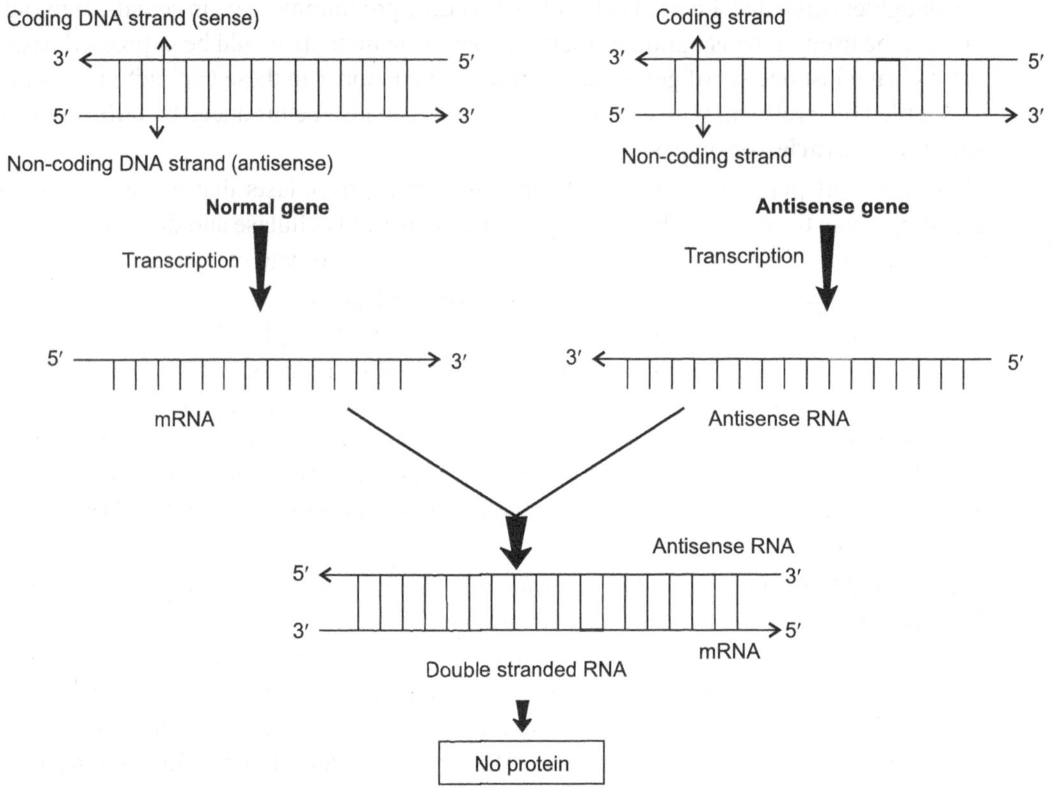

Fig. 19.2 Introduction of antisense gene for polygalacturonase enzyme for the development of Flavr Savr tomato.

This method of using *Agrobacterium tumefaciens* to manipulate the genome is unsuitable for monocotyledonous plants, like wheat, rice, maize, etc. In **monocotyledonous** plants gene transfer is mostly achieved by gene gun by bombarding the foreign DNA-coated gold or tungsten particles (micro projectiles) into the host plant cells that penetrate the cell wall and deliver the DNA into the nucleus. The recombinant cells can be regenerated into whole plants.

The widespread use of insecticides, fungicides, herbicides (to kill the plant weeds) and pesticides has the damaging effect on the environment and human health. Many crop plants have not only been improved with respect to the pest and disease control but also to increase the yield and quality of the product by genetic engineering technology.

(*i*) ***Resistance to Insect Pests***: Proteinacious toxin genes (*cry* genes, where *cry* means crystalline toxins) from *Bacillus thruringiensis* (Bt), that is toxic to larvae of certain insects, have been introduced into several crops, so that they become resistant to that particular insects, *e.g.*, cotton, tomato etc. In India also, some insect resistant *cry* genes containing cotton varieties are grown in the fields.

(*ii*) ***Resistance to Microbial Diseases***: Microbial diseases, particularly fungal and viral, cause huge losses and limit the plant productivity worldwide. Resistance to many viruses by integrating the genes for viral coat proteins have been achieved in several crop plants, *e.g.*, rice.

(*iii*) ***Resistance to Herbicides***: Herbicide-tolerant crop plants have been produced by genetically engineering the crop plant genomes that would reduce the use of herbicides, *e.g.*, potato, rapeseed etc. This method of weed control is more effective, less costly and more environment friendly.

(*iv*) ***Delayed Ripening Fruits***: Losses during storage and transportation of some crops is very high due to over-ripening and softening of fruits and vegetables. The ripening is due to some endogenous enzyme activities that can be genetically stopped or slowed down, *e.g.*, in tomato, the polygalacturonase breaks down the cell wall, leading to softening of fruits during ripening. 'Flavr Savr tomato' has been engineered in USA by blocking the expression of polygalactarunase enzyme by introducing the antisense gene for this enzyme. This tomato which has improved flavour and shelf life and is now marketed in USA. Such techniques will now be used in a wide variety of soft fruits. Other studies are considering inhibition of ethylene synthesis for controlling fruit ripening (ethylene is a plant hormone; which induces fruit ripening).

(*v*) ***Stress Resistance***: Cold resistant tobacco plant has been produced by introducing the gene for glycerol-1-phosphate acyl transferase from the plant *Arabidopsis*.

Despite their beneficial effects, it has been argued that transgenic plants may be harmful to the environment as:

(*i*) There may be possibility of transfer of transgene (foreign gene) from transgenic crops to their wild relatives through pollens, making them more persistent and damaging. To prevent this, transgenics should not be grown near the occurance of their wild relatives.

(*ii*) Transgenic crops may themselves become persistent weeds.

(*iii*) They may damage the environment in some unknown manner.

BIOLOGICAL CONTROL

Though the use of chemical pesticides has improved the production in agriculture and forestry it has also damaged the environment and contaminated the food by accumulation of pesticide residues. Therefore, alternatives of these detrimental chemical pesticides are the direct use of some microorganisms, like bacteria, fungi, viruses, called biocontrol agents, to control mostly the insects, pests and occasionally the disease causing pathogens or weeds in agriculture. These are basically naturally occurring disease causing microorganisms or predators of the weeds or pests.

(*i*) ***Bioinsecticides***: The most successful biocontrol agent is the spores (or crystalline proteins) of *Bacillus thuringiensis* that produce toxic proteins against the insect pests and are highly specific in their activity. The commercially available preparation contains a mixture of spores, crystal proteins and a carrier material. The toxic protein activity does not persist in the environment long and disappears after 1–2 days of application. Moreover, these toxic proteins are highly specific and do not affect the non-target organisms. Certain genetically modified bacteria, *e.g., Clavibacter* containing *B. thuringiensis cry* gene, are used to protect the crop plants against insects.

Many viral (like, bacculoviruses) and fungal (*e.g., Hirsutella* sp. for *Citrus* mites) biopesticides are available in the market.

(*ii*) **Bioherbicides:** Fungal pathogens of the weeds are most attractive commercially used weed control agents, *e.g., Phytophthora* sp., *Collectotrichum* sp. etc.

(*iii*) **Disease Control:** Both bacteria (*e.g., Agrobacterium radiobacter*) and fungi (*e.g., Trichoderma* sp. to control soil borne fungal pathogens) are commercially used to control many plant diseases. Certain genetically modified bacteria, (*e.g., Pseudomonas* containing chitinase gene from *Serratia*) are used against some fungal diseases (fungal cell wall is made of chitin, which gets dissolved by the enzyme chitinase).

Unlike chemical pesticides, biopesticides are **cheaper** and do **not** leave harmful **residues** to contaminate food and fodder and to pollute the environment and do not destroy non-target species, as they are highly **specific.** Moreover, insects are not develop resistance to biopesticides.

BIOFERTILIZERS

Microorganisms (bacteria, cyanobacteria and fungi) used to enhance the availability of nutrients, like nitrogen and phosphorus for the crops are called biofertilizers. Chemical fertilizers are short in supply and the production of (nitrogenous) these fertilizers needs use of fuels. Unlike chemical fertilizers, biofertilizers do not give quick responses, have to compete for the native microorganisms and substitute for only a part of the chemical fertilizer. Still, biofertilizers are preferred as they are **cheaper, ecofriendly** and have long-term **sustainability.** For optimum crop yields biofertilizers are used in combination with the chemical fertilizers. Some important microorganisms with practical or potential application as biofertilizer are:

(*i*) ***Rhizobium* Sp.:** They are G^- symbiotic N_2^- fixing bacteria forming pink coloured root nodules in leguminous plants, like gram, mung, pea, pigeonpea (arhar dal) etc. (Fig. 19.3). Inside the nodules, many bacteria change into non-dividing bacteroids which fix the atmospheric nitrogen into ammonia with the help of the enzyme nitrogenase. Nitrogenase is highly sensitive to O_2 and the pink coloured **leghaemoglobin** binds with O_2 to maintain a correct concentration of O_2 to protect the nitrogenase. The estimated N_2 fixed by *Rhizobium* ranges from 50–150 kg/ha/year and increases about 10–15% yield. The culture of *Rhizobium* is grown in flasks or fermenters and mixed with a carrier material like peat. This *Rhizobium* biofertilizer is used for coating the seeds.

(*ii*) ***Azotobacter*:** It is a free-living N_2–fixing bacterium utilizing the organic matter present in the soil and is capable of fixing 15–30 kg N/ha/year. Low O_2 concentration in the cells is maintained by high rate of respiration.It is used for crops, like wheat, rice, and vegetables.

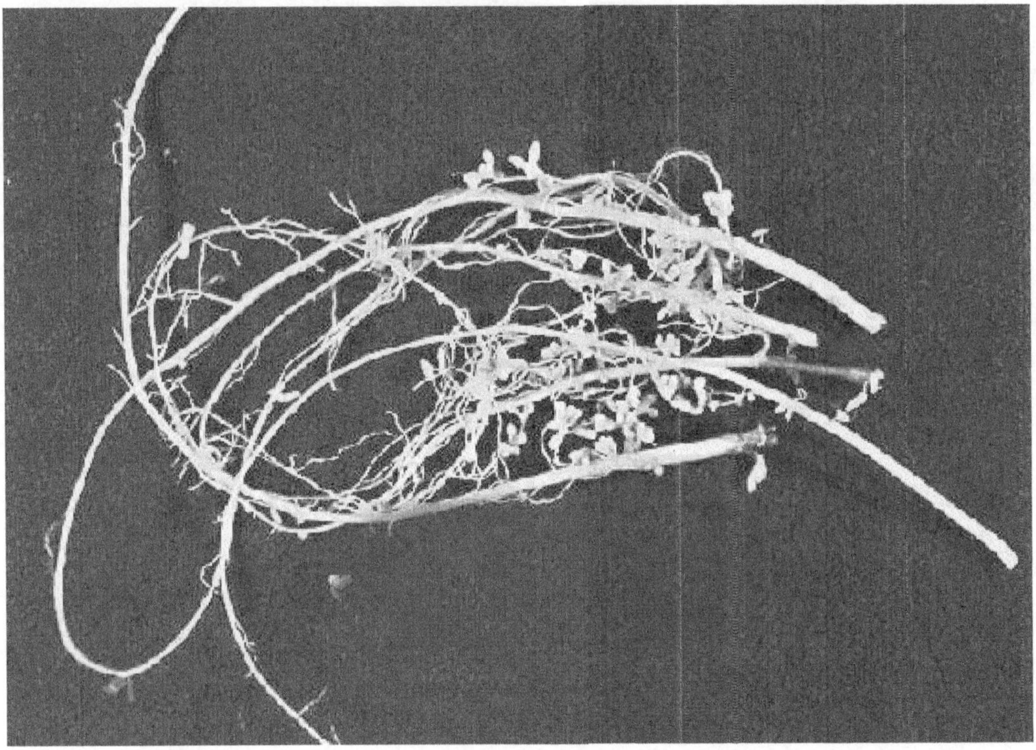

Fig. 19.3 Soybean root nodules.

(*iii*) ***Azospirillum*:** It is found in association with roots of many plants of grass family (associative symbiosis) and is used for crops like jowar, wheat, bajra, etc. and fix about 15–30 kg N/ha/year.

(*iv*) **Cyanobacteria (blue-green algae):** They are photosynthetic, prokaryotic organisms, many of which fix N_2 asymbiotically, *e.g., Nostoc, Anabaena, Aulosira, Tolypothrix*, etc. (Fig. 19.4). Generally, nitrogen fixation occurs in specialized cells called heterocysts where there are mechanisms (*e.g.,* impermeable cell wall to inhibit the diffusion of gases, absence of O_2-evolving Photosystem II, high rate of respiration, etc.) to make the environment anaerobic in order to protect the enzyme nitrogenase from O_2 damage. They are preferably used as biofertilizer, containing a mixture of 2–4 cyanobacterial cultures and a carrier material (generally soil), in tropical rice fields. Cyanobacterial biofertilizer not only contributes for 20–30 kg N/ha/year but also provides organic carbon, various growth promoting substances and improves the physical properties of the soil. After application for 3–4 successive years in the field, there is no need to apply this biofertilizer again as sufficient population builds up in the soil. They are also useful in reclamation (curing) of alkaline and saline soils. Cyanobacteria are grown in small tanks or ponds in the sunlight and after harvesting the biomass, it is dried in the sun and used to sprinkle the rice field as biofertilizer.

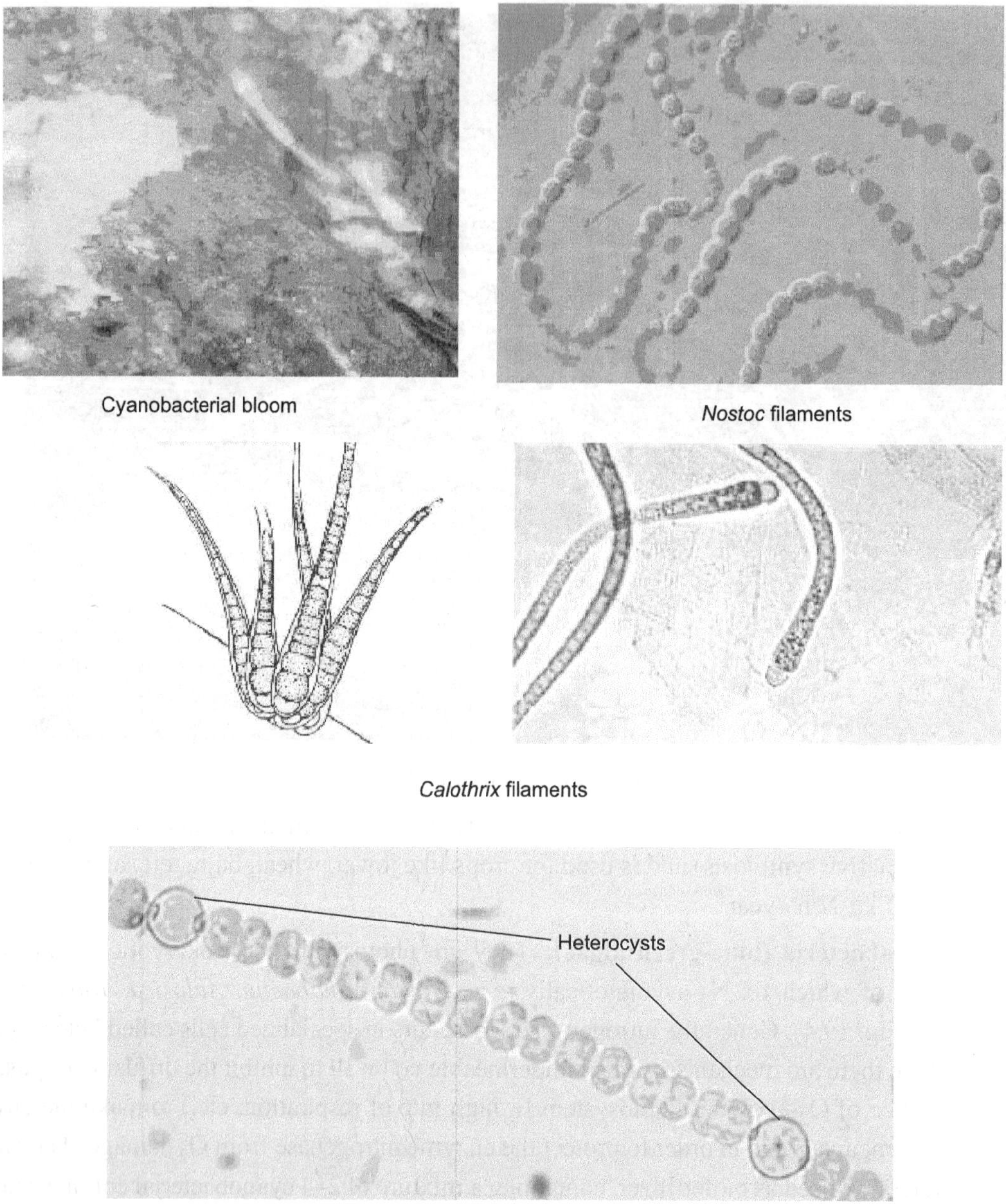

Fig. 19.4 Cyanobacterial bloom in the tank and some biofertilizer strains of cyanobacteria.

(*v*) ***Azolla*:** *Azolla* is a fern and the symbiotic cyanobacterium, *Anabaena azollae,* as observed under the microscope found in the cavity on the underside of its leaves. It is widely used as biofertilizer in rice fields of temperate countries, like Vietnam (Fig. 19.5).

(*vi*) **Phosphate Solubilizing Bacteria (PSB):** Some bacteria, *e.g., Pseudomonas, Bacillus, Thiobacillus,* etc. convert insoluble non-available inorganic phosphorus (rock phosphate) present in the soil into soluble form which is utilizable by the crop plants. These bacteria also protect the plants from harmful microorganisms by producing siderophores which chelate the iron present in the root zone, thus, iron becomes unavailable for the growth of the harmful microorganisms.

Fig. 19.5 Cultivation of *Azolla* in cement tank.

(*vii*) **Vesicular Arbuscular Mycorrhizal (VAM) Fungi:** Certain fungi, *e.g., Glomus*, are found in association with plant roots and are called mycorrhizae. These fungi may be located at the root surface (ectomycorrhizae) or inside the roots (endomycorrhizae) and convert non-available phosphorus into the available form, produce growth promoting substances and also protect crop plants against soil pathogens. These VAM fungi are yet to be exploited on a commercial scale. In USA these are commercially applied in *Citrus,* where seedlings are inoculated with VAM.

QUESTIONS

1. **Discuss the applications of biotechnology in the field of agriculture industry.**
2. Write short notes on the following:
   (*i*) **Micropropagation** in crop improvement
   (*ii*) Somaclonal variation
   (*iii*) Haploid plants in crop improvement
   (*iv*) Somatic hybrids in crop improvement
   (*v*) Artificial seeds
3. Briefly describe the various approaches to large scale culture for plant cells and discuss their advantages and limitations.
4. Define micropropagation. Briefly describe the various approaches for micropropagation and discuss their advantages and limitations.
5. **Define somaclonal variation. Briefly descibe their isolation, characterization, molecular basis and applications.**
6. Briefly describe the mode of production of haploid plants and their various applications for crop improvement.
7. **Define somatic hybrids and their possible applications.**

Chapter 20 BIOTECHNOLOGY AND MEDICINE

About 200 years ago the life was very harsh and less than 2% people were able to live beyond the age of 65 years. With research, it became possible to control various diseases by using different antiseptics, vaccines, antibiotics, etc. Life expectancy over the last 150 years rapidly increased from 35 to almost 80 years. Today, the infectious diseases are no longer the main threat to life and only the chronic diseases, like cancer, cardiovascular diseases, etc. are dangerous and can be solved through genetic manipulation that has lead to the development of new strains with higher productivity and new medicines. New techniques of protoplast fusion and gene transfer technologies gave products, which resulted in the decrease in cost of production.

Pharmaceutical drugs in the market are either synthetic chemicals or derived from biological sources. The various disease treatment agents are either obtained from:

(*a*) **Non-recombinant organisms,** *e.g.*, antibiotics and vitamins from microorganisms, taxol (for the treatment of breast and ovarian cancers) from cell culture of *Taxus*, interferons from animal cell cultures, etc.

(*b*) **Recombinant organisms,** *e.g.*, insulin, human growth hormone, interferons, blood clotting factor, etc. produced by genetically engineered microbial cells or cultured mammalian cells.

Antibiotics: are the antimicrobial compounds produced by living microorganisms to control infectious diseases and are extensively used in medicine since about 1945. The first antibiotic, **penicillin** which inactivates a wide range of bacteria, was discovered by **Alexander Fleming** (1928) and was produced by the fungus *Penicillium notatum.* This antibiotic was used for the treatment of deadly diseases, like pneumonia, tuberculosis, cholera, leprosy, etc., but today most of the pathogenic bacteria have developed resistance against this antibiotic.

Though, over 4,000 antibiotics have been isolated mostly from actinomycetes (*e.g., Streptomyces*) and mould fungi (*e.g., Aspergillus, Penicillium*), only 50 are widely used in fields, lke:

(*i*) Human and veterinary medicine
(*ii*) In animal farming to increase the weight of livestock and poultry
(*iii*) To control plant diseases
(*iv*) As insecticides
(*v*) In food and feed preservation

Some antibiotics are broad spectrum and affect a wide range of microorganisms, *e.g.*, chloramphenicol, tetracyclines, etc., whereas, others are narrow spectrum and effective against a few bacterial species, *e.g.*, streptomycin, penicillin, etc.

Biopharmaceuticals: are the recombinant drugs, recombinant vaccines and monoclonal antibodies developed through the techniques of molecular biology or genetic engineering. Earlier, they were derived from organs of cadavers and blood banks. But through genetic engineering, some of these scarce molecules can be obtained in unrestricted quantities. In this, the human-derived gene is inserted into a suitable microorganism or mammalian cell culture, which produces the desired therapeutic protein identical to the normal human protein or with meaningful improvements. Moreover, in this method they are free from dangerous contaminants.

Certain human proteins of pharmaceutical potential, *e.g.*, blood clotting factor, have been expressed in the lactating glands of some transgenic animals, like mouse, sheep, cow, pig, and the products are secreted in the milk from which they can be easily extracted and purified. This method of production of valuable products in the milk, blood or urine of the transgenic animals (equivalent to bioreactors) is called **molecular farming,** that is at the experimental stage and yet no commercial product is available in the market. This does not cause any adverse effect on the animals. An American company is producing human haemoglobin in the blood of transgenic pigs, which could serve as a human blood substitute that would be free from pathogen like AIDS virus.

Some of the important biopharmaceuticals are:

(A) RECOMBINANT DRUGS

(*i*) **Insulin:** is used for the treatment of diabetes and was earlier extracted from pigs and cattle which had some side effects due to additional contaminating compounds, whereas, the recombinant insulin, introduced into the market in 1982, produced by recombinant *E. coli* has no such problems. Banting and Best discovered and isolated insulin from the dog's pancreas. **F. Sanger** (1955) (fourth person in the world who was awarded Nobel Prize two times; the first one in chemistry, for giving the structure of insulin in 1958 and the second for 'di-deoxy chain termination method of DNA sequencing' in 1980) gave the amino acid sequence of insulin and was awarded Noble Prize (Fig. 20.1). It is made up of 2 polypeptides — one is 21 amino acids long and the other is 30 amino acids long (total 51 amino acids).

Fig. 20.1 Frederick Sanger

(*ii*) **Somatostatin:** is a growth hormone and is a protein which was introduced into the market in 1985. It is extremely difficult to isolate from the animals (about 0.005 g hormone is isolated from 0.5 million sheep brains). After cloning of the human growth hormone gene in *E. coli,* the yield has been increased many times (0.005 g/L bacterial fermentation medium). It is used for the treatment of hypopituitary dwarfism which is caused due to growth hormone deficiency (about 1 child in 5000). The hormone improves the muscle growth and is exploited by some athletes.

(*iii*) **Interferons:** They were discovered by two British researchers, Issacs and Lindenmann in 1957. They have been named interferons as they interfere with the virus infection by inducing the degradation of viral mRNA in infected cells by the enzyme ribonuclease. Human interferons are glycoproteins (protein+sugar) and are produced by virus infected cells in order to protect healthy cells. They control a large number of viral diseases, *e.g.*, common cold, cancer etc. They have potential application in cancer therapy as they attack cancer cells by inhibiting their growth (division). They also check the growth of cancer virus, AIDS virus, hepatitis-B virus, etc. The protection conferred by them is non-specific (interferons induced by one virus can provide protection against other viruses also). They have not become the antibiotics of the viral infections because minute amounts of interferons are produced in the cells and, moreover, their isolation (extraction) from other cellular proteins is complicated. Conventionally, they were obtained from human blood. Currently they are produced from cultured mouse fibroblasts or human leucocytes infected with Sendai virus and genetically engineered bacteria (*E. coli*) or yeast with inserted gene for human fibroblast interferon, which are more pure and much cheaper (about 1/150 cost of the conventional method).

(B) RECOMBINANT VACCINES

When a foreign substance (antigen) enters the body, it triggers the immune system and, if successful, it results in the inactivation of the invading organism and, if unsuccessful, creates illness or death. Generally, antigens are proteins. Vaccines are living attenuated (weakened) microorganisms (viral vaccines) or dead microorganisms (bacterial vaccines) or their fractions which mimic the infectious agents without pathogenic consequences. In the development of vaccines the antigens or the components that actually induce immune response are identified and characterized. Some vaccines are very effective, *e.g.,* polio vaccines, whereas others are unsatisfactory, *e.g.*, vaccines of typhoid, cholera, etc. In some cases, fully effective vaccines are not available, *e.g.*, malaria. Massive efforts are in progress to develop a vaccine against AIDS. Through conventional methods, the production of vaccines is costly and occurs in low amounts and, moreover, there is risk of disease development due to occasional presence of active pathogen or reversion of virulence in vaccinated individuals. Moreover, some viruses are difficult to grow in cultured cells, *e.g.*, hepatitis-β virus. Through biotechnology using animal or human cell cultures, safer vaccine production would increase in near future. Some important types of biotechnological vaccines are:

(*i*) **Purified antigen vaccines:** They are **non-recombinant** vaccines and are prepared by the separation of the antigen from the pathogen, *e.g.*, polysaccharide antigen against pneumonia from *Streptococcus pneumoniae*, endotoxin (toxoid) for tetanus, etc. Monoclonal antibodies are used to purify antigens with certainty.

(*ii*) **Recombinant vaccines:** They are produced through recombinant DNA technology by integrating the gene for the immunogenic proteins in *E coli*, yeast, cultured animal cells or transgenic plants. Here, DNA segment, necessary for the production of either

(*a*) **whole protein** (*e.g.,* coat protein gene for hepatitis-B has been cloned in yeast)or

(*b*) **a part of a protein** (*e.g.,* amino acid 114–160 of coat protein for foot and mouth disease virus) necessary for immunogenicity, is cloned.

But these protein vaccines (second revolution in vaccines) are very costly and need to be stored at low temperature.

(*c*) A third kind of vaccine is the **DNA vaccine** (third revolution in vaccines or genetic immunization) in which the isolated gene for immunogenic protein is cloned in a suitable vector; after getting injection, it expresses in the vaccinated individual by the formation of particular immunogenic protein. Influenza vaccine is at advanced stage of testing. They are more safer, more specific, give more potent immune response and are easier to prepare.

(C) MONOCLONAL ANTIBODIES

Conventional serum contains heterogeneous (a mixture of many kinds of antibodies) antibodies produced in the blood in response to an antigen, whereas, monoclonal antibodies are homogenous (of one kind) and specific (Fig 20.2). They were produced for the first time by G. Kohler (1975) by hybridoma technology and got Nobel Prize for this important achievement. In this technology, he first immunized mouse by injecting a specific antigen to develop specific antibody in the lymphocyte cells of its blood. He fused this mouse lymphocyte cell with a myeloma (bone marrow tumour) cell to produce hybridoma that has the property to multiply indefinitely (myeloma cell character) as well as to produce large quantity of that specific antibody (lymphocyte character) through cell culture (Fig. 20.3).

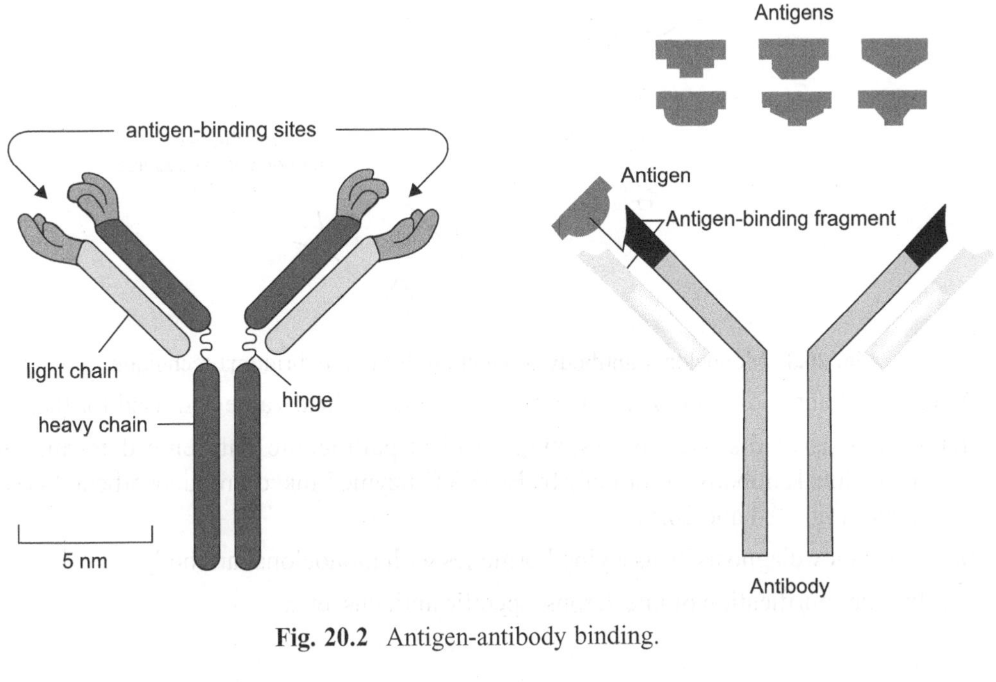

Fig. 20.2 Antigen-antibody binding.

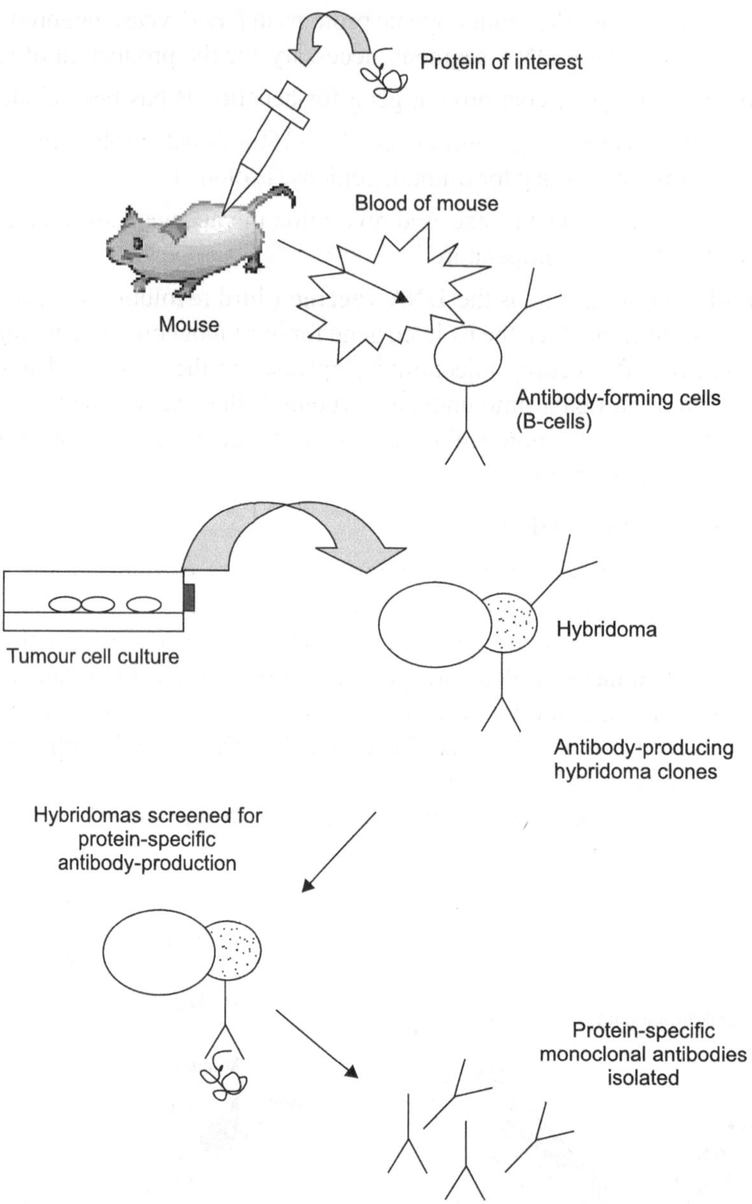

Fig. 20.3 Monoclonal antibody production through hybridoma technique.

Monoclonal antibodies are of great commercial and medical value and used for the:

(*i*) Diagnosis of diseases by assaying specific pathogenic antigenic determinant with monoclonal antibody, commonly by ELISA (Enzyme Linked Immunosorbent Assay) technique (Fig. 20.4 and 20.5).

(*ii*) Pregnancy diagnosis by assaying hormones with monoclonal antibody,

(*iii*) Immunopurification of interferons, specific antigens, etc., (*iv*) Treatment of cancer by delivering toxins specifically to the tumour cells by developing specific monoclonal antibodies against tumour cells and then linking these antibodies with the toxins or radioactive substances that kill only tumour cells.

(*v*) Identification of blood groups (A,B,O,Rh).

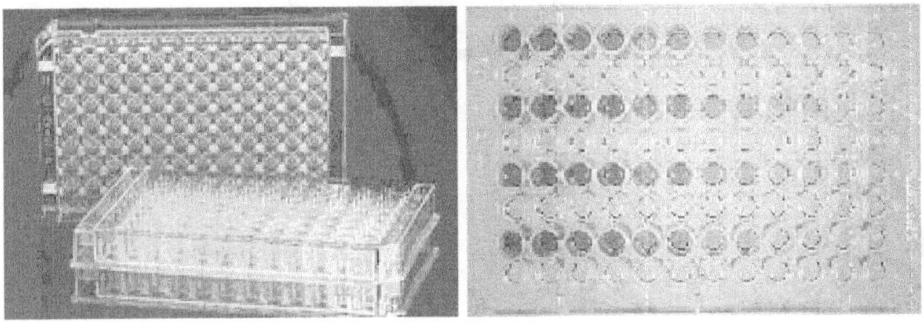

ELISA plates

ELISA plate well is coated with a capture antibody

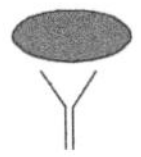

Sample is added, and any antigen present binds to capture antibody

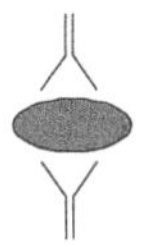

Detecting antibody is added, and binds to antigen

Enzyme-linked secondary antibody is added, and binds to detecting antibody

Substrate is added, and is converted by enzyme to detectable form

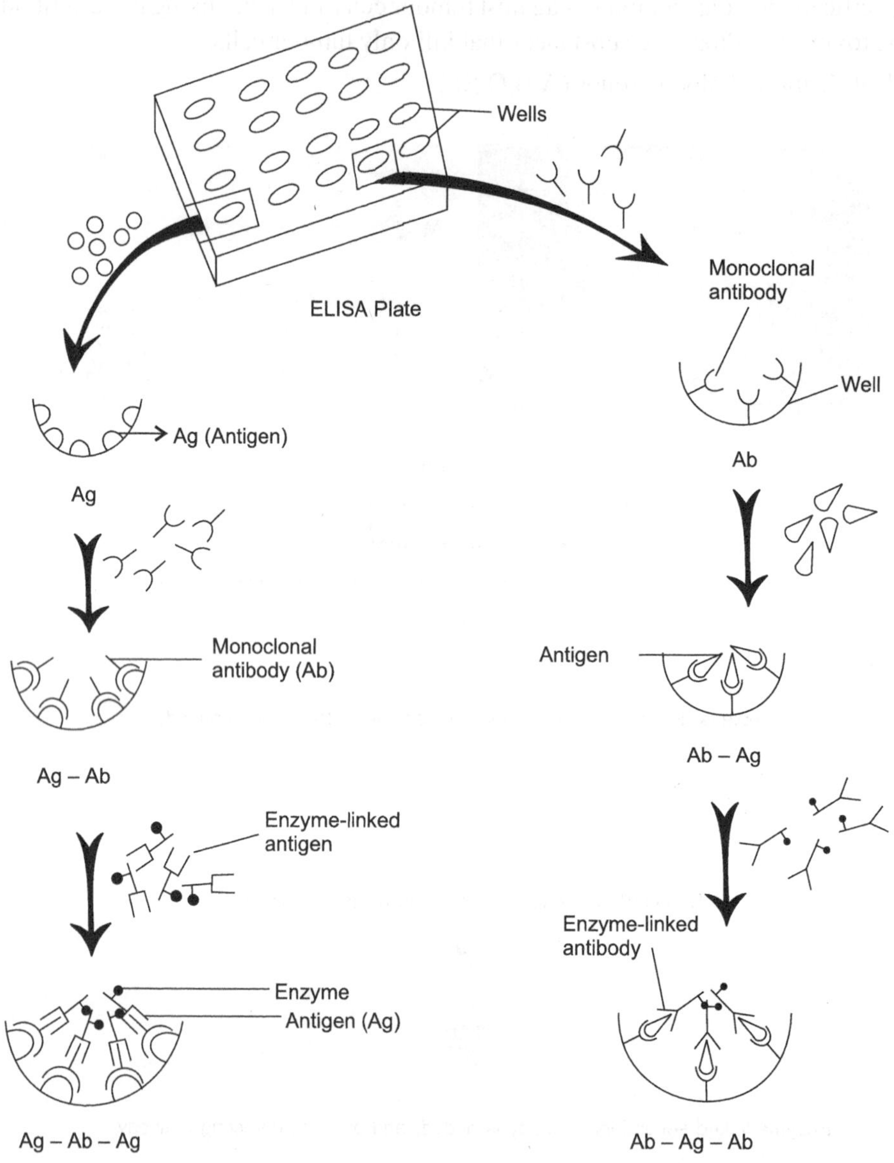

Fig.20.4 Use of monoclonal antibodies in ELISA technique.
Ab = Antibody, Ag = Antigen

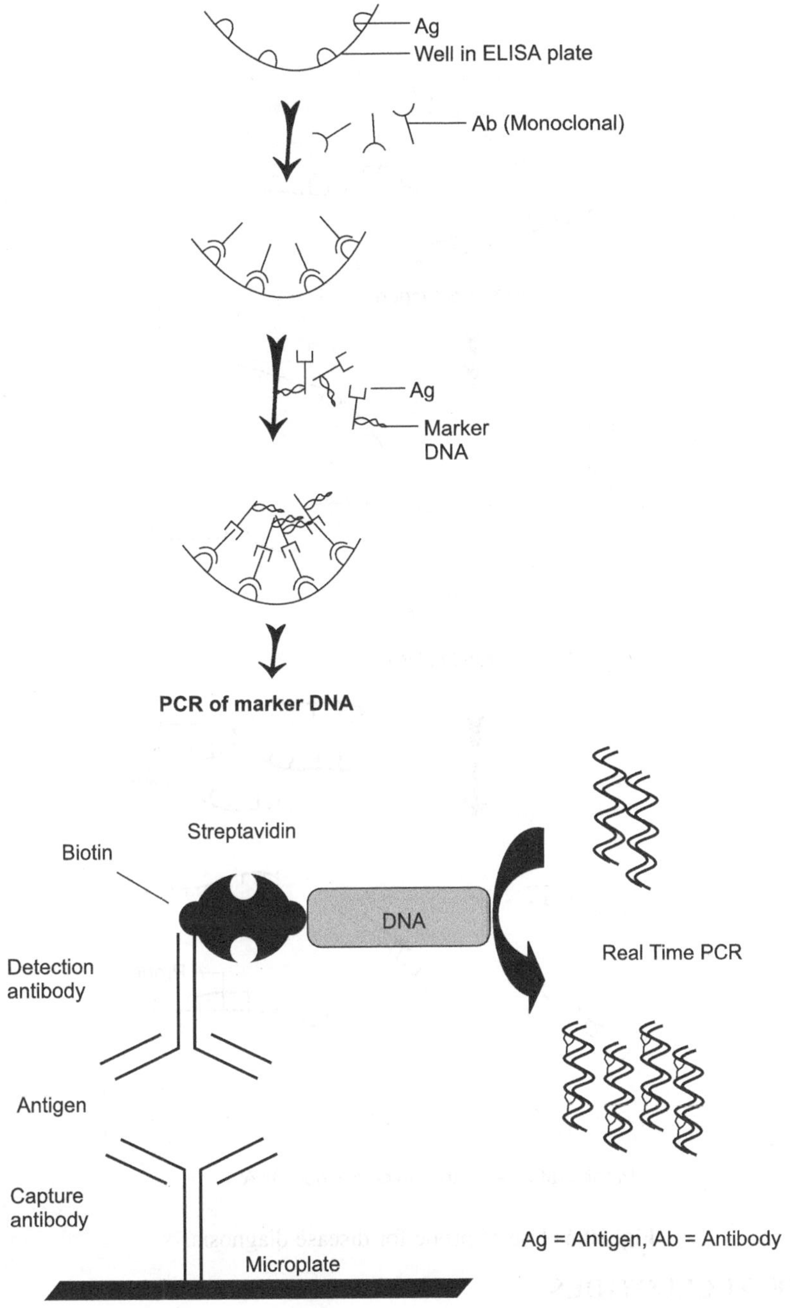

Fig. 20.5 Immuno-PCR technique.

(D) PROBES

Probes are radioactively (^{32}P) or non-radioactively (*e.g.*, biotin, fluorescent molecules) labeled, 15–30 bases long segments of DNA or RNA and can detect complementary bases in nucleic acid samples of pathogen by autoradiography (Fig. 20.6). Molecular probes for malaria, filaria, tuberculosis, etc., are available for disease diagnosis.

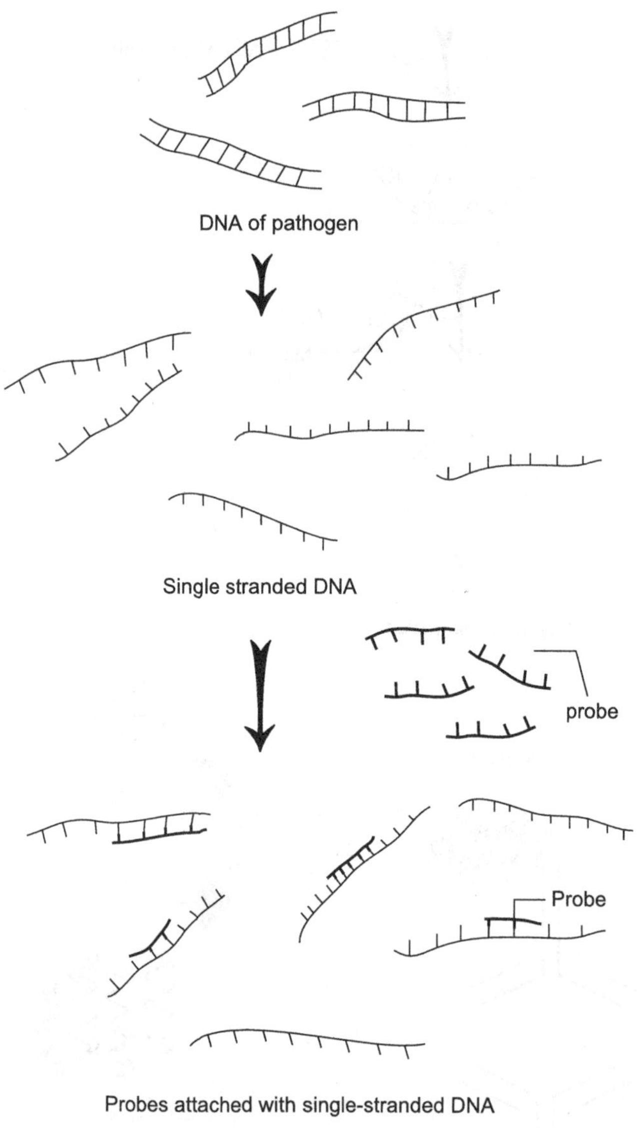

Fig. 20.6 Use of probe for disease diagnosis.

(E) ANTISENSE NUCLEOTIDES

25–35 bases long oligonucleotides, complementary to 5' end of parasitic mRNA molecules can be used as therapeutic agents to kill the pathogen or for treatment of cancer (Fig. 20.7). The work is at various evaluation stages.

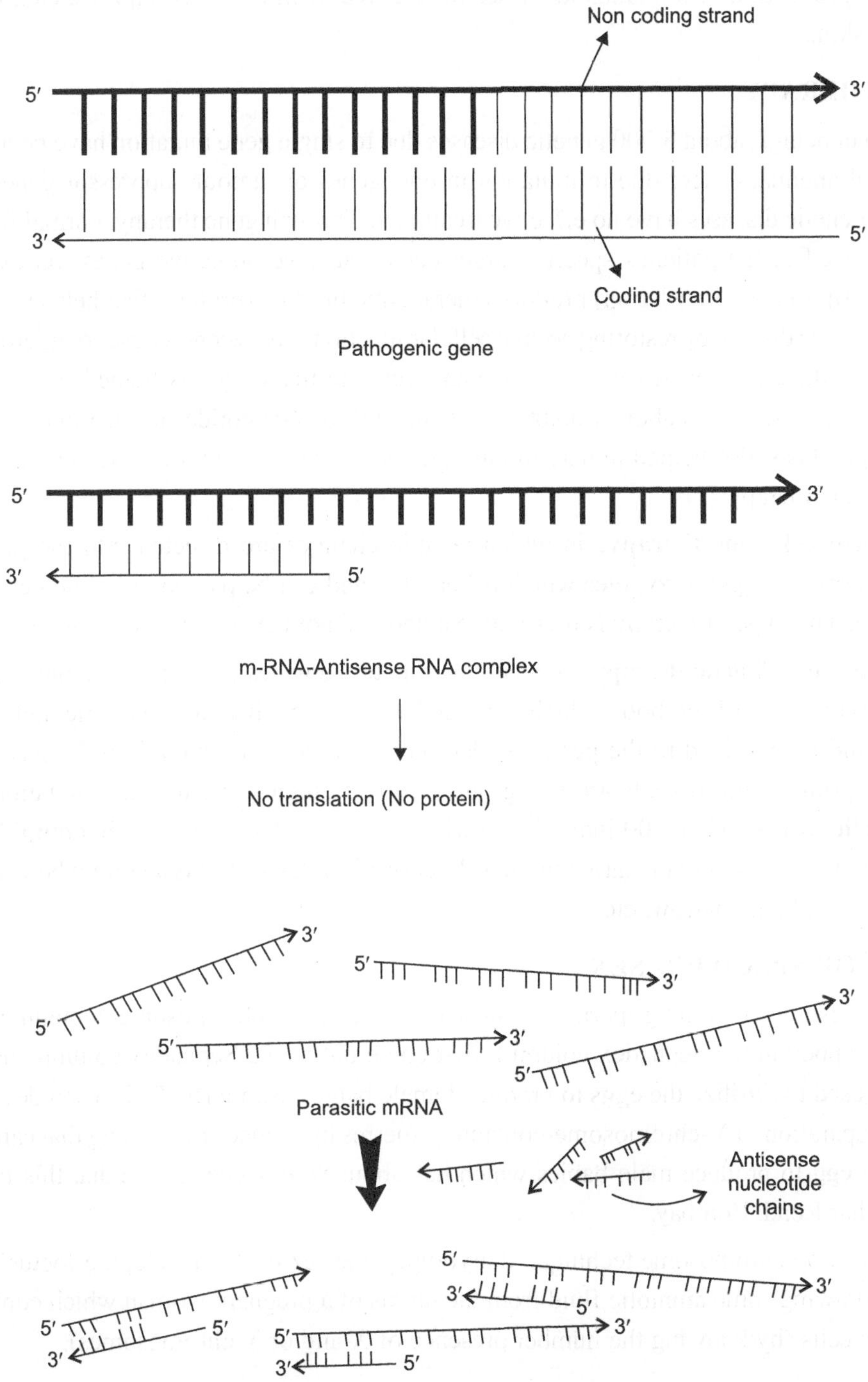

Fig. 20.7 Use of antisense nucleotides for the treatment of disease.

(F) ARTIFICIAL TISSUES

In vitro production of implantable tissues for the treatment of burns, injuries, etc., is in use, *e.g.,* artificial skin.

(G) GENE THERAPY

In human beings, about 5,000 genetic diseases due to single gene mutation have been known, *e.g.,* sickle cell anemia, cancer due to mutation in oncogenes or tumour suppressor genes, and at present, most genetic diseases have no effective treatment. Through gene therapy, normal functional gene is transferred in the patient's specific target cells that have defective genes. Its expression (through transcription and translation) produces therapeutic product (protein) that helps in the treatment of such genetic disease by restoring normal cellular function. Live adeno viruses or microinjection are used to introduce the functional genes. Human genome mapping has helped in locating the relative and actual positions of about 1,00,000 genes in 3 billion nucleotides in a haploid set of chromosomes (23) and has also helped in determining the sequences of many of these genes. There are two ways of gene therapy:

(*i*) **Germ cell gene therapy,** in which genetic changes are directed into the germ cells (sperms or eggs) or zygotes which is heritable and can be passed on to the next generation. This type of therapy is presently banned and not permitted in any country.

(*ii*) **Somatic cell gene therapy,** in which the functional genes are introduced only into those somatic cells of the body which have defective genes. It is not heritable and its effect would be confined to the person undergoing treatment and would not be passed to its offspring. Clinical trials are going on to correct single-gene defects (mutations), *e.g.,* **cystic fibrosis** (1 in 500 individuals) where the target tissue is liver, **haemophilia** (1 in 6,000 males) where the target tissue is liver or fibroblasts, **thalassaemia** where the target tissue is bone marrow, etc.

(H) BABIES OF SPECIFIED SEX

X-chromosome-containing sperms, being heavier than the Y-chromosome-containing sperms (held up and trapped in the Sephadex column), first come out of the Sephadex column and can be preferentially used to fertilize the eggs to produce female babies. Similarly, R. Ericson developed a technique of separation of Y-chromosome-containing sperms by quinacrine staining that can be used to fertilize the egg to produce male babies which has about 80% success. In India, this facility is available in Khar Road, Bombay.

Moreover, by chromosome technique (karyotype), the sex of the developing foetus can also be known by drawing some amniotic fluid from the uterus of a pregnant woman which contains the floating foetus cells (by knowing the number presence of X and/or Y chromosomes).

QUESTIONS

1. **What is biotechnology? Describe its role in medical science giving suitable examples.**
2. **Describe monoclonal antibody and its various applications.**
3. **Discuss the contributions of biotechnology in the control of diseases.**
4. What is gene therapy? Briefly describe its various approaches.
5. Discuss briefly the contributions of biotechnology in:
   (*i*) **Prevention of diseases**
   (*ii*) **Disease diagnosis or early detection of diseases**
   (*iii*) **Disease treatment**
   (*iv*) **Forensic medicine (or DNA fingerprinting)**
   (*v*) Fertility control
6. Write short notes on:
   (*i*) **Probes in disease diagnosis**
   (*ii*) Test tube babies
   (*iii*) ELISA and immuno PCR in disease diagnosis
   (*iv*) **Therapeutic agents from genetically engineered organisms**
   (*v*) **Recombinant vaccines**
   (*vi*) Interferons
   (*vii*) Babies of specified sex
   (*viii*) *In vitro* fertilization and embryo transfer

Index

D

E

F

I

J

K

L

M

N

O

P

R

S

T

APPENDIX

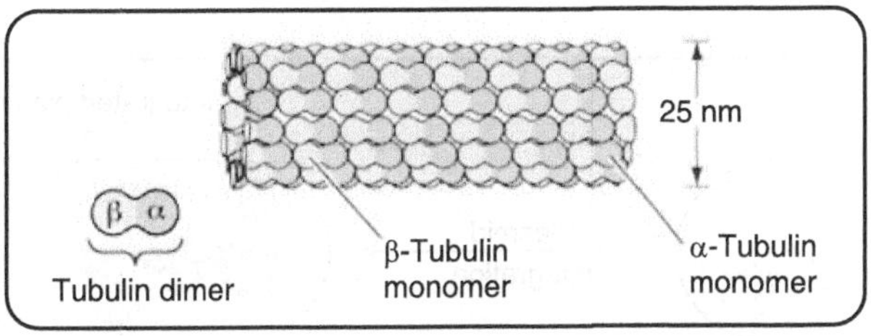

Fig. 1.28 Structure of a microtubule.

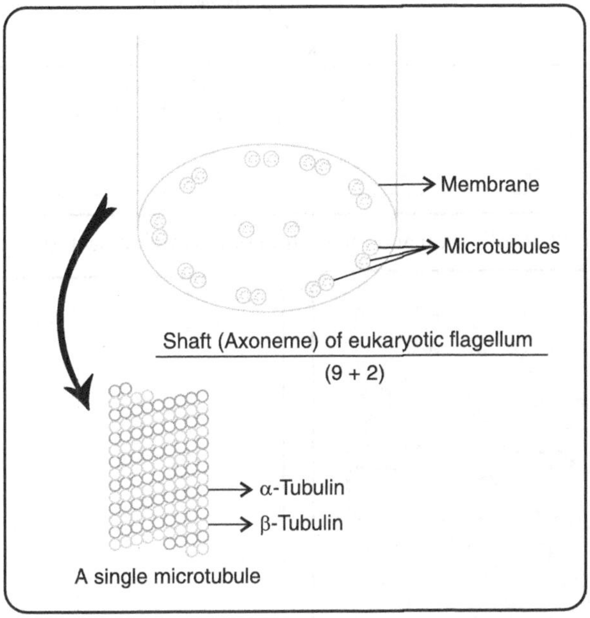

Fig. 1.34 Cross section of an eukaryotic flagellum.

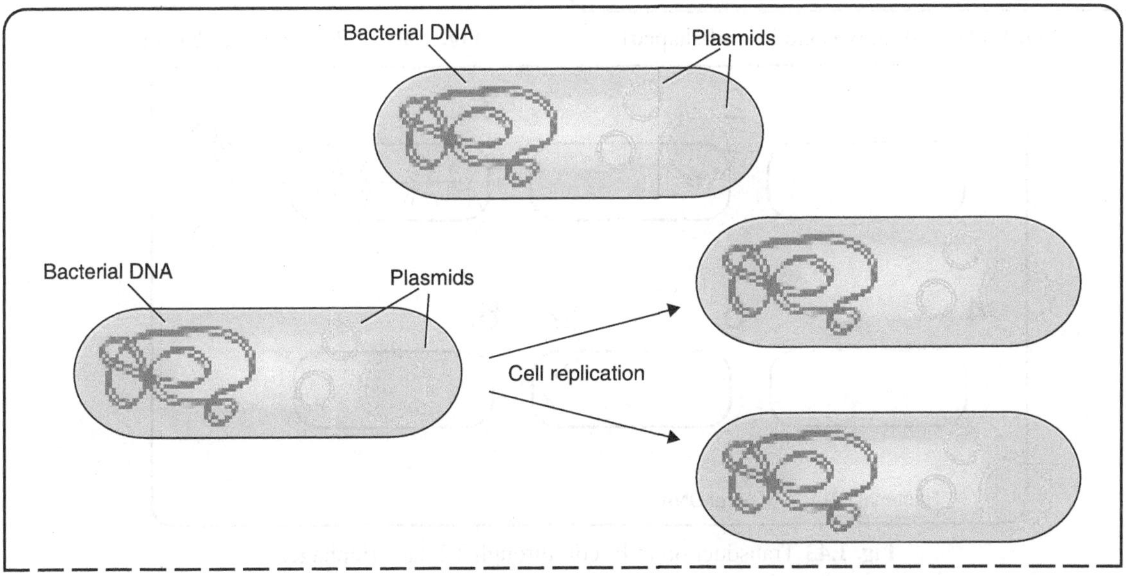

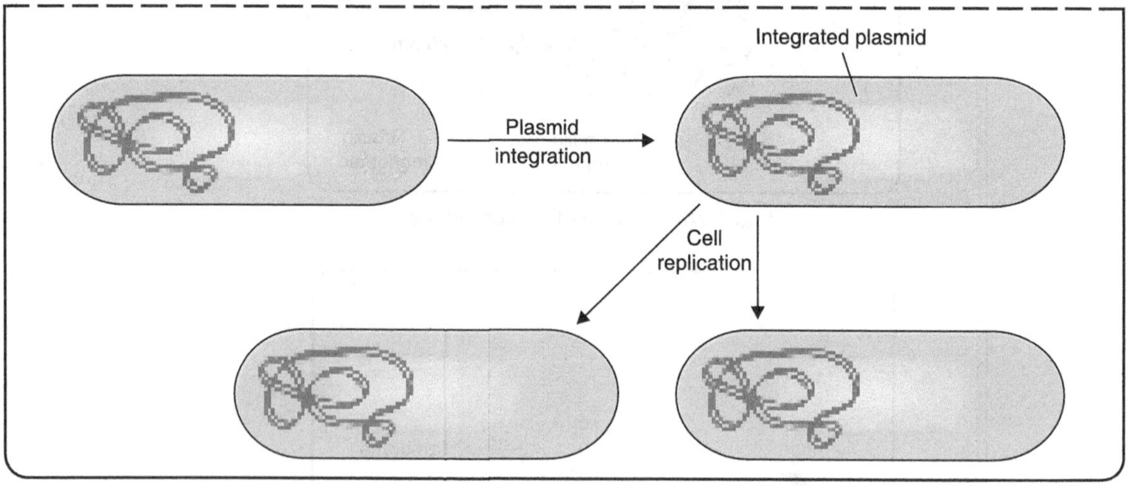

Fig. 1.38 Replication and integration of plasmid in bacteria.

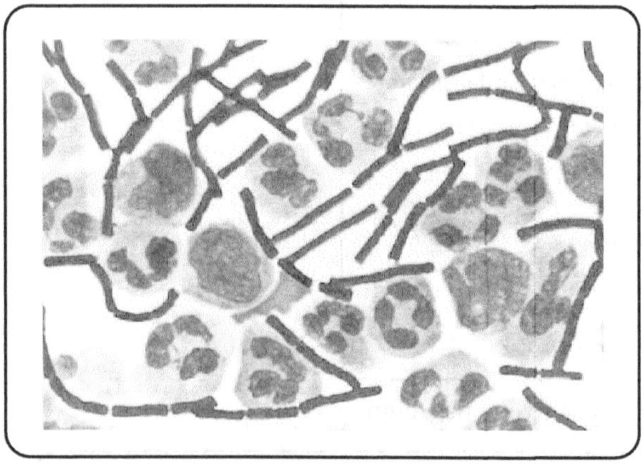

Fig. 1.39 G^+ Anthrax bacteria (rod-shaped).

Fig. 1.40 G^- Pseudomonas bacteria.

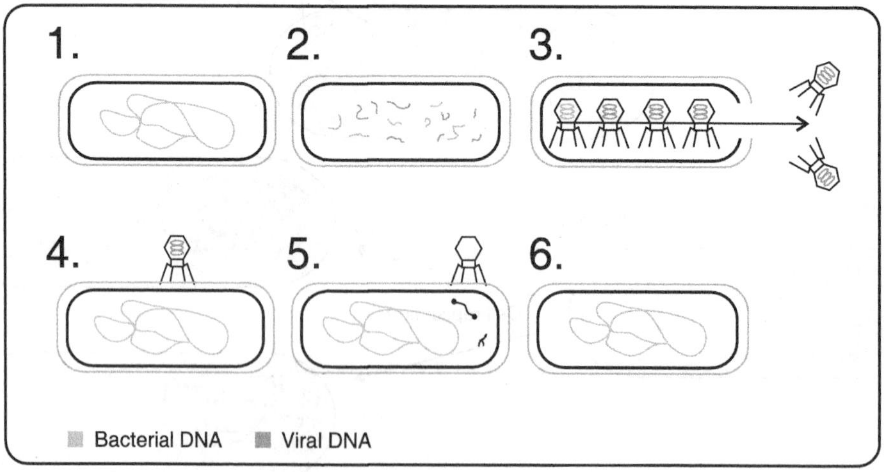

Fig. 1.43 Transduction in E. coli through T2 bacteriophage.

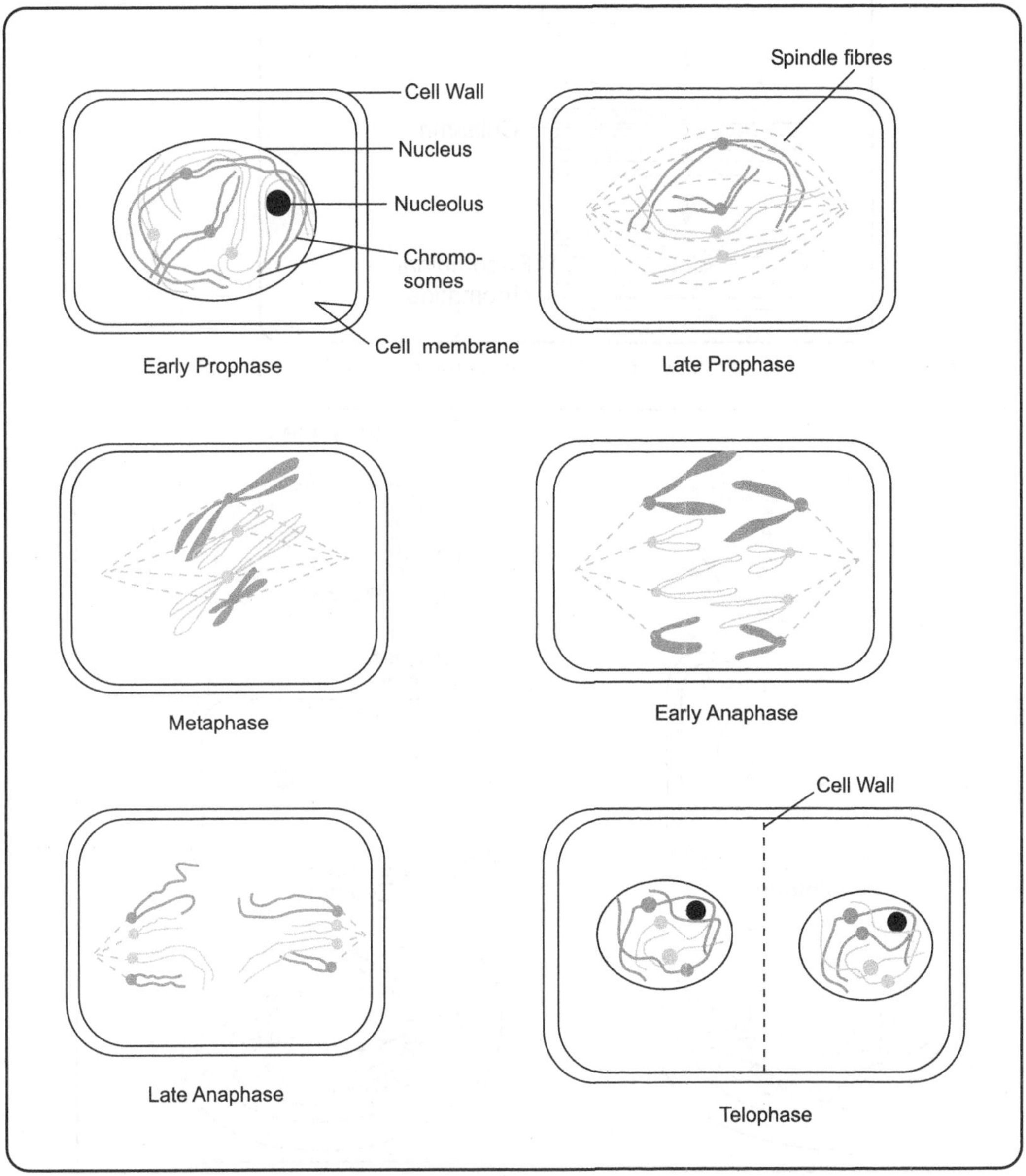

Fig. 5.7 Various stages of mitosis.

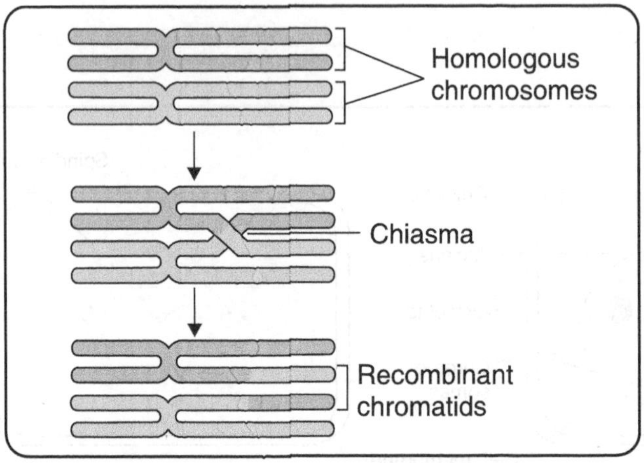

Fig. 5.11 Crossing over between homologous chromosomes during Prophase I.

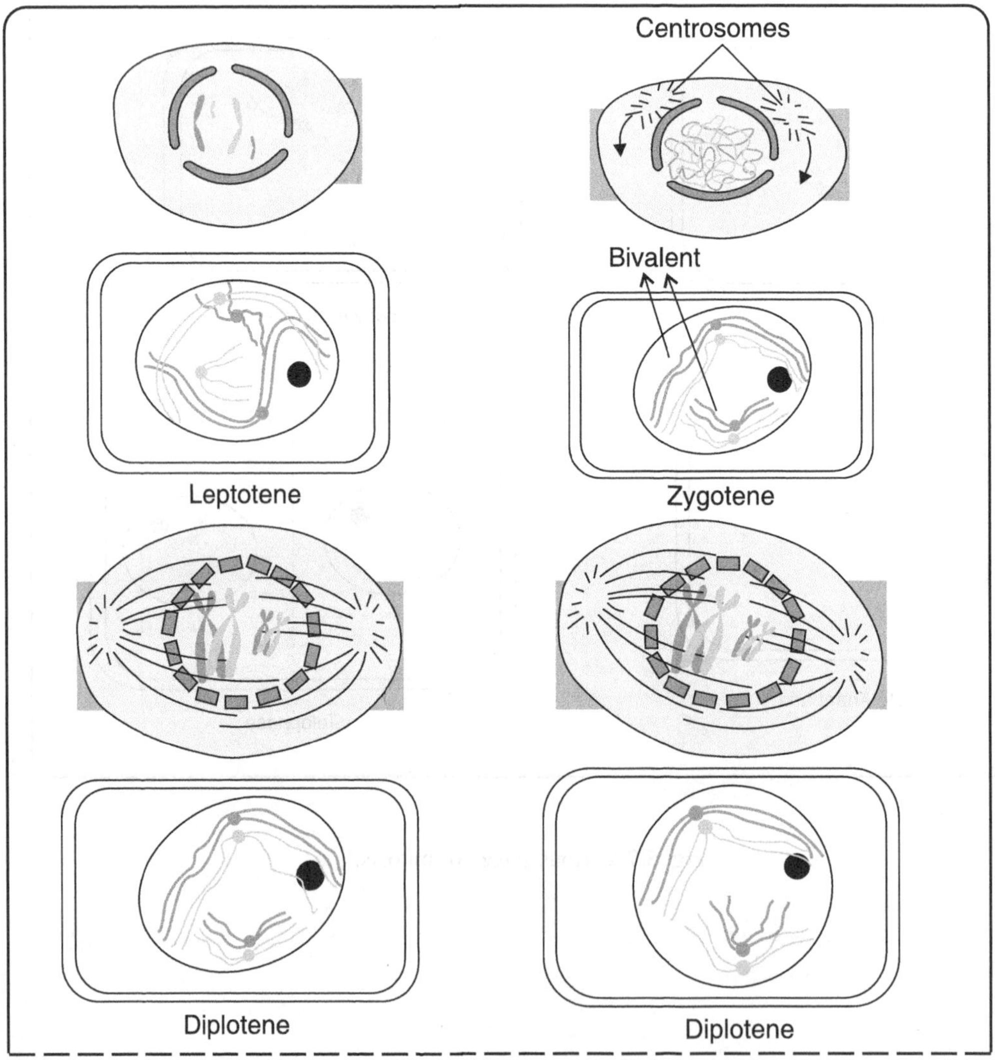

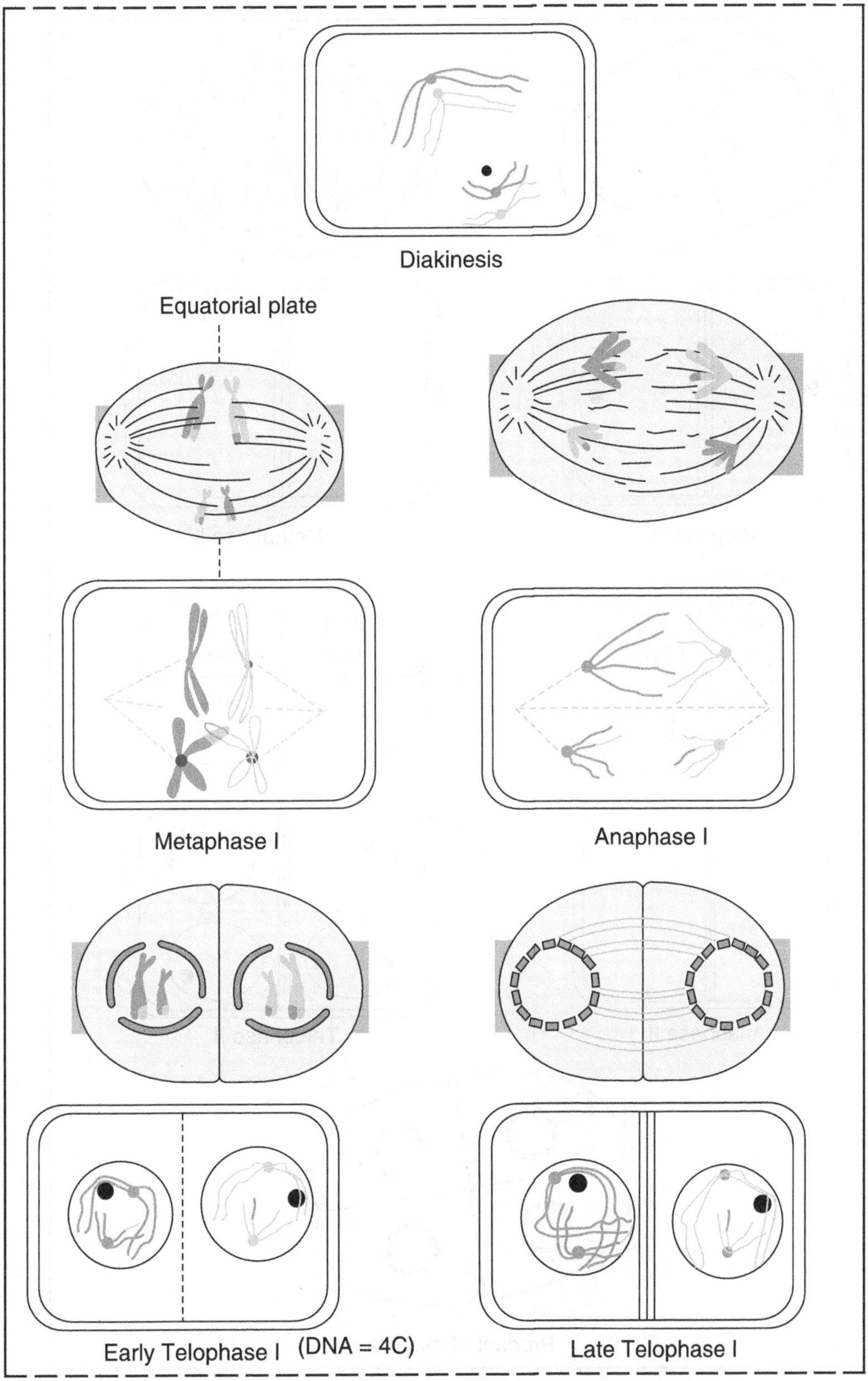
Diakinesis
Equatorial plate
Metaphase I
Anaphase I
Early Telophase I
(DNA = 4C)
Late Telophase I

(*vii*)

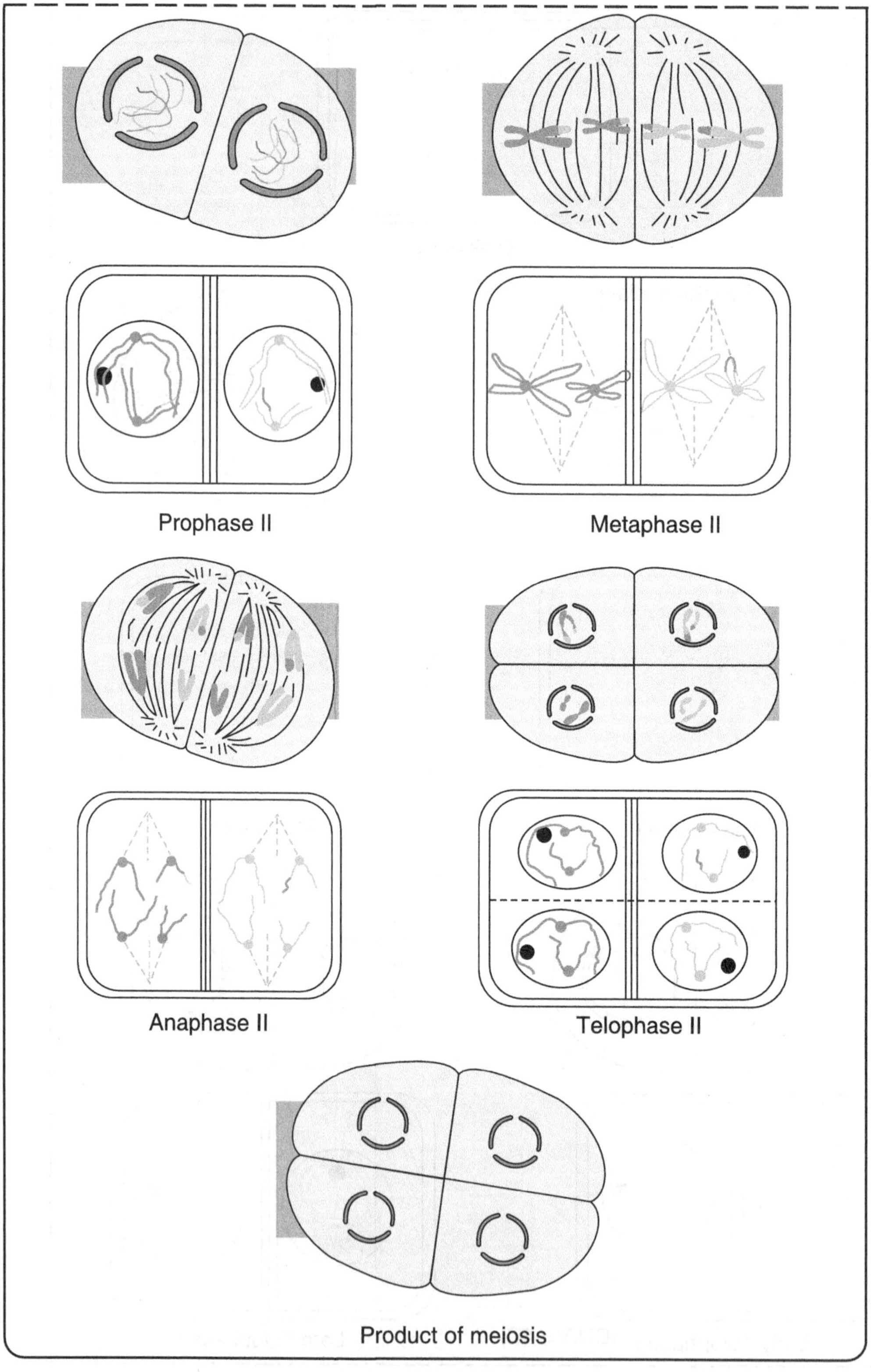

Fig. 5.12 Meiosis.

(viii)

Fig. 6.2 Garden pea plant with flowers and pods.

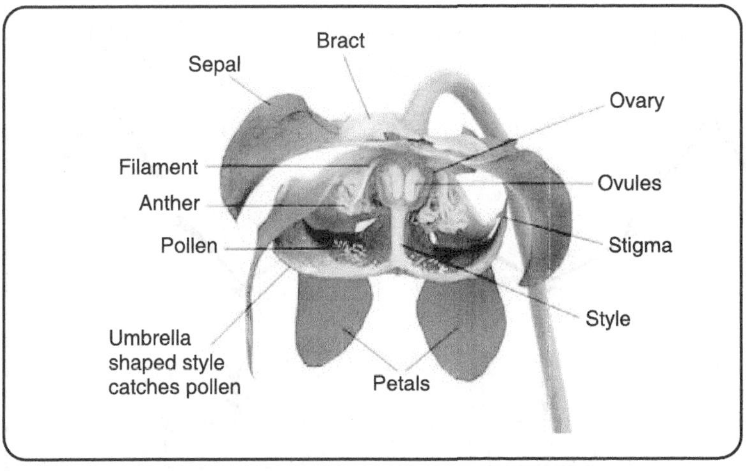

Fig. 6.3 Parts of a flower.

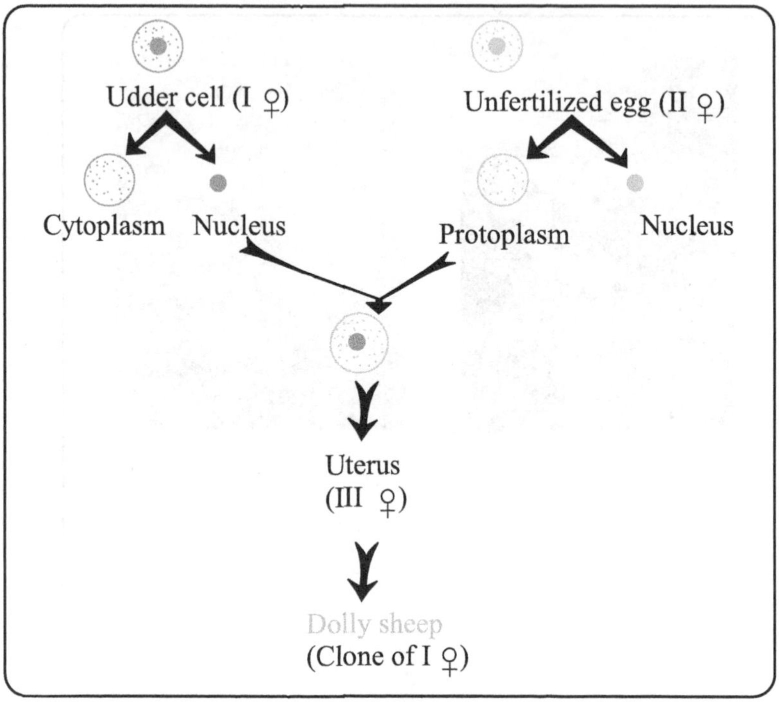

Fig. 10.12 Role of nucleus in controlling differentiation and development in the clone, Dolly sheep.

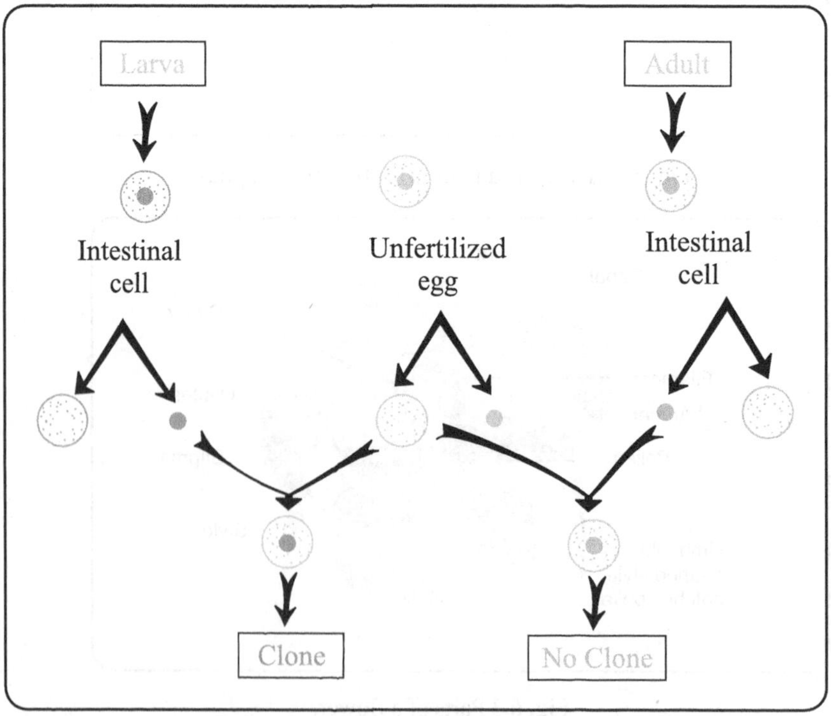

Fig. 10.13 Role of regulation of gene expression in controlling differentiation and development of frog.

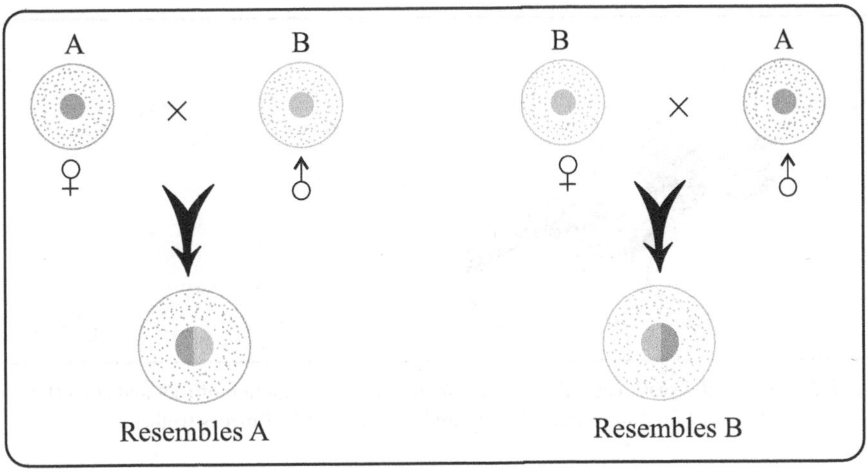

Fig. 10.15 Maternal effect in sexually-reproducing organisms.

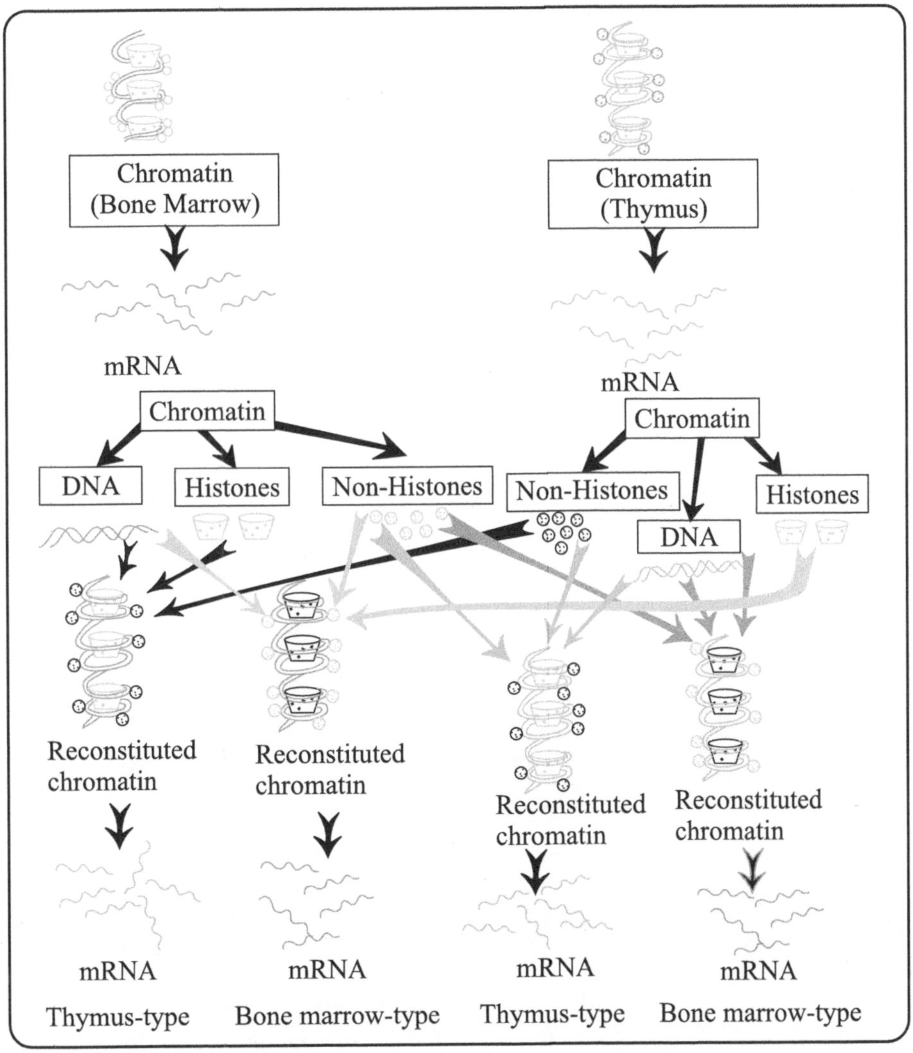

Fig. 11.11 Chromatin reconstitution experiment of Gilmour and Paul.

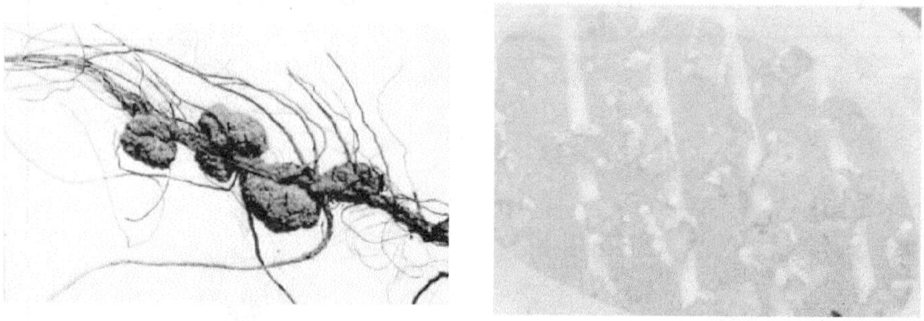

Fig. 12.7 Crown gall disease caused by *A. tumefaciens* near the junction of root and stem (left) and callus formation from the cultured transformed-leaf pieces (right).